The New Peasantries

The New Peasantries

Struggles for Autonomy and Sustainability
in an Era of Empire and Globalization

Jan Douwe van der Ploeg

from Routledge

First published in 2008 by Earthscan
2 Park Square, Milton Park, Abingdon, Oxon OX14 4RN
Simultaneously published in the USA and Canada by Earthscan
711 Third Avenue, New York NY 10017

Earthscan is an imprint of the Taylor & Francis Group, an informa business

Paperback edition first published in 2009

Copyright © Jan Douwe van der Ploeg, 2008

All rights reserved

ISBN 978-1-84407-558-4 (hardback)
ISBN 978-1-84407-882-0 (paperback)

Typeset by FiSH Books, Enfield
Cover design by Susanne Harris

Earthscan publishes in association with the International Institute for Environment and Development

A catalogue record for this book is available from the British Library

Library of Congress Cataloging-in-Publication Data

Ploeg, Jan Douwe van der, 1950-
  The new peasantries : struggles for autonomy and sustainability in an era of empire and globalization / Jan Douwe van der Ploeg.
    p. cm.
  ISBN 978-1-84407-558-4 (hardback)
  1. Peasantry. 2. Agricultural systems. I. Title. II. Title: Struggles for autonomy and sustainability in an era of empire and globalization.
  HD1521.P56 2008
  305.5'633—dc22
                                                                    2008011072

Contents

List of Figures, Tables and Boxes	ix
Preface	xiii
List of Acronyms and Abbreviations	xix

1 Setting the Scene — 1
  Introduction — 1
  Industrialization — 5
  Repeasantization — 6
  Deactivation — 7
  Interrelations between constellations and processes — 8
  The coming crisis — 10
  The methodological basis — 12
  Contents and organization of the book — 14

2 What, Then, Is the Peasantry? — 17
  Introduction — 17
  The 'awkward' science — 18
  A comprehensive definition of the peasant condition — 23
  On commonalities, differentiation and change — 35
  From peasant condition to the peasant mode of farming — 42
  Labour-driven intensification — 45
  Multilevel distantiation and its relevance in the 'modern' world — 49

3 Catacaos: Repeasantization in Latin America — 53
  Introduction — 53
  Repeasantization — 54
  Mechanisms of repeasantization — 59
  The effects of repeasantization: Intensification of production — 62
  Spurred intensification — 63
  New modalities of repeasantization — 65
  Meanwhile: The rise of Empire — 69
  The peasant community and Empire — 80

4	**Parmalat: A European Example of a Food Empire**	87
	Introduction	87
	The mechanics of global expansion	87
	Parmalat as a three-tiered network	93
	Did Parmalat ever produce value?	96
	The last resort: Fresh blue milk	101
	The distorted development of food production and consumption	105
	The non-exceptional nature of food degradation: The rise of 'lookalikes'	106
	Empire compared with a contrasting mode of patterning: Regressive centralization versus redistributive growth	109

5	**Peasants and Entrepreneurs (Parma Revisited)**	113
	Introduction	113
	The multiple contrasts between peasant and entrepreneurial farming	113
	From deviation to modernization: The historical roots of agrarian entrepreneurship	125
	The political economy of entrepreneurial farming	128
	Heterogeneity reconsidered	136
	The moral economy of the agricultural entrepreneurs	140
	The fragility of entrepreneurial farming in the epoch of globalization and liberalization	142

6	**Rural Development: European Expressions of Repeasantization**	151
	Introduction	151
	Mechanisms of repeasantization	152
	Magnitude and impact	157
	The quality of life in rural areas	160
	Newly emerging peasant types of technology	167
	Repeasantization as social struggle	178

7	**Striving for Autonomy at Higher Levels of Aggregation: Territorial Co-operatives**	181
	Introduction	181
	What are territorial co-operatives?	182
	A brief history of the North Frisian Woodlands	185
	Novelty production	192
	Dimensions of strategic niche management	201
	Design principles	204
	The construction of movability	206

8	**Tamed Hedgerows, a Global Cow and a 'Bug': The Creation and Demolition of Controllability**	**211**
	Introduction	211
	Taming hedgerows	211
	The global cow	214
	State apparatuses as important ingredients of Empire	218
	Science as a Janus-faced phenomenon	220
	The creation of a bug	226
	Postscript	230
9	**Empire, Food and Farming: A Synthesis**	**233**
	Introduction	233
	From the Spanish to the current Empire	235
	On railway systems and corporations	243
	The third level	245
	The central but contradictory role of information and communication technology	247
	State, markets and institutions	252
	The role of science	253
	Synthesis	255
10	**The Peasant Principle**	**261**
	Introduction	261
	Empire and the peasantry	262
	Resistance	265
	Reconstituting the peasantry	271
	The 'peasant principle'	273
	The peasant principle and agrarian crisis	278
	Some notes on rural and agrarian policies	282

Notes   289
References   319
Index   347

List of Figures, Tables and Boxes

Figures

1.1	Different but interlinked modes of farming	3
1.2	Patterns of connectivity	5
1.3	Transitional processes	9
1.4	An outline of the coming agrarian crisis	10
2.1	The contours of the theoretical impasse	18
2.2	Choreography of the peasant condition	24
2.3	The basic flows entailed in farming	29
2.4	Border zones, degrees and movements	37
2.5	The relatively autonomous, historically guaranteed scheme of reproduction	44
2.6	Market-dependent reproduction	45
3.1	Land distribution in Catacaos, Castilla and Piura (1995)	56
3.2	The symbolic organization of the agricultural process of production in Catacaos	67
3.3	'Clever geography'	71
3.4	Available but de-linked resources	72
3.5	Modelling the world according to Empire	73
3.6	Representations of Empire: Barbed wire	74
3.7	Representations of Empire: Armed guards	74
3.8	Representations of Empire: Machinery	75
3.9	Representations of Empire: Artificial lakes	75
3.10	Representations of Empire: Water scarcity	76
3.11	Fresh Peruvian asparagus sold on European markets	77
3.12	An alternative patterning	84
4.1	The mechanics of expansion through mortgaging	89
4.2	Parmalat as a socio-technical network	94
4.3	Value flows	96
4.4	The making of *latte fresco blu*	103
4.5	Relative shares in world food market	110
5.1	The *contadini* (peasant) logic	118
5.2	The logic of the *imprenditori agricoli* (agricultural entrepreneurs)	119

5.3	Differential farm development trajectories in Emilia Romagna, 1970–1979	120
5.4	Added value for main industrial branches in Italy	129
5.5	The double squeeze on agriculture	130
5.6	The evolution of production per cow over time	132
5.7	The changing biophysics of production	133
5.8	Outcomes of a scenario study that compared different development trajectories (dairy farming in Friesland, The Netherlands)	136
5.9	Space for manoeuvre and different degrees of 'peasantness'	137
5.10	International comparison of investment levels in dairy farming	141
6.1	The choreography of repeasantization	153
6.2	Newly emerging expressions of repeasantization	158
6.3	Differentiation of rural and semi-rural areas in Italy	161
6.4	Where do people move to?	162
6.5	The theoretical model underlying the enquiry into quality of life in rural areas	164
6.6	Explaining the quality of life (overall path diagram)	166
6.7	Re-patterning resource use in Zwiggelte: An illustration of peasant inventiveness	169
6.8	Centrifugal filtering of olive oil	171
6.9	New technological devices	175
6.10	Rural development as a contested and fragmented process	180
7.1	An overview of the hedgerow landscape	182
7.2	The anatomy of a hedgerow	183
7.3	Distribution of nitrogen surpluses among VEL/VANLA member farms	187
7.4	Nitrogen surpluses on VEL/VANLA member farms compared to the regional average	187
7.5	The outline of the new North Frisian Woodlands plan	189
7.6	The cattle–manure–soil–fodder balance	193
7.7	Development of margins per 100kg of milk for several groups	195
7.8	A web of interconnected novelties	199
7.9	A second web relating to the management of nature and landscape	200
7.10	Dimensions of strategic niche management	202
7.11	Improved connectivity suggested by Landscape IMAGES	207
7.12	Nature and economy as mutually exclusive categories	208
7.13	Pareto optimization	208
8.1	Calculating the nitrogen excretion of the 'global cow'	215
8.2	Understanding the local groundings of manure production	217
8.3	Nitrogen delivery of sand, clay and peat soils	222

8.4	Nitrogen delivery of the soil (empirical observations)	223
8.5	The wider effects of legally prescribed slurry injection	225
8.6	Empirical levels of nitrogen excretion in relation to milk production per cow	231
10.1	Lost or carving out new pathways?	266
10.2	Humility or pride?	266
10.3	Going beyond the agrarian crisis	280

Tables

2.1	Different degrees of market dependency in The Netherlands, Italy and Peru (1983)	40
2.2	The variability in interrelations between dairy farms and the markets (The Netherlands, 1990)	41
3.1	Rates of repeasantization	56
3.2	The development of agricultural employment in Catacaos	58
3.3	Cotton yields in the community of Catacaos compared with neighbouring districts	63
4.1	Contrasting value chains	98
4.2	Prices for farmers in relation to prices paid by consumers	99
5.1	The main differences between the peasant and entrepreneurial modes of farming	114
5.2	Contrasting degrees of commoditization in Emilia Romagna, 1980	116
5.3	Differentiated growth patterns of production and value added (dairy farming in Parma province at current prices)	123
5.4	Comparison between a peasant and an entrepreneurial approach in Dutch dairy farming	139
5.5	Comparative analysis of Dutch dairy farms (2005)	148
7.1	Some quantitative data on the management of nature and landscape	188
10.1	Evolution of the agricultural labour force in Latin America (1970–2000)	273

Boxes

2.1	Mechanisms of distantiation	50
3.1	The shared values of the peasant community of Catacaos	61
5.1	The entrepreneurial condition	128
6.1	Features of the peasant mode of energy production	178
7.1	Commonly shared values as specified by the North Frisian Woodlands in its mission statement	190

8.1	The 'global algorithm'	216
10.1	An expression of the peasant principle	275
10.2	A fragment from the Taormina policy document	282

Preface

For many centuries, the peasantry was as omnipresent as it was self-evident. There was little need to enquire into it, let alone ask why it existed. It clearly did so through a wide variety of time- and place-bound expressions, the main contrasts of which might be summarized by referring to the Greek and Roman cradles of European agriculture. In Greek culture, the peasant was a free man, farming in a proud and independent way. The Greek γεωργός (*gheorgos*) represented the sublime. In contrast, in the Roman tradition the peasant was subordinated, a condition still echoed in the current Italian word for peasants: *contadini*, which literally means the 'men of the lord' – subordinated, mean, ugly and unable to control their own destiny. Of course, in every specific location the struggle for freedom and the danger of subordination go together, the one never far from the other. Probably the most telling expression of this intimate connection is that elaborated by Bertolucci in his seminal film *Novecento*. In a poignant scene, we see the ugly peasant and facing him the landlord, *il padrone*, who explains that the wages are to be lowered or the rents raised. To express his opposition, the peasant takes out his knife and cuts off – in one violent stroke – his own ear. This is meant to make clear to the landlord that he will no longer listen to him or accept his explanations. Surrounding the two, there is the peasant family: the wife and young children who are crying, suffering as they are from hunger. Then, in a heartbreaking moment, the now mutilated man seems to reach again for his knife, which leads us to believe that he might kill one of his children, maybe to end their suffering. But, instead, he takes up his flute and starts to play a sweet melody to comfort them.

Subordination and disobedience, humility and the longing for freedom, the ugly and the sublime are closely interwoven and, thus, present an undeniable combination of opposite elements, a combination in and through which the one provokes the other and vice versa. That is precisely what Bertolucci demonstrates in such a masterly way. It is also one of the central themes of this book.

In today's world, the peasantry is no longer a self-evident reality and the tensions inherent in the concept no longer seem relevant. In the modern world there is, apparently, no place or attention for this strange Janus-faced phenomenon. Much attention was given to the peasantry during the grand transformations of the last two centuries, and many of the resulting theories centred on the peasant as an obstacle to change and, thus, as a social figure that should disappear or be actively removed. Theoretically, peasants have been cut off from the land, their place taken by 'agricultural entrepreneurs' – well equipped to listen to the logic of the market. Such a view might just admit that some peasants may still exist in remote places, typically in developing world countries; but that they will, for sure, disappear as progress marches on.

In this book, I argue that behind this manufactured invisibility, which is greatly strengthened by the negative connotation that the word peasant has in the language of everyday life, there is an empirical reality in which there are far more peasants than ever before. Worldwide, there are now some 1.2 billion peasants (*Ecologiste*, 2004; Charvet, 2005). 'Small-farm households, after all, still constitute nearly two-fifths of humanity' (Weis, 2007, p25). Among them are millions of European farmers who are far more peasant than most of us know or want to admit.

In view of the uneasy combination of invisibility and omnipresence, this book pursues three interconnected lines of reasoning. The first centres on the contradictory nature of the peasant condition by defining it as the ongoing struggle for autonomy and progress in a context characterized by multiple patterns of dependency and associated processes of exploitation and marginalization. The basic mechanisms through which such struggles unfold go beyond the specificities of time and space. However, farming might also deviate from these basic mechanisms by tending, for example, towards system integration instead of autonomy. Then, new forms, patterns and identities emerge, such as those of the agricultural entrepreneur.

The second line of reasoning contextualizes the first by claiming that there is a critical role for peasants in modern societies and that there are millions who have no alternative to such an existence. In many developing world countries, millions fight to escape misery (including urban misery) by turning themselves into peasants – the Brazilian movement of landless people, the MST (*Movimento dos Sem Terra*), being the most outspoken, though far from the only expression of this tendency. And in the so-called 'civilized' parts of the world we will probably come to the conclusion that a world with peasants is a better place than one without them. As I will show, their presence often relates positively to the quality of life in the countryside, to the quality of our food and to the need to make sustainable and efficient use of water, energy and fertile land.

The third line of reasoning concerns the opposite: it shows how the dominant mode of ordering – I refer to this new mode using the notion of Empire – tends to marginalize and destroy the peasantry along with the values that it carries and produces.

Thus we have a first arena, one that is located in the real world and which will be, in several respects, decisive for our futures. It is the arena in which Empire and the peasantry, wherever located, engage in multilayered and multidimensional contradictions and clashes. Then there is a second arena, which intersects with the first – that of science, knowledge, theory and, more generally, the battle of ideas. In this arena there are basically two opposing approaches. One I have already referred to – namely, the approach (or, I should say, a broad range of somehow interlocking approaches) that has made the peasantry invisible and which is unable to conceptualize a world in which peasants are 'possible'. Opposed to this dominant approach is a new 'post-modern' approach[1] that is being developed worldwide by many researchers, which argues that a proper understanding of the rise and expansion of what are essentially global markets[2] is crucial for post-modern peasant studies. While for many centuries there have been worldwide transactions of agricultural products, present-day global markets for agricultural and food products represent a new phenomenon that strongly impacts upon agriculture wherever it is located. The strategic importance of these global markets has stimulated a range of new studies that enquire into the patterns that now govern these markets. Within this enquiry, the notion of 'Empire' operates as a heuristic device for characterizing the new 'superstructure' of the globalizing markets (see especially the work of Hardt and Negri, 2000; Holloway, 2002; Negri, 2003, 2006; Friedmann, 2004; Weis, 2007).

Empire, as I will show throughout this book, is a new and powerful mode of ordering. It increasingly reorders large domains of the social and natural worlds, subjecting them to new forms of centralized control and massive appropriation. However, the places, forms, expressions, mechanisms and grammar of Empire are, as yet, insufficiently explored, documented and critically elaborated upon, especially in so far as farming, food processing and the newly emerging food empires are concerned.

Along with many others, I have been engaged in the exploration of Empire. Through the analysis of a wide range of changes in agricultural production, the processing and consumption of food and the 'management' of nature, I have probed the mechanics and characteristics of Empire and the new order that it entails. The analysis shows that currently emerging food empires, which constitute a crucial feature of Empire, generally, share several features, such as expansionism, hierarchical control and the creation of new, material and symbolic orders. There is imperial conquest with respect to the integrity of food, the craft of farming, the dynamics of nature, and the resources and prospects of many agricultural producers. This conquest proceeds as the ongoing deconstruction and subsequent reassembling of many interrelations and connections that characterize the domains of farming, food and nature. New technologies and a widespread reliance on expert systems play a strategic role in this imperial reassembling.

New peasantries play a central role in this book and I think it important from the outset to stress that in the following chapters the peasantry is not treated as a remnant from the past, but as an integral part of our time and societies. The peasantry cannot be explained by mere reference to the past; it is rooted in the realities of today and is, therefore, to be explained by the relations and contradictions that characterize the present. Nor does the peasantry figure in this book as representing only a problem since it also offers promising, albeit as yet somewhat hidden, prospects and solutions. There are, thus, several reasons for reconsidering the peasantry and its future.

Current patterns of accumulation produce high levels of both urban and rural unemployment. Lack of income and prospects, hunger and other forms of deprivation are among the many results that together might be summarized as the condition of marginality. In my view, it would seem that, in most continents, there is only one adequate mechanism for tackling and superseding this condition of marginality and that is by enlarging the ranks of the peasantry and providing for peasant-managed forms of rural and agricultural development.[3] I am more than aware that such a statement will be perceived – especially among the 'experts of development'– as cursing in front of the Pope. However, in practice, there simply is no alternative and politically the need for certain levels of integration can no longer be neglected.

In Europe, the imperial restructuring of the natural and social worlds implies an overall degradation of landscapes, biodiversity, rural livelihoods, labour processes and the quality of food, all of which are outcomes that are triggering widespread opposition among a large number of the population, including urban residents. At the same time, the farming population is confronted with an increased squeeze on agriculture. Prices are stagnating, costs are soaring and many farming families are pushed into conditions of marginality. It is intriguing, at least at first sight, that within this panorama growing segments of the farming population in Europe are reconstituting themselves as peasants. They face and fight the condition of marginality imposed upon them through actively creating new responses that definitely deviate from the prescriptions and logic of Empire, while simultaneously creating and strengthening new interrelations with society at large through the care they invest in landscape, biodiversity, the quality of food, etc. In fact, the grassroots processes of rural development that are transforming the European countryside might best be understood as ever so many expressions of repeasantization.

From a socio-political point of view, today's peasantries constitute many 'multitudes', from which resistance, countervailing pressure, novelties, alternatives and new fields of action (Long, 2007) are continuously emerging. Maybe there is even more to this – namely, that by simply being there, these peasantries remind us constantly that the countryside, agriculture and the processing of food are not necessarily to be ordered as part of Empire. The peasantry presents, in this respect, a materialized and often highly visible critique of today's world and how it is organized.

Alongside the foregoing observations, important steps are being made within rural studies worldwide to rethink and redefine the concept of the peasantry. That is, new and probably decisive efforts are being made to go beyond peasant theories as they were developed and formulated during the late 19th century and first eight decades of the 20th century. I will discuss these new theoretical insights – which are evidently inspired by a range of fundamental new trends at the empirical level – in terms of the emergence of post-modern peasant studies. During the modernization period (that materially embraced the 1950s to the 1990s), the perception and interpretation of different practices and policies, the social definition of interests by farmers and the elaboration of programmes by social and political movements were all encapsulated, if not entrapped in and governed, by the modernization paradigm. Now, at the beginning of the 21st century, it is clear that this modernization project has run counter to its own self-produced limits – not only materially but also intellectually. Hence, a new approach is needed – one that definitively goes beyond modernization as a theoretical (and practical) framework. I refer to this new approach that is beginning to emerge from many sources as post-modern peasant studies.

In the aftermath of modernization it is increasingly recognized that the peasantry will remain with us, in many new and unexpected forms, and that we need to come to grips with this both in practice and in theory. This 'discovery', that constitutes the backbone of newly emerging post-modern peasant studies, is not always easy to digest, as becomes apparent from many international debates. It runs, that is, counter to the core of both Marxist and modernization approaches, which interpret the peasant as disappearing and which neglect, to a large degree, the empirical development trajectories of agricultural sectors in both the centre and the periphery.

In this book I attempt to summarize this newly emerging reconceptualization of the peasantry and its role in societies located at the beginning of the third millennium. I am happy to have been part of some of the 'laboratories' located at the interfaces between peasant struggles, scientific analysis and political debates. I feel fortunate to be able to draw upon the many experiences and theoretical insights achieved in those laboratories. Many people have helped me along the way and I have, where possible, indicated their creative contributions in the first note of each of the following chapters. Here I limit myself to Ann and Norman Long. This book is dedicated to them for their presence and involvement in the 'battlefields of knowledge'.

List of Acronyms and Abbreviations

AID	General Inspection Service (The Netherlands)
BSE	bovine spongiform encephalopathy
CAP	Common Agricultural Policy
CCP	capitalist commodity production
CIDA	Comite Interamericano de Desarrollo Agricola
CLA	conjugated linoleic acid
cm	centimetre
CRPA	Research Centre for Animal Production (Italy)
DES	diethylstilbestrol
DLG	National Service for the Rural Areas (The Netherlands)
DOP	Protected Denomination of Origin
ECN	Petten Research Institution (The Netherlands)
EEAC	European Environmental and Agricultural Councils
ESRS	European Society for Rural Sociology
EU	European Union
EU-15	15 European Union member states before the expansion on 1 May 2004 (Austria, Belgium, Denmark, Finland, France, Germany, Greece, Republic of Ireland, Italy, Luxembourg, The Netherlands, Portugal, Spain, Sweden and the UK)
FEDECAP	Federación Departamental Campesina de Piura (Departmental Peasant Federation of Piura) (Peru)
FPCM	fat and protein corrected milk
GMO	genetically modified organism
GPS	global positioning system
GVA	gross value added
GVP	gross value of production
ha	hectare
HACCP	Hazard Analysis and Critical Control Point
HYV	high-yielding variety
ICT	information and communication technology
IEP	Instituto de Estudios Peruanos (Lima)
IGP	Protected Geographical Indication
ISMEA	Service Institute for the Agricultural and Food Market (Italy)
kg	kilogram
km	kilometre
LDC	less developed country

LEI	Farm Accountancy Institute (The Netherlands)
LFA	less favoured area
LTO	Land- en Tuinbouw Organisatie Nederland (Dutch Farmers' Union)
M − s	mean minus standard deviation
M + s	mean plus standard deviation
mg	milligram
MINAS	management of nutrient accountancy systems
MP	member of parliament
MPA	medroxyprogesterone acetate
MST	*Movimento dos Sem Terra* (Brazil)
n	total sample population size
NCBTB	Dutch Christian Federation of Farmers and Horticulturists
NFW	North Frisian Woodlands (The Netherlands)
NGO	non-governmental organization
NWO	Netherlands Scientific Research Organization
OECD	Organisation for Economic Co-operation and Development
PCP	petty commodity production
RD	rural development
RLG	Council for the Rural Areas (The Netherlands)
RSA	Cooperating Register Accountants (The Netherlands)
SBNL	Society for the Protection of Nature and Landscape (The Netherlands)
SCP	simple commodity production
SIDEA	Italian Society for Agrarian Economics
SME	small- and medium-sized enterprise
SPN	Regional Products Netherlands
STS	Sociology of Technology and Science
TATE	technological–administrative task environment
UCP	*unidades comunales de producción* (communal units of production)
UHT	ultra-high temperature
UK	United Kingdom
US	United States
VA	value added
VANLA	Vereniging Agrarisch Natuur en Landschapsonderhoud Achtkarspelen (Achtkarspelen Society for Agrarian Management of Landscape and Nature) (The Netherlands)
VAT	value-added tax
VEL	Eastermars Lânsdouwe (The Netherlands)
WALIR	Water Law and Indigenous Rights
WRR	Scientific Council for Advice to Government
WTO	World Trade Organization

1
Setting the Scene[1]

Introduction

As chaotic and disordered as it may appear at first sight, worldwide agriculture is today clearly characterized by three basic and mutually contrasting development trajectories: a strong tendency towards *industrialization*; a widespread, though often hidden, process of *repeasantization*; and, third, an emerging process of *deactivation*, especially in Africa. These three processes each affect, albeit in highly contrasting ways, the nature of agricultural production processes. By doing so, they place a specific imprint upon employment levels, the total amount of produced value, ecology, landscape and biodiversity, and the quantity and quality of food. They interact in many different ways and at several levels, thus contributing to the overwhelming impression of chaos and disorganization that currently seems to characterize world agriculture (Charvet, 1987; Uvin, 1994; Brun, 1996; Weis, 2007).

These development trajectories interlink with a certain segmentation of agriculture, which I argue may be conceptualized as three unequal but interrelated constellations (see Figure 1.1). The first is that of peasant agriculture, which is basically built upon the sustained use of ecological capital and oriented towards defending and improving peasant livelihoods. Multifunctionality is often a major feature. Labour is basically provided by the family (or mobilized within the rural community through relations of reciprocity), and land and the other major means of production are family owned. Production is oriented towards the market as well as towards the reproduction of the farm unit and the family.

In the second constellation an entrepreneurial type of agriculture may be distinguished. It is mainly (though not exclusively) built upon financial and industrial capital (embodied in credit, industrial inputs and technologies), while ongoing expansion, basically through scale enlargement, is a crucial and necessary feature. Production is highly specialized and completely oriented towards markets. Entrepreneurial farmers actively engage in market dependency (especially in markets on the input side of the farm), whereas peasants try to distance their farming practices from such markets through a multitude of often very clever mechanisms. Forms of entrepreneurial farming often arise from state-driven programmes for 'modernization' of agriculture. They entail a partial industrialization of the labour process and many entrepreneurs aim at a further unfolding along this pathway.

Third, there is the constellation composed of large-scale corporate (or capitalist) farming. Once having nearly disappeared, among other things through the many land reform processes that swept the world, it is now re-emerging everywhere under the aegis of the agro-export model. The corporate farming sector comprises a widely extended web of mobile farm enterprises in which the labour force is mainly or even exclusively based on salaried workers. Production is geared towards and organized as a function of profit maximization. This third constellation increasingly conditions major segments of food and agricultural markets, although sharp differences can be noted between different sectors and countries.

It is often thought that the main differences between these three constellations reside in the dimension of scale. Peasant agriculture then would represent the tiny and vulnerable units of production, the relevance of which is only of secondary importance. Opposed to this would be corporate farming: large, strong and important – at least, that is what is generally assumed to be the case. The in-between situation is represented by entrepreneurial farming, moving along the scale dimension from small to larger units. If entrepreneurial farmers are successful, they might, it is argued, join the ranks of corporate farmers – which is precisely what some of them dream of achieving.

There are undoubtedly empirical correlations between size and scale of farming and the different modes of farming. The point is, though, that the *essence* of the difference resides somewhere else (i.e. in *the different ways in which the social and the material are patterned*). Peasants, for instance, create fields and breed cows that differ from those created by entrepreneurs and corporate farmers. Also, the *mode* of construction differs between the three categories. And beyond this, peasants *relate* in a different way to the process of production than do the other two categories, just as they relate in a contrasting way to the outside world. Regardless of size, they constitute themselves as a social category that differs in many respects from those of corporate farmers and entrepreneurs.

As I will show throughout this book, these different modes of patterning deeply affect the magnitude of value added and its redistribution, as well as the nature, quality and sustainability of the production process and the food resulting from it.

Equally important is the time dimension. Normally, it is assumed that the peasantry and peasant farming belong to the past, while entrepreneurial and corporate farming represents the future. Here again, in essence, it is all about *patterning*. Within the peasant mode of production, the past, present and future are linked in a way that sharply contrasts with the social organization of time entailed in entrepreneurial and corporate farming (Mendras, 1970).

Setting the Scene 3

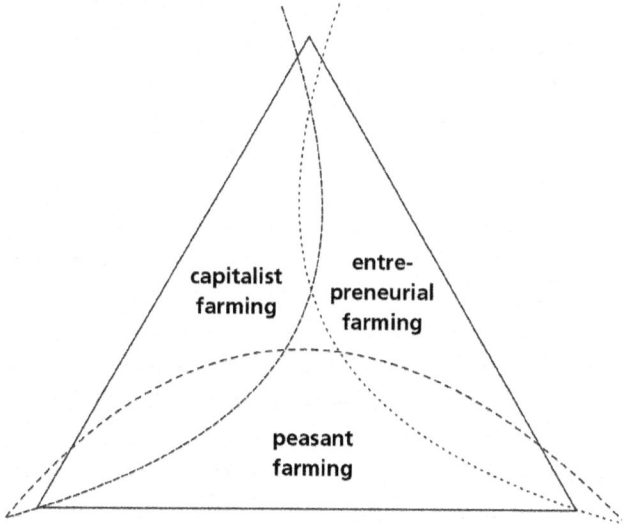

Figure 1.1 *Different but interlinked modes of farming*
Source: Original material for this book

Although the differences between the three constellations are manifold and often quite articulated, there are no clear-cut lines of demarcation. At the interfaces there is considerable overlap and ambiguity, and 'borderlines' are crossed through complex moves both backwards and forwards. Several of these border crossings (e.g. from peasant to entrepreneurial farming and vice versa) will be discussed at some length in this book. The 'outer borders' of the constellation summarized in Figure 1.1 are, likewise, far from being sharp and clear. Peasant farming flows through a range of shades and nuances, frequently summarized as pluriactivity (compare this notion with that of *polybians*[2] discussed by Kearney, 1996; see also Harriss, 1997), to the situation of the landless and the many urban workers who cultivate plots for self-consumption. Industrial entrepreneurs might also invest in agriculture (and vice versa), thus constituting themselves as a kind of 'hybrid' capitalist farmer. Hence, confusion is, it seems, intrinsic to all these borderlines.

The interconnections between the three agrarian constellations and society at large are patterned in many different ways. However, we can distinguish two dominant patterns. One pattern is centred on the construction and reproduction of *short and decentralized circuits* that link the production and consumption of food, and, more generally, farming and regional society. The other highly centralized pattern is constituted by large food processing and trading companies that increasingly operate on a world scale. Throughout the book I refer to this pattern as *Empire*. Empire is understood here as a mode of ordering that tends to become dominant. At the same time, Empire is embodied in a wide range of specific expressions: agribusiness groups, large retailers, state apparatuses, but also in laws, scientific models, technologies, etc. Together these expressions (which I refer to in the plural as *food empires*) compose a regime: 'a grammar or rule set comprised in the coherent complex of scientific knowledge, engineering practices, production process technologies, product characteristics, [enterprise interests, planning and control cycles, financial engineering, patterns of expansion, and] ways of defining problems – all of them embedded in institutions and infrastructures' (Rip and Kemp, 1998; see also Ploeg et al, 2004b).[3] On the one hand, this regime is, indeed, continuously made coherent, while, on the other, it is equally an arena in which internal struggles and contradictions are omnipresent. Authoritative hubs of control mutually contest for hegemony, while specific carriers of Empire as an ordering principle might emerge, become seemingly powerful, then erode or even collapse. Hence, Empire is not only an emergent and internally differentiated phenomenon; it is, above all, the *interweaving* and *mutual strengthening* of a wide range of different elements, relations, interests and patterns. This *interweaving* increasingly relates in a *coercive* way to society: single projects (of individual and collective actors) become aligned, at whatever level, to the grammar entailed in Empire. Indeed, to a degree, Empire is a disembodied mode of ordering: it goes beyond the many sources from which it is emerging; it also goes beyond the many carriers and expressions into which it is currently materializing. These carriers might crack or collapse (later I describe and analyse several cases of this); however, through such episodes, Empire as a mode of ordering might even be strengthened.

I am aware that the representation of Empire as a disembodied whole implies a considerable danger of reification. I also think that there is no semantic solution for such danger: it is only when resistance, struggles and the creation of alternatives are systematically included in the analysis that such a danger of reification can be avoided.

The creation of *disconnections* is a key word for understanding the *modus operandi* of Empire. Through Empire, the production and consumption of food are increasingly disconnected from each other, both in time and in space. Likewise, agricultural production is decontextualized: it is disconnected from the specificities of local ecosystems and regional societies. Currently, Empire is engaged, as it were, in a fierce endeavour to conquer and control increasing parts of food production and consumption on a world scale (although it should not be forgotten that some 85 per cent of food production in the world is channelled through short and decentralized circuits).[4]

There are no simple one-to-one clear-cut relations between these two mutually contrasting patterns of connectivity and the three agrarian constellations. All three constellations interact with and are, in a way, constituted through the different mechanisms that link them to wider society. However, corporate and entrepreneurial farming are mainly linked (as illustrated in Figure 1.2) through large-scale food processing and trading companies to world consumption, while peasant agriculture is basically, though far from exclusively, grounded in short and decentralized circuits that at least escape from *direct* control by capital (though indirect control is, of course, considerable and far reaching).

Industrialization

Corporate farming is the main laboratory and Empire the main driver of the process of industrialization, although parts of the entrepreneurial segment also provide significant contributions. In the first place, then, industrialization represents a definitive disconnection of the production and consumption of

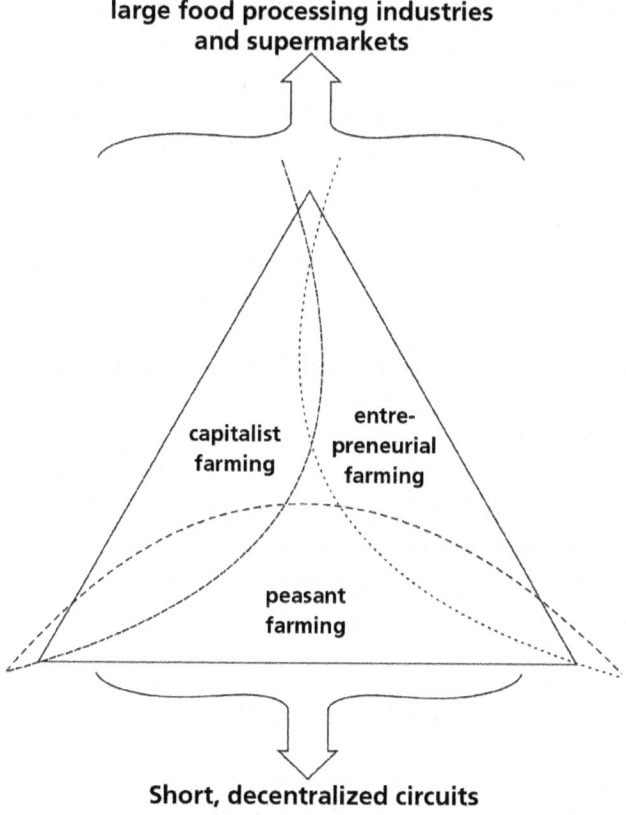

Figure 1.2 *Patterns of connectivity*

Source: Original material for this book

food from the particularities (and boundaries) of time and space. Spaces of production and consumption (understood as specific localities) no longer matter. Nor do the interrelations between the two. In this respect food empires may be said to create 'non-places' (Hardt and Negri, 2000, p343; see also Ritzer, 2004, for a provocative discussion).

Second, the industrialization of agriculture represents an ongoing move away from 'integrity'. This is a triple-layered process of disintegration and recomposition. Agricultural production is 'moved' away from local ecosystems. Industrialization implies, in this respect, the superimposition of artificial growth factors over nature and a consequent marginalization and, in the end, probably a complete elimination of the latter. Beyond that, the once organic unity that characterized the agricultural production process is broken down into isolated elements and tasks that are recombined through complex and centrally controlled divisions of labour, space and time. The well-known 'global chicken' (Bonnano et al, 1994) is, in this respect, a telling metaphor. And, finally, there is the disintegration and recomposition of food products as such. Food is no longer produced and processed – it is engineered. The once existing lines between fields, grain and pasta, or, for that matter, gardens, tomatoes and tomato sauce to be poured over the pasta, are being broken. This has given rise to what we now know as the 'food wars' (Lang and Heasman, 2004).

Third, industrialization coincides with (and is an expression of) an increased and direct 'imperial' control over food production and consumption. The search for elevated levels of profitability, the associated conquest and the imposition of an overarching control become new and dominant features that reshape agricultural production, processing and food consumption on a global scale.

The current process of industrialization of food production and consumption is expressed in and carried forward by a well-defined agenda: globalization, liberalization, a fully fledged distribution of genetically modified organisms (GMOs), and the claim that the world has never had safer food at its disposal than now are key elements of that agenda. It is equally claimed that this same agenda contains promising prospects for poor peasants in the developing world. In fact, the industrialization agenda claims that there is no alternative except further industrialization.

Repeasantization

Throughout the world the process of agricultural industrialization introduces strong downward pressures on local and regional food production systems, whatever their specific nature. A dramatic strengthening of the already existing squeeze on agriculture is one of the most visible consequences: although we see temporary upheavals, off-farm prices are, on the whole, nearly everywhere under pressure. This introduces strong trends towards marginalization and new patterns of dependency, which, in turn, trigger considerable repeasantization – whether in the developing world or in industrialized countries. Repeasantization is, in essence, a modern expression of the *fight for autonomy and survival in a context of deprivation and dependency*. The peasant condition is definitively not static. It represents a flow through time, with upward as well as downward movements. Just as corporate farming is continuously evolving (expanding and simultaneously changing in a qualitative sense – that is, through a further industrialization of the processes of production and labour), so peasant farming is also changing. And one of the many changes is *repeasantization*.

Repeasantization implies a double movement. It entails a quantitative increase in numbers. Through an inflow from outside and/or through a re-conversion of, for instance, entrepreneurial farmers into peasants, the ranks of the latter are enlarged. In addition, it entails a qualitative shift: autonomy is increased, while the logic that governs the organization and development of productive activities is further distanced from the markets.[5] Several of the time- and place-bound mechanisms through which repeasantization occurs are discussed in this book. In the same discussion I will make clear that repeasantization occurs as much in Europe as it does in developing world countries.

Deactivation

Deactivation implies that levels of agricultural production are actively contained or even reduced. In several instances, deactivation translates into an associated sub-process: the resources entailed in agriculture are released (i.e. converted into financial capital and invested in other economic sectors and activities). Equally, the necessary labour may flow, permanently or temporarily, out of agriculture. Deactivation (which is not to be confused with de-peasantization)[6] knows many specific causes, mechanisms and outcomes. A dramatic expression is presented by sub-Saharan Africa. While throughout history, demographic and agricultural growth went together – the former being the driver of the latter – contemporary Africa has already shown for decades an ongoing and dramatic decline in agricultural production per capita. Deactivation translates here directly into widespread de-agrarianization (Bryceson and Jamal, 1997). Hebinck and Monde (2007) and Ontita (2007) provide an empirically grounded critique of the assumptions of de-agrarianization.

So far, deactivation has occurred in Europe only on a minor scale. While Eastern European agriculture was temporarily deactivated (due to the demise of the socialist regime and the transition to a neo-liberal market economy), this was followed by widespread repeasantization and a surge of entrepreneurial and corporate farming (the latter two mostly based on migration from Western Europe). Close by large and expanding cities there is often deactivation: speculation in land becomes more attractive than agricultural production. There is also deactivation imposed by state apparatuses and the European Union. Set-aside programmes, the McSharry reforms (that introduced a deliberate extensification of agricultural production), quota systems, as well as several spatial and environmental programmes all contain or even reduce agricultural production. It is to be expected, however, that in the years to come, deactivation will go far beyond the levels realized so far. Globalization and liberalization (and the associated shifts in the international division of agricultural production) will introduce new forms of deactivation that will no longer depend upon state interventions, but which will be directly triggered by the farmers involved. In Chapter 5 (when discussing the major trends in Italian dairy farming) I offer evidence of such deliberate deactivation. Within entrepreneurial farming, in particular, deactivation might become a 'logical' response. When price levels decrease so much that profitability becomes illusionary, opting out and reorienting invested capital elsewhere become evident expressions of entrepreneurial behaviour. Processes of suburbanization, development of recreational facilities, the creation of 'nature reserves' and new forms of water management will further accelerate this movement.

Interrelations between constellations and processes

It is my impression that, at this moment, the two main developmental processes are industrialization and repeasantization. Deactivation has been, so far, a less prominent process; but it might in the future be triggered and thus also provoke a considerable imprint upon rural areas. The three processes are evidently interlinked. Since industrialization, for instance, proceeds as the takeover of market shares, entrepreneurial economies will enter (slowly or abruptly) into crisis, their reproduction possibilities being reduced through deteriorating terms of trade. Hence, new degrees, forms and spaces for autonomy are sought and constructed. This is how repeasantization is triggered. In order to reduce cost levels, a part of entrepreneurial farming will be re-patterned into more resistant peasant-like forms of production. However, it is equally possible that deteriorating terms of trade will be countered from within the entrepreneurial constellation through a further industrialization and/or through deactivation. Within peasant agriculture itself, further repeasantization might also emerge. The 'peasant condition' is not static. 'Like every social entity, the peasantry exists only as a process (i.e. in its change)' (Shanin, 1971, p16).

There are many other interlinkages between the developmental trajectories mentioned above, of which several are explored in this book. Together they compose a highly complex panorama. We are confronted with the simultaneity of three mutually opposed, but interlinked, transitional processes. Within this panorama, at least one of the three is explicitly searching for hegemony – in this case, the industrialization process rooted in corporate farming and Empire. At the same time, its fragility is omnipresent, although highly camouflaged.

The three transitional processes are located in a complex and changing way in the three constellations outlined earlier (see Figure 1.3). The practice and prospects for further industrialization are clearly located in corporate farming and – to a lesser degree – in entrepreneurial farming. Through industrialization, parts of the entrepreneurial constellation are moving towards and becoming reconstituted as integral parts of the corporate sector.

Deactivation basically stems from and resides in the domain of entrepreneurial farming, although it could be argued that engagement in pluriactivity – a frequent feature of peasant agriculture – also represents a kind of deactivation. Repeasantization, in turn, appears within Figure 1.3 in a manifold form: it occurs through an inflow of, for example, urban people into agriculture as represented by the impressive case of the landless people's movement in Brazil, the *Movimento dos Sem Terra* (MST) (see Long and Roberts, 2005, for a convincing specification of the theoretical significance of this case). It likewise occurs through the less visible creation of new microscopic units in Pakistan, Bangladesh and India. It may also arise as an important reorientation within entrepreneurial farming itself: in order to face the squeeze imposed by falling

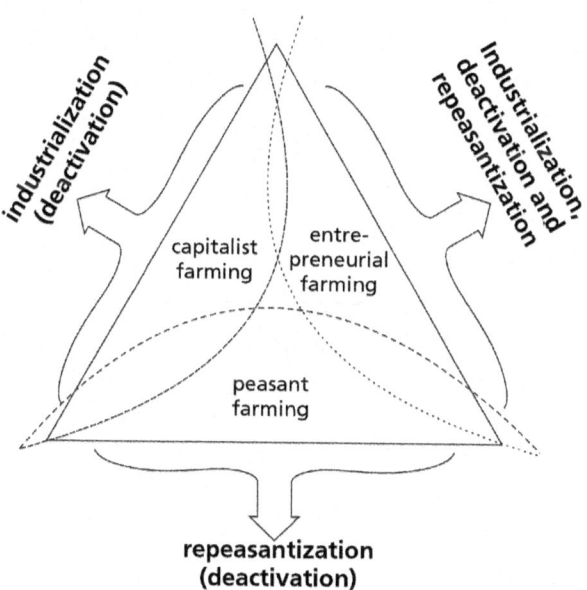

Figure 1.3 *Transitional processes*

Source: Original material for this book

prices and rising costs, such farming increasingly switches over to peasant-like modes of organization. And, finally, repeasantization occurs within the peasant sector itself, which often shows a *further unfolding* of the peasant mode of farming.[7]

These transitional processes also connect up with Empire. Empire triggers and reproduces corporate farming, especially in the current conjuncture. Empire also builds on entrepreneurial farming since it subjects agriculture, wherever located, to an 'external squeeze' that translates, especially in respect to entrepreneurial farming, into an 'internal squeeze' (the details of which I discuss in Chapter 5). Peasant agriculture is also submitted to Empire, albeit partly through other mechanisms, although, at the same time, the peasantry represents resistance to it, sometimes in an overt and massive way, but mostly in hidden, tangible ways of escaping from or even overcoming the pressures. In this respect, (re)assessing short and decentralized circuits that connect producers and consumers independently from Empire frequently play a decisive role.

The coming crisis[8]

Whatever its location in time and space, agriculture always articulates with nature, society and the prospects and interests of those directly involved in farming (see Figure 1.4). If a more or less chronic disarticulation emerges in one of the defined axes, then one is faced with an agrarian crisis.

The 'classical' idea of agrarian crisis centres upon the interrelations between the organization of agricultural production and the interests and

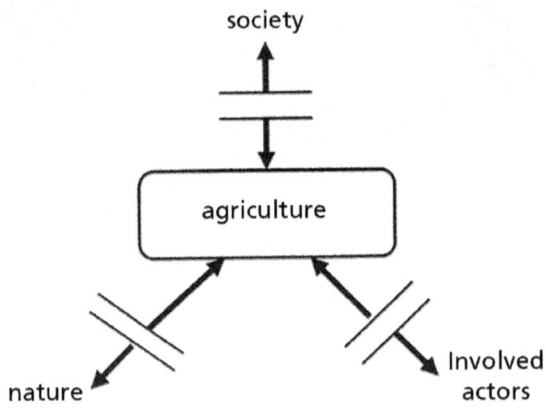

Figure 1.4 *An outline of the coming agrarian crisis*

Source: Adapted from Ploeg (2006a, p259)

prospects of those directly engaged in it.⁹ This is the form of crisis that throughout history has triggered massive peasant struggles and often land reform. However, mankind has also witnessed (especially in recent times) agrarian crises that concern how agricultural and livelihood practices interrelate with nature. When agriculture becomes organized and develops through a systematic destruction of the ecosystems upon which it is based and/or increasingly contaminates the wider environment, an 'agro-environmental crisis' is born. And, finally, there is the relation with society at large in which the quality of food is an important, though not the only, relevant feature. The current range of food scandals (notably BSE, or mad cow disease, and the public outcry following containment of animal diseases such as foot and mouth, avian influenza, swine fever and blue tongue disease) are expressions of the crises emerging on the axis that links agriculture with society at large.

Currently, a crisis is looming that for the first time in history concerns:

- All three axes contained in Figure 1.4: it concerns the quality of food and the security of food delivery; it concerns the sustainability of agricultural production; and it is associated with a far-reaching negation of the emancipatory aspirations of those involved in primary production.
- For the first time, it represents a global crisis whose effects are felt throughout the world.
- Finally, this many-faceted and internationalized agrarian crisis increasingly represents a Gordian knot in the sense that alleviation of one aspect at any one particular moment and place only aggravates the crisis elsewhere at other moments and/or transfers to other dimensions.

Thus, the thesis I present in this book argues that it is the rise of Empire as an ordering principle that increasingly governs the production, processing, distribution and consumption of food, and in so doing contributes to the advance of what seems like an inevitable agrarian crisis. This is also because Empire proceeds as a brutal ecological and socio-economic exploitation, if not degradation of nature, farmers, food and culture. Industrialization implies the destruction of ecological, social and cultural capital. Moreover, the very forms of production and organization that are introduced turn out to be highly fragile and are scarcely adequate in confronting the very conditions intrinsic to globalization and liberalization. Thus, new, immanent contradictions emerge (Friedmann, 2004, 2006).

It is, I think, only through the widespread and possibly renewed repeasantization that this international and multidimensional crisis might be redressed and averted. In Chapter 10 I come back to repeasantization as a way out of the global agrarian crisis.

The methodological basis

Throughout I argue that peasant, entrepreneurial and corporate ways of farming are (interrelated) movements through time. Hence, the methodological grounding of the book consists of *longitudinal* studies. It is especially through such studies that movements over time may be grasped – that is, the study of long-term trends enables one to comprehend the nature, dynamics and impact of different modes of ordering. These longitudinal studies focus, in the first place, on the peasant community of Catacaos in the north of Peru, where during the early 1970s I witnessed the disappearance of corporate agriculture, partly due to the implementation of a state-organized process of land reform, but especially to the impressive struggles undertaken by the Catacaos community itself. Thirty years later (my last long stay in Catacaos was in the second half of 2004), corporate agriculture was again omnipresent, now as an expression of Empire; at the same time, processes of repeasantization had stretched far beyond what one imagined possible. This is precisely what makes longitudinal studies so important, stimulating and difficult: they underline that the many contradictions that characterize everyday life scarcely have easy, unilinear and predictable outcomes. At the same time, the Catacaos case shows how particular contradictions are reproduced over time, resulting in an evolving agenda that urges one to reflect upon the interrelations between past, present and future.

My second longitudinal study focuses on dairy farming in the area where milk is transformed into *Parmigiano-Reggiano*, or, as it is internationally known, Parmesan cheese. During the period of 1979 to 1983, together with a team of colleagues, I studied in detail a sample of cheese- and milk-producing farms in the area. Later, in 2000, I had the opportunity to carry out a restudy of exactly the same farms. On a personal level, this was as heart warming as being back in Catacaos. This revisit, however, also confronted me with considerable perplexity. What we had initially diagnosed as continuously expanding farms (i.e. those characterized as typical entrepreneurial farms) turned out, at the beginning of the 2000s, to be involved in a process of deactivation, while peasant-type farms found themselves far better placed to face and respond to the processes of globalization and liberalization *avant la lettre* that the area was confronted with. This apparent contradiction once again called for a more thorough theorization of what, in the end, the peasant, entrepreneurial and corporate modes of farming really amount to.

I was intellectually shaped in an epoch (i.e. during the 1960s and 1970s) in which the demise of the peasantry was predicted and heralded everywhere and from virtually all theoretical perspectives. I never felt comfortable with this prospect, but did not have, at that time, the elements and tools to really argue against it. Now, more than 30 years later, I understand somewhat better the *mystery of farming*. Mystery is, in this context, an intriguing concept. In the English language 'mystery' refers to both 'the enigma' or 'secret' of farming and to the tasks required. In this sense it is like the Italian word *mestiere*, which equally refers to a job or – to put it more precisely – the capacity to realize a specific work or task in a well-executed fashion. Every job contains its secrets. Doing a job well implies knowledge, insight and experience not available to others – or, at least, better knowledge, superior insight and more extensive experience (MacIntyre, 1981, p175; Keat, 2000).

This same mystery of farming underlies my third longitudinal study, which looks at dairy farming in the Northern Frisian Woodlands of The Netherlands. Due to its particular history, this area has been, and still is, characterized by relatively small farms that operate in a beautiful man-made hedgerow landscape rich in biodiversity. During the 1970s and 1980s, the main expert systems considered that farming here was doomed to disappear. The structure of the landscape (many small to very small plots) and the relatively small-scale nature of most farms seemed to exclude any competitiveness (a concept that became very fashionable from those times onwards). However, farming did not disappear. Many farms closed down or moved to other locations; but, simultaneously, many farms remained and developed further along a highly interesting track that started to unfold from the second half of the 1980s. At farm level, a style of farming 'economically' (Ploeg, 2000) was optimized and, at the level of the area as a whole, a new territorial co-operative was created that turned the maintenance of landscape, biodiversity and the regional ecosystem by farmers into a new, solid pillar that now sustains the economy of both the farm units involved and the region as a whole. Apart from having been born there myself, I also came to know the area through a range of multidisciplinary studies in which I was involved. These studies commenced in the mid 1980s and still continue. In Chapter 7, I detail some of the outcomes.

The availability of these three longitudinal studies allows for a comparative analysis that attempts to grasp regularities that go beyond time- and place-bound specificities.[10] Are there any commonalities in the way in which farming is organized? And, if so, to what do they refer? What responses are emerging vis-à-vis the restructuring of agriculture that follows the current processes of globalization and liberalization? And, again, are there common patterns underlying these new responses and associated practices and trajectories? At the same time as identifying similarities, this comparative approach allows us to specify the uniqueness of every constellation encountered. Thus, step by step, both the general and the specific can be assessed in what otherwise remains, indeed, a confusing 'chaos'.

Contents and organization of the book

Following this introductory chapter, Chapter 2 discusses the 'peasant condition' as an ongoing struggle for autonomy and progress in a world characterized by often harsh dependency relations and (often high levels of) deprivation. To counter dependency and deprivation, autonomy is sought. Such a *condition*, of course, is basic to all simple commodity producers. It also characterizes, for instance, independent producers and artisans in the urban economy.[11] Specific to the peasantry, then, is that autonomy and progress are created through the co-production of man and living nature. Nature – that is, land, animals, plants, water, soil biology and ecological cycles – is used to create and develop a resource base, which is complemented by labour, labour investments (buildings, irrigation works, drainage systems, terraces, etc. – in short: objectified labour), knowledge, networks, access to markets and so forth. Thus, departing from the peasant *condition*, a peasant *mode* of farming can be specified. Other modes of farming evidently also require resources. However, as I will specify (especially in Chapter 5), the way in which resources are created, developed, combined, used and reproduced within the peasant mode of farming is highly distinctive, with sustainability being an important feature. Following Martinez-Alier (2002, pviii), I do not claim 'that poor people [and, more specifically, peasants] are always and everywhere environmentalists, since this is patent nonsense. But I would argue that, in ecological distribution conflicts, the poor are often on the side of resource conservation and a clean environment.'

The struggle for autonomy and progress is, of course, not limited to developing world conditions. European farmers are equally involved in such struggles, although the immediate conditions under which it occurs are often strikingly different, just as outcomes may also differ. Chapter 3 then looks at the repeasantization process that has taken place during the last three decennia in the peasant community of Catacaos in the north of Peru. I show how this process increasingly runs counter to emerging forms of Empire. Chapter 4 focuses on a dramatic expression of Empire in Europe: the Parmalat case.

Dealing with agriculture does not imply, of course, that we are talking about peasants alone. In Chapter 5 I focus on the differences between the peasant and the entrepreneurial modes of farming, using both Italian and Dutch data. Chapter 6 introduces and discusses processes of repeasantization that are currently taking place within Europe. The chapter also presents the results of Italian research on the quality of life in rural areas. This is followed by Chapter 7, which focuses on new forms of creating autonomy at higher levels of aggregation. The example analysed concerns the creation of a territorial co-operative in the north of The Netherlands. It is, as it were, about the creation of a new 'Catacaos' – albeit far from Peru where the original Catacaos is located. Here special consideration is given to newly emerging moral economies (Scott, 1976). Then, in Chapter 8, attention shifts to the 'global cow' – a metaphor that refers to the schemes that state apparatuses build for implementing prescription and control in the agricultural sector. The chapter also discusses the role of science in the elaboration of such schemes. Chapter 9 attempts to knit together the different storylines that characterize Empire as a new mode of ordering. In the tenth and final chapter, I discuss the relevance of the peasant principle vis-à-vis this new imperial framework.

2
What, Then, Is the Peasantry?[1]

Introduction

Science generates both knowledge and ignorance and one of the black holes it has created systematically obscures the ways in which peasants operate within the modern world. Thus, the phenomenon of the peasant has been delegated to remote places hidden in history and the periphery. What science did was to create an image and model of the agricultural entrepreneur – a model that posits the farmer, his practices and the relations in which he is engaged *as they are supposed to be* (Jollivet, 2001; Ploeg, 2003a). This model – realized through extended and far-reaching processes of modernization – represented the opposite of what Shanin (1972) designated the 'awkward' class of peasants. It heralded *'la fin des paysans'* (Mendras, 1967). Silvia Pérez-Vitoria (2005), in her discussion of the relations between modernization and the peasants, signals that *'personne ne voulait les entendre; on était trop ocupés à se modernizer'* ('nobody wanted to understand them; everybody was too busy becoming modern').

The agricultural entrepreneur develops, it is assumed, a farm enterprise that is highly, if not completely, integrated within markets on both the input and output sides. In other words, the degree of commoditization is high. The farm is managed in an entrepreneurial way: it follows the logic of the market. Classical beacons, such as autonomy, self-sufficiency and the demographic cycle contained within the farm family (Chayanov, 1966) are no longer considered relevant. The farm enterprise is completely specialized and through strategic choices is oriented towards the most profitable activities, with other activities externalized. Both long- and short-term objectives centre on the search for, and maximization of, profits. The entrepreneur not only behaves as *Homo economicus*; he (or she) also operates as an 'early adopter' of new technologies compared with others who are 'laggards' (Rogers and Shoemaker, 1971). Hence, it is assumed that agricultural entrepreneurs have at their disposal considerable competitive advantage, which they use to invest in expansion.

There is no point in discussing whether this model is true or not. The crux of the matter is that such a model has been *made* true, albeit to different degrees and with contrasting outcomes during the 1950 to 1990 period when big modernization projects dominated worldwide agriculture. And although the modernization paradigm is now theoretically discredited, it still persists as a central model in policy, albeit often under cover. Consequently, it is generally assumed, especially in those spaces where the modernization project has been successful, that the peasantry has *de facto* disappeared. As both 'modernists' and Marxists see it, they have either been converted into entrepreneurs or into proletarians.

The 'awkward' science[2]

As outlined in Chapter 1, most agrarian constellations today are made up of a confusing and highly diversified mix of different modes of farming, some peasant like, others entailing a completely different logic. At the same time, there is, as yet, no adequate theory to understand and unravel these new agrarian constellations. Together, the highly diversified, empirical constellations and weak theoretical approaches present a confusing and contradictory set of relations. As indicated in Figure 2.1, there is, at the level of empirical reality, a range of expressions that go from the more entrepreneurial to the more peasant-like ways of farming. At the same time, at the level of theory, we have the modernization approach (that focuses on entrepreneurship) and the tradition of peasant studies that hardly provides a place for peasants in the modern world.

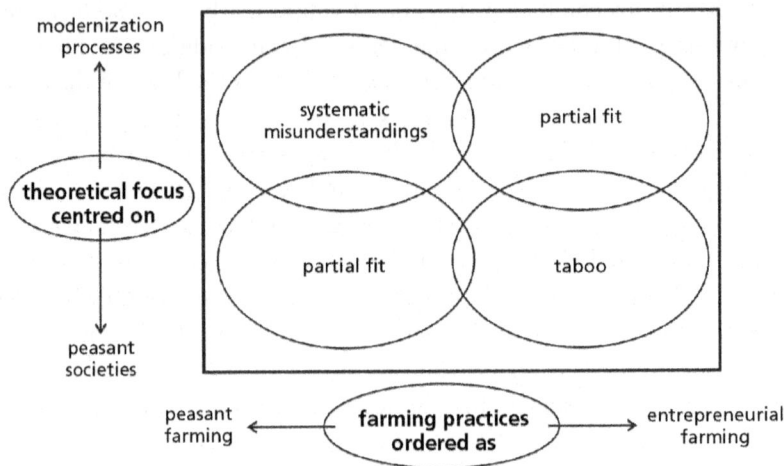

Figure 2.1 *The contours of the theoretical impasse*

Source: Original material for this book

The problems that relate to these remarkable, albeit understandable, relations between practice and theory are manifold. First, it follows that peasant-like ways of farming often exist as *practices without theoretical representation*. This is especially the case in developed countries. Hence, they cannot be properly understood, which normally fuels the conclusion that they do not exist or that they are, at best, some irrelevant anomaly. And even when their existence is recognized (as in developing countries), such peasant realities are perceived as a hindrance to change[3] – a hindrance that can only be removed by reshaping peasants into entrepreneurs (or into fully fledged 'simple commodity producers').[4]

Second, in so far as there have been effective transitions towards entrepreneurial farming, the continuities and commonalities entailed in such processes are mostly missed, especially since entrepreneurial farming as practice and the entrepreneur as social identity are thought to be completely opposite to the peasant and the way in which he or she farms. Thus, misunderstood changes enter the panorama.[5]

Third, wherever entrepreneurial farming deviates from the model as specified in modernization theories, such deviations are seen as temporary imperfections having no theoretical significance whatsoever. Thus, *virtual realities* are created, which are inadequate for policy preparation, nor very helpful for farm development (for further discussion, see Ploeg, 2003a).

Such problems contribute considerably to some of the dramas that the world is currently witnessing. The first problem translates into a denial of the typical way in which peasant agriculture unfolds – that is, as labour-driven intensification. It is a promising trajectory for tackling unemployment, food shortages and poverty; yet, it is absent on political agendas and in the international forums that discuss issues of agriculture and development. The lack of adequate theoretical conceptualization has also impacted – in a tragic way – upon land reform processes, often described, in retrospect, as 'broken promises' (Thiessenhuisen, 1995). Thus, they increasingly became a vehicle for a further, and unnecessary, marginalization of 'those who till the Earth' (Ploeg, 1977, 1998, 2006d).

In turn, the problem of misunderstood changes blinds many of those involved (whether they are scientists, politicians, farmers or farm union leaders). Since these changes (often actively organized as modernization) were, by definition, understood as an *adieu* to the assumed economic irrationality and backwardness of the peasant, *current* patterns of behaviour (individual or collective) can only be understood in terms of 'rational decision-making' – which evidently leads to chains of interrelated misunderstandings and fictions.

Finally, there is the drama relating to newly created virtual realities. Since farming is now widely conceptualized and understood as the expression of entrepreneurial activity, agriculture is consequently perceived as being an economic sector that differs in no significant way from other economic sectors. Hence, agriculture can and must be aligned and governed by markets. Within such virtual realities things cannot but be perceived in this way. Thus, a major danger is introduced, since, as Polanyi (1957) indicates: 'leaving the fate of soil and people to the market would be tantamount to annihilating them'. What started as 'the demise of the peasantry' (I am slightly paraphrasing the title of Gudeman's 1978 book) could thus very well end up as the demise of considerable parts of agriculture as we know it today.

There is evidently no point in discussing whether or not this partly virtual, partly real image of the agricultural entrepreneur has an opposite. I refer to this counterpoint with the concepts of peasant and peasantry. In so doing I am not revoking the peasant of the past, but am explicitly referring to peasants of the 21st century. This raises the question summarized in the title of this chapter: what, then, is the peasantry?[6] And within what theoretical framework should it be elaborated upon?

I am more than aware of the richness, amplitude and reach of the tradition of peasant studies (well-documented and interesting overviews are found in Bernstein and Byres, 2001; Sevilla Guzman, 2006; and, more generally, Buttel, 2001). However, despite its many virtues, I consider the outcomes of this multifaceted tradition as inadequate for fully comprehending *today's* contradictions, potentials and constraints.

The shortcomings of available literature can be summarized in four points. First, it separates the world into two parts and then applies different theories and different concepts to each part (i.e. to the developed centre and the underdeveloped periphery). Thus, highly contrasting images emerge and are reproduced – images of different worlds inhabited by different people.[7] Although it was seldom made explicit, the decisive borderline between the two is that of being 'developed' as against being 'underdeveloped'. Peasants were understood, in the main body of peasant studies, as a 'hindrance to development' (see Byres, 1991), and as an obstacle to industrialization as 'the route-way from backwardness' (Harriss, 1982, in a critical introduction to Byres). Thus, in the underdeveloped parts of the globe, the peasantry dominated (the one even implicitly defining the other): 'peasant populations occupy the *margins* of the modern world economy' (Ellis, 1993, p3; emphasis added). On the other side of the divide that separates the 'precarious from the prosperous' (Ellis, 1993), there could, logically, no longer be any peasants. Thus, different concepts were thought to be required in research, analysis and theory. It goes without saying that the phenomenological manifestations of today's peasantries are manifold and often highly contrasting. This does not exclude, however, that analytically they build upon the same mode of ordering.[8] Throughout this book I try to specify the implied commonalities. I will do this also because I firmly believe (following Hofstee, 1985b) that only when the commonalities are well understood can the relevant dissimilarities be assessed.

A second troubling aspect of the peasant studies tradition is that the peasant's *way of farming* has largely been neglected: the emphasis has simply been on involvement in agriculture as one of the defining elements. That a peasant was involved in agriculture was taken for granted; but *how* peasants were involved, *how* they practised agriculture and whether or not this was *distinctive* vis-à-vis other modes of practising agriculture has hardly been touched upon – leaving aside exceptions such as the rich empirical studies by Comite Interamericano de Desarrollo Agricola (CIDA) realized in Latin America during the 1960s and early 1970s (for a summary, see CIDA, 1973).[9] Hence, the distinctiveness of peasants has principally been sought in terms of unequal power relations and/or their socio-cultural characteristics. Of course, denying the presence of such phenomena is not my intention. However, many questions remain, among them, how inequality in power relations *translates* into a specific ordering of the many activities and relations in which peasants are engaged. Peasants, wherever located, relate to nature in ways that sharply differ from the relations entailed in other modes of farming; likewise, they shape and reshape the processes of agricultural production into realities that contrast significantly with those created by entrepreneurs and capitalist farmers; and, finally, they mould and develop their resources, both the natural and the social, in distinctive ways.[10]

Third, peasant studies have generally been weak in acknowledging *agency*, which evidently is an (unintended) consequence of their epistemological stance. Thus, peasants often figure as 'passive victims'. Shanin (1971) even uses 'the underdog position, [i.e.] the domination of the peasantry by outsiders' as one of the basic facets that define and delimit peasant societies. Hence, the 'subordinated' position of peasants is central to Shanin's (1971, p15) conceptualization: 'Peasants, as a rule, have been kept at arms' length from the social sources of power. Their political subjection interlinks with cultural subordination and economic exploitation through tax, corvee, rent, interest and terms of trade unfavourable to the peasant.' And Wolf (1966, p11) argues that it is 'only when ... the cultivator becomes subject to the demands and sanctions of power-holders outside his social stratum that we can appropriately speak of peasantry'. Evidently, in itself, such a description is not entirely wrong. For example, such elements can easily be encountered in present-day Dutch agriculture. The point is that such views are *incomplete*. They stress only one side of the equation. As Long and Long (1992, pp22–23) observe: 'Agency attributes to the individual actor the capacity to process social experience and *to devise ways of coping with life*' (emphasis added). This point has been reaffirmed in a wide array of studies on farming styles.[11] Recently, Long (2007) has extended this position into a finely tuned conceptual and methodological framework for studying resistance. I will return to this approach in Chapter 10.

Fourth, I would argue that even when peasant studies has paid considerable attention to its immediate expressions (such as the Green revolution, small farmer credit programmes and land reform), it has significantly missed the key point of the enormous modernization wave that has rolled over developing world agriculture, just as it did in Europe and elsewhere. Regardless of its overall degree of success or failure, the modernization project has transmuted into new patterns of politico-economic differentiation. This has occurred as much in the periphery as at the centres of the world economy. Alongside the already well-known peasants, modernization processes created agricultural entrepreneurs and entrepreneurial farming in the agricultural sectors of the developing world just as they did elsewhere. The theoretical implication of this was that classical dualism (peasants versus capitalist farmers) suddenly became inadequate for reflecting theoretically on the situation in the countryside. There are no longer just two delineations that define the peasantry (namely, peasant versus proletarian and peasant versus capitalist farmer). We now need a third – that is, a strategic way of distinguishing the peasant from the agricultural entrepreneur (see also Figure 1.1). If not, no *theoretical* difference may be perceived nor attributed, for example, to a Brazilian *poseiro* family composed of father, three sons and two uncles, owning and working 1500 hectares (ha) of highly mechanized *soya,* and another, probably neighbouring, *sem terra* family of father, mother and three children who work 15ha of poor land with fruits, vegetables and some cows in a settlement on recently occupied land (see, for example, Caballo Norder, 2004, and, for a more general discussion, Schneider, 2006, and Otsuki, 2007).

In order to go beyond these shortcomings, one must develop concepts that meet certain crucial requirements. First, they must embrace both the centre and the periphery. They must be applicable to present-day constellations, as well as historical settings, and *a priori* segmentation must be excluded. Second, they must go beyond the divide created between socio-economic and agronomic approaches. Third, they must be grounded in the recognition that, since the heyday of modernization theories and peasant studies, *agriculture has been materially reshaped* in accordance with the large modernization processes that took place on a world scale during the period of 1950 to 1990. Fourth, the concepts must go beyond the simplicity of black-and-white schemes in order to allow for degrees, nuances, heterogeneity and specificity. And since the peasantry and the peasant way of farming represent, above all, dynamic processes that unfold through time – in many different, sometimes even diametrically opposed, directions – the concepts must facilitate the exploration and analysis of the different outcomes of such processes. This implies that relative differences and degrees will be central to the analysis (Toledo, 1995, translates this into 'degrees of peasantness'). Fifth, the concepts need to be elaborated upon in such a way that comparative analysis becomes feasible. Sixth, they should reflect the multidimensional, multilevel and multi-actor nature of peasant realities (Paz, 1999, 2006a). And, finally, whatever the constellation, the concepts must be based on positive and substantive definitions. Thus, peasants should be defined, as Palerm (1980) argues, according to what they are, not as a negation of what they definitely are not. Likewise, characterizing the peasant as not yet possessing the traits of an entrepreneur or as a disappearing category is clearly deficient.

A comprehensive definition of the peasant condition

I think it is possible and urgent to go beyond these shortcomings; but in order to do so we must attempt a re-theorization of the peasantry in terms of the 'peasant condition' that places the peasantry firmly in its present-day context, while simultaneously acknowledging the agency contained within it – not as an additional attribute, but as the central characteristic.[12] Following a definition of the peasant condition, I will also specify the 'peasant mode of farming', which focuses on the multiple and internally coherent ways in which peasants actively pattern the agricultural process of production. The two concepts are intimately related: the peasant mode of farming is embedded in the peasant condition and originates from it. These two concepts meet the theoretical requirements I formulated above and together allow for a theoretical enrichment of peasant studies while, at the same time, augmenting its usefulness in practice.

Central to the peasant condition, then, is the *struggle for autonomy* that takes place in *a context characterized by dependency relations, marginalization and deprivation*. It aims at and materializes as the *creation and development of a self-controlled and self-managed resource base*, which in turn allows for *those forms of co-production of man and living nature* that *interact with the market, allow for survival and for further prospects* and *feed back into and strengthen the resource base, improve the process of co-production, enlarge autonomy* and, thus, *reduce dependency*. Depending upon the particularities of the prevailing socio-economic conjuncture, both survival and the development of one's own resource base might be *strengthened through engagement in other non-agrarian activities*. Finally, *patterns of cooperation* are present which regulate and strengthen these interrelations. Figure 2.2 (overleaf) illustrates the 'choreography' of these elements of the peasant condition.

I will first briefly discuss the different elements that together constitute this comprehensive definition of the peasant condition, after which I will identify the dynamics that it entails and through which it materializes into different constellations embedded in specific and often contrasting time–space relations.

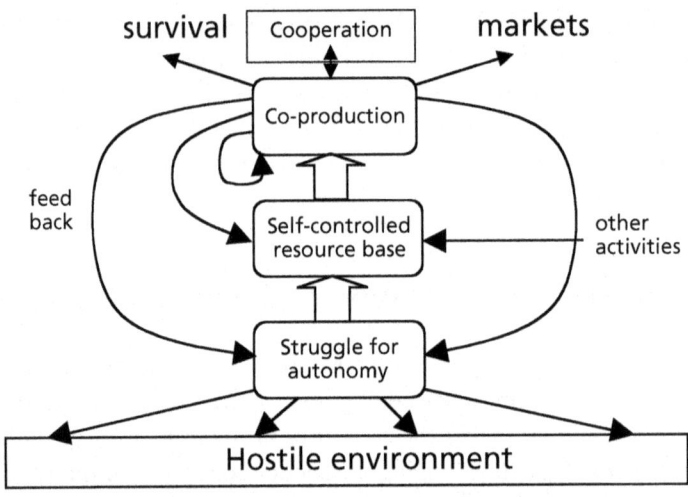

Figure 2.2 *Choreography of the peasant condition*
Source: Original material for this book

Co-production

Co-production, one of the important defining elements of the peasantry, concerns the ongoing interaction and mutual transformation of man and living nature (Toledo, 1981, 1990, 1994; González de Molina and Guzmán Casado, 2006). Both social and natural resources are constantly moulded and remoulded, thus continually generating new levels of co-production (Guzmán Casado et al, 2000; Gerritsen, 2002; Ploeg, 2003a). Agriculture, animal husbandry, horticulture, forestry, hunting and fishing, but also the further transformation of products obtained into other, more refined, ones (e.g. dung and straw into manure, raw milk into cheese, meat into ham), as well as new phenomena such as agro-tourism, all constitute expressions of co-production. Here the interaction between man and *living* nature is decisive – it delineates the rural from the urban (Ploeg, 1997b). The interaction with living nature also shapes the social into specific forms: the artisan nature of the process of production, the centrality of craftsmanship and the predominance of family farms are closely interrelated with co-production and the co-evolution of man and living nature. Theoretically important is the point that it is through co-production that progress is wrought. Thus, endogenous forms of development emerge (Ploeg and Long, 1994; Ploeg and Dijk, 1995). Overviewing and assessing the production process as a whole enables different subtasks to be coordinated and aligned in a more productive way; and in the longer run natural and social resources might be improved. Through a (re)moulding of resources (such as changing a degraded field into a more fertile field), as well as through the construction of new resource combinations, higher levels of productivity might be obtained (see, for a general discussion, Ploeg et al, 2004b; an empirical case is presented in Verhoeven et al, 2003). Thus, an important difference is established between the comprehensive definition of the peasant condition proposed here and earlier ones, since the former systematically integrates the agricultural process of production as *a potentially dynamic praxis*.

Every definition of the peasantry refers to its involvement in agricultural activities. The point is, however, that in most studies agriculture merely functions as decor. It is just there, like wallpaper on the wall. And even when agricultural activities are described at length, it is the *routine* that is stressed (the stable organization of space, the agrarian calendar that governs time, the painstaking tasks that arise out of the labour process and their fixed distribution according to age and gender). The dynamics and malleability of agricultural production are, like those associated with the process of reproduction, hardly explored. Hence, agricultural production as organized by peasants is basically seen as being stagnant, which is frequently translated as the general and intrinsic 'backwardness' of the peasantry as a whole.[13]

The main argument of this book contests this point of view and maintains that peasant agriculture is far from stagnant and intrinsically backward (for both historical and contemporary evidence, see, for example, Richards, 1985; Bieleman, 1992; Osti, 1991; and Wartena, 2006). In and through agricultural production, progress can be wrought. By slowly improving the quality and productivity of the key resources – land, animals, crops, buildings, irrigation infrastructure, knowledge, etc. – and by means of a meticulous fine-tuning of the process of production and a continuous re-patterning of relations with the outside world, peasants strive for and eventually obtain the means of enlarging their autonomy and improving the resource base of their farm units.

Resource base

The construction and maintenance of a self-controlled resource base is another defining element of strategic importance. The creation and growth of such a resource base allows a degree of freedom from economic exchange; it is built, at least partly, *upon an exchange with nature* (Toledo, 1990, 1992). The creation and development of a resource base is a crucial and indispensable condition for co-production.[14] Simultaneously, an evolving resource base is one of the major (and non-commoditized) outcomes of co-production. Through co-production, resources are not only converted into a range of goods and services, they are at the same time reproduced as resources. Hence, co-production always refers to two neatly interwoven processes: one of production and the other of reproduction. Although the required resource base will differ from place to place and from one specific conjuncture in time to another, it generally holds that without an adequate resource base, co-production and its self-propelled development become difficult, if not impossible. Successful co-production feeds back into the needed reproduction (and further development) of the resource base. It also feeds into the survival, standard of living and improved prospects of the farming families involved. *Thus, the development of agriculture and the resource base on which it is grounded coincide with and translate into the emancipation of the peasantry.*

Together, a self-controlled resource base and peasant-managed co-production constitute a specific labour process, which is, for those involved in it, far from being an endless and, in the end, terribly boring repetition of more or less simple tasks and subtasks. The labour process is, in the first place, the locus where man and living nature meet and where different cycles are integrated within a coherent and therefore often aesthetic whole.[15] Since living nature cannot be planned and controlled completely, there will always be surprises – for the better or the worse. The art of mastering these surprises and turning them into novel practices is often a key element of the labour process (Remmers, 1998; Swagemakers, 2002; Wiskerke and Ploeg, 2004; Wolleswinkel et al, 2004). This indicates a second crucial aspect of the labour process: it is the place where learning takes place and where novel ways of doing things are designed. A third and probably decisive aspect is that the agricultural process of production is a social process through which not only end-products (such as milk, potatoes, meat, etc.) are constructed. During the labour process the actors involved also construct, reconstruct and develop a specific finely tuned and well-balanced resource combination – that is, they construct a *style of farming* and link it in a specific way to the outside world.

Farming is all about *actively making* things, resources, relations and symbols. From this there follows a fourth key feature. It is in and through the labour process that progress can be wrought. This implies that *the labour process is a very important arena of social struggle for the peasantry*. Social struggle not only takes place in the streets, in land occupations, factories or big supermarkets (i.e. outside the domain of agricultural production), nor does it necessarily require banners or rousing speeches of whatever colour. Social struggle is also to be seen in the sturdy striving to improve available resources, making small adaptations which together contribute to the creation of better well-being, improved incomes and brighter prospects. Cooperation is often a key mechanism in this respect.

The importance of the labour process as one of the places where progress is constructed explains the tenacity with which peasants defend their autonomy. At whatever level of development, the possibility of designing, controlling, constructing and reconstructing the labour process (and the many resources, cycles, tasks and relations that it entails) is strategic.

Patterning relations with markets in ways that allow for autonomy

A third defining element concerns the specific relations established with the markets. These relations are part of a wider set of relations that connect the peasantry with the surrounding world, and which peasants pattern in such a way as to allow for maximum flexibility, fluidity and autonomy. External relations are ordered in order to allow for contraction or expansion at moments deemed appropriate: becoming entrapped will be avoided as much as is possible. Relations with the outside world – whether with markets, market agencies, political authorities, bandits or priests – are constructed, maintained and changed according to local cultural repertoires (or moral economies) that centre on the issue of distrust and consequently translate into the construction of autonomy (Pérez-Vittoria, 2005, pp132ff and especially p227). Here distrust is clearly a reflection of, as well as a response to, hostile environments. Entering dependency relations, even if this might help to construct something that looks impressive, *macho* and powerful, is deeply distrusted. And related to this is a mistrust of immediacy and its inherent temptations. Immediacy is suspect in nearly all peasant cultures, whether in the developing world or highly developed countries. Immediacy means that things must be taken at face value. In the peasant world, though, questions are continually asked about what underlies immediate appearances. Does a high-yielding cow indicate highly successful breeding strategies and competencies within the farm and keenly maintained networks with other farmers who are providing 'new blood'? Or is it about the costly acquisition of cattle bred elsewhere, high input levels of expensive concentrates, high veterinarian expenses and short longevity? When one sees an impressive-looking large farm, questions to ask are: is it built upon large debts and thus facing huge financial costs, or is it what Dutch farmers refer to as a 'free farm'? Immediacy is treacherous (this evidently translates into the anti-cyclical economic behaviour of peasants). Things cannot be taken at face value. As such they are not transparent. What is often decisive is their location in carefully constructed relations that link the past, present and future[16] – forwards and backwards – and which simultaneously locate the farming unit in specific divisions of labour and space.

All of this points to the centrality of local cultural repertoires and their associated narratives. Whatever phenomenon or image is presented, it will be considered meaningless, if not dangerous, if it is not embedded within its own specific history: where does it come from,[17] what will it lead to? What might be its costs and what its benefits? And what are the conditions under which such benefits might emerge and materialize? And who is going to reap the associated fruits? Such questions reflect deeply rooted distrust (which noticeably helps to contain transaction costs). In the context of modernity, such institutionalized distrust will admittedly be perceived as an anachronism. However, in a world increasingly dominated by Empire and the associated invasion of daily life with all kinds of virtual images, such institutionalized distrust (and its associated stubbornness) is probably not out of place.

At the same time, distrust is combined with trust as far as the local, social and material resources embedded in it are concerned. Nearly all local repertoires stress the virtues entailed in labour and especially the values of the objects and relations created in and through the (self-controlled) labour process. Thus, the *art* of making good manure, breeding good cows and creating a horse of good character are all central elements of local repertoires that refer to farming as a socially constructed process. Connected with this is the importance attached to hard work, dedication, passion and knowledge – as strategic sources of the values created. Even in highly modernized ('secularized') societies as, for instance, The Netherlands, where only economic reasoning (of the neo-classical type) seems permitted, most farmers carefully delineate what they refer to as their 'hobby'. Such a hobby (e.g. on-farm cattle improvement) is the *grey zone* (compare with 'zones of deliberate incertitudes' as Crozier, 1964, puts it) where one's own labour, knowledge, experience and desires are the guiding beacons (where external prescription introduced through dependency relations is not accepted) and where things containing superiority and beauty are constructed that offer satisfaction and pride. The balance of trust and distrust translates into a specific patterning of economic relations in which farming is embedded. As illustrated in Figure 2.3, farming consists, from an analytical point of view, of three interrelated and mutually adapted processes:

1 the *mobilization* of resources;
2 the *conversion* of resources into (end-)products; and
3 the *marketing* and *reuse* of the end-products.

The first and the third processes, and, increasingly, the second, assume and *de facto* imply relations with markets. However, they might be patterned in completely different ways.

Resources might be mobilized through different markets; they may also be produced and reproduced within the farm. This applies to all relevant social and material resources: to cows, feed and fodder, fertilizers, seed, labour, knowledge, working capital, buildings, etc. These might be obtained through market transactions and consequently enter the process of production as

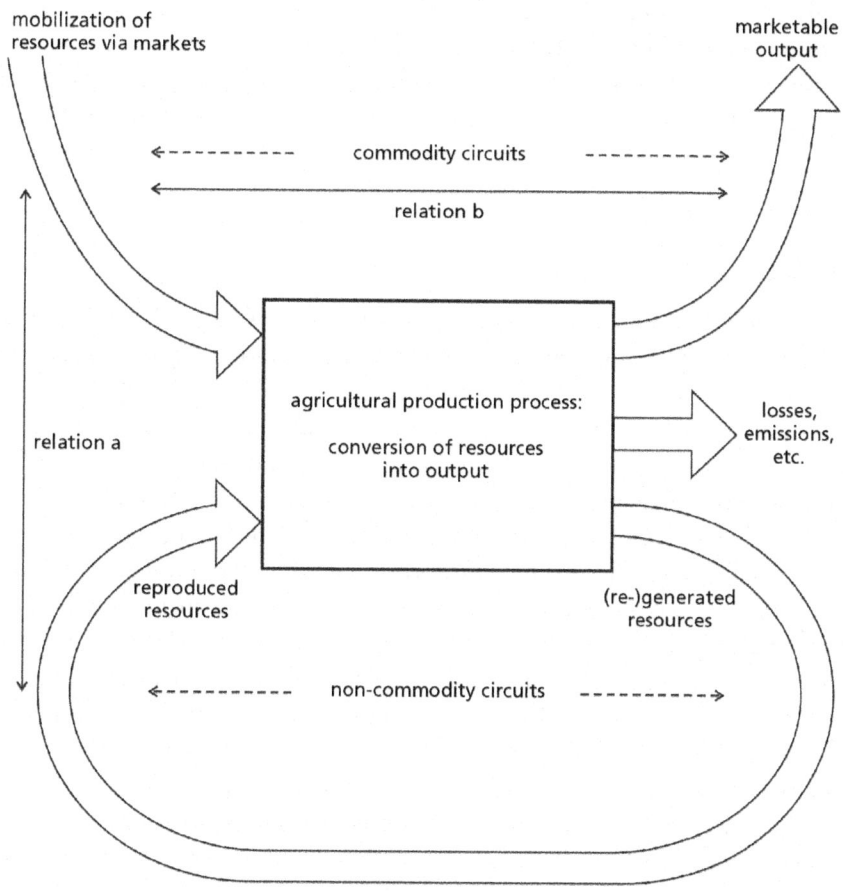

Figure 2.3 *The basic flows entailed in farming*
Source: Ploeg (2003a, p56)

commodities, or be produced and reproduced within the farm unit itself (or obtained through socially regulated exchange). Even resources that cannot be physically produced on the farm (such as heavy machinery) might be obtained by *converting* one's own resources (e.g. savings) into the required ones – in contrast to borrowing in order to buy them. Thus, it is the specific 'social history' (Appadurai, 1986) that makes the difference.

Peasant farming is mainly, though not exclusively, built upon a relatively autonomous flow of resources produced and reproduced within the farm unit itself. From an analytical point of view, relations *a* and *b*, as indicated in Figure 2.3, are strategic. They refer to degrees of 'peasantness'. Through reproduction the solid and finely tuned resource base is further developed. Total production is only partly sold; a part (which will, of course, be variable in time and space) is reused in the farm itself. It flows back into the coming cycles – thus creating a form of self-sufficiency (or self-provisioning) that is not related (as is still assumed in many theories) to the family consumption of food, but to the operation of the farm unit *as a whole*. I will come back to this typical *peasant way of patterning* when I discuss the peasant mode of farming.

Survival

Survival (or the 'pursuit of livelihood', as Pearse, 1975, p42, phrased it) is another element of the comprehensive characterization of the peasantry. It refers to the reproduction and, hopefully, to the improvement of one's existence. Survival is, as it were, the metaphor that refers to the 'symbiotic unity' (Tepicht, 1973) of the unit of production and the unit of consumption entailed in the peasantry. The nature and level of survival evidently depend upon location in time and space – that is, upon relations with the state, capital groups, other social groups, classes and institutions, as well as the internal relations within the peasantry itself. Friesian farmers 'survive' on an income level of 35,000 Euros per year (or, rather, they survive due to the salary obtained by their wives). Small mixed farms in the Gelder Valley (also located in The Netherlands) survive on 4000 Euros (Bruin et al, 1991) and Peruvian potato growers in the Andes survive on a few dollars a day. However, not only the level, but also the notion, changes. In some situations, self-sufficiency implies that production, first and foremost, is for meeting the nutritional needs of the farming family. In other situations, the notion basically refers to the level of obtained income. In yet another set of situations it is the capacity to meet the requirements imposed by banks, agro-industrial groups and the state that becomes decisive for survival. Survival is, in synthesis, a time- and space-bound notion.

It is important to avoid identifying or limiting the concept of survival (and, for that matter, the concept of the peasantry more generally) to that of 'subsistence' (or self-provisioning of food). Such self-provisioning might be one expression of survival, but not necessarily the only one (Salazar, 1996, p27). By the 17th century, Dutch peasants had already stopped producing grain for home consumption. It was imported from the Baltic area far more cheaply than it could ever be produced in The Netherlands. In addition, the supply was stable and trustworthy. Thus, Dutch peasants specialized in other ('high value') crops, especially breeding and dairy production (Hoppenbrouwers and Zanden, 2001). Peasants constantly adapt to particular conjunctures and so the specificities of survival are equally adapted to these – without implying any *basic shift* in peasant conditions as such.

Further strengthening of the resource base

Co-production not only feeds into survival, it also strengthens the resource base. This can take many forms. It might occur as an extension of the resource base, but it often materializes as a qualitative improvement of the available resources and/or as a redefinition of its composition. The quality of land, cattle, plant varieties, labour, irrigation systems, buildings, instruments, etc. are improved, thus allowing for higher productive results. The relations between objects of labour, instruments and labour force might be reshuffled, as well, by combining, for example, the same quantity of material resources with more labour inputs, thus spurring a process of intensification. In practice, the extension and consolidation of the resource base is also regarded as the creation of a patrimony – a *patrimonialização*, as Portela and Caldas (2003) described it for the peasant area of Tras-os-Montes in northern Portugal. It is also associated with pride (Lanner, 1996). Strengthening the resource base often implies the use of extensive social networks through which promising genetic material circulates (Badstue, 2006), or collective action such as fighting landlords for control over water. Evidently, strengthening the resource base is not only about resources as such, but also about the relations and networks that govern their mobilization, use and valorization (Schneider, 2006).

Reducing dependency

In so far as the definition of the peasant condition refers to a situation of dependency and deprivation, it points to the general tendency (especially in current globalizing economies) of unequal and worsening terms of trade. This occurs through decreasing prices, deteriorating conditions of sale, rising costs, taxation, (partial) expropriation, diminished access to essential goods and services, rising costs of living, and the imposition of regulatory schemes that increase costs, diminish the efficiency of production and/or close specific ways forward. Due to the mechanics of the general process of capital accumulation, dependency relations and associated levels of deprivation are constantly (re)introduced into the peasant condition.[18] This is not immanent to that condition; it is due to its embedding in globalizing capitalist economies (and, for that matter, in forms of state socialism as well). Confronting dependency and deprivation is thus not a one-step process, nor situated only at the beginning of the journey through time called farm development. It is repeated endlessly.

A reduction in dependency might be realized through the survival and strengthening of the resource base (here the cyclical and self-sustaining nature of the peasant condition comes to the fore). Rural livelihoods can be improved, both in the short and in the long run, through ongoing and renewed endeavours to reduce dependency. It is important here to stress that, with some important exceptions, the reduction of dependency does not refer to the politico-economic context as such – but, above all, to the *interrelations* between the unit of consumption and production and its context. It refers, that is, to the question of how, by whom, with which means, and through which encounters and contradictions such interrelations are *patterned*.

Striving for autonomy

Thus emerges the common denominator of the defining elements presented so far. The peasantry basically represents a *constant striving for autonomy* or pursuance of, as Slicher van Bath (1948, 1978) phrases it, 'farmers' freedom'. As he made clear, such freedom entails two sets of relations: one that secures at least some relative freedom *from* harsh relations of exploitation and submission; and the other (evidently linked to, and conditioned by, the first) freedom *to* act in such a way that farming is aligned with the interests and prospects of the involved producers.[19] While dependency relations are located within the social formation as such, the search for and construction of autonomy again focuses on the *interrelations* between the farm and its context (see also Robertson, 1912). Further on in this text I probe deeper into this issue. At this introductory level it is important, though, to note that we are dealing here with *degrees of 'systemness'* (Gouldner, 1978) that run from high levels of system integration and dependency via all kinds of in-between situations towards elevated levels of relative autonomy. Such differences partly concern the possibility of creating *room for manoeuvre* (Long, 1985) at micro and meso levels.

The struggle for autonomy takes many, often interrelated, forms. It may occur through classic 'peasant wars' (Wolf, 1969; Paige, 1975) or with the less visible 'weapons of the weak' (Scott, 1985). And more often, almost continuously, it passes through the fields, barn yards and cowsheds, through the many decisions over cattle breeding, seed selection, irrigation and labour input. The struggle for autonomy also articulates at higher levels of aggregation (see, for example, Haar, 2001, and Sandt, 2007).

Finally, I wish to add that autonomy as discussed here is not to be interpreted as a negative category, as a 'state of not being conditioned by anybody'. Instead, I refer to relative autonomy – the room for manoeuvre as defined by Long (1985) that consists as a constellation in which responsibility and agency are manifested. The relevance of this specification is highlighted when I discuss Empire.

Pluriactivity

Peasants engage more often than not in pluriactivity – not only in the periphery (Ellis, 2000a, 2000b; Schneider, 2003), but also at the centre (Gorgoni,

1980; Bryden et al, 1992; Wilson et al, 2002). They mostly do so to supplement their income, but also to obtain funds that allow them to invest in farming, to buy diesel, irrigation pumps, seeds, fertilizer, oxen, a tractor and/or to feed the family. Through engaging in pluriactivity, dependence upon banking circuits and moneylenders can be avoided. In a superficial analysis it might seem that in this way one pattern of dependency is simply replaced by another. There is, however, a strategic difference. When seeds, fertilizer, etc. have been paid for with money earned elsewhere, they have, indeed, 'been paid for'. They are bought as commodities, but then they enter the farm production process as use values. They are no longer to be strictly assessed in terms of exchange value. The particular social history of these resources gives the peasant the *freedom to* do with them what he thinks best (he might lend them to a neighbour, or sell them again in order to pay a hospital bill for his wife; I am using this second example on purpose since such conduct is considered 'delinquent' within credit schemes). If, however, such products or services are bought on credit, they are to be repaid with interest from the results to be generated in the *coming* cycle of production, which often implies a re-patterning of the process of production (e.g. to avoid risks; see Ploeg, 1990a, Chapter 3). And if bad weather should lead to the loss of the harvest, the peasant will probably lose his land.

Several observers (compare Kearney, 1996; but also Bryceson and Jamal, 1997; Bryceson et al, 2000) have systematically misinterpreted the phenomena of pluriactivity and migrant labour.[20] They analyse it as just another (and probably definitive) stage in the disappearance of the peasantry. Yet, if they were to go beyond the immediacies of time and place, they would see that migrants return home with earnings precisely in order to invest in a renewed and strengthened farm. Such processes explain, for instance, the current blossoming of farming in the Tras-os-Montes area in northern Portugal and the strong agricultural boom in the South of Poland.

In The Netherlands, some 70 to 75 per cent of all farming families are engaged in pluriactivity (Vries, 1995). Either the husband or wife (or both) earn a considerable part of the farm family income in places located outside the farm. On professional (i.e. 'full-time') dairy farms some 30 per cent of available income is derived from pluriactivity (Ploeg, 2003a). On arable farms this reaches more than 50 per cent (Wiskerke, 1997). Generally speaking, family income levels are higher on pluriactive farms than on so-called full-time farms. An Irish enquiry into pluriactivity concluded that it is not an expression of poverty; it is, instead, associated with well-being (Kinsella, et al, 2000). This does not rule out, of course, that in other social circumstances pluriactivity may acquire a drastically different significance (see, for example, Hebinck and Averbeke, 2007).

Patterns of cooperation

Facing a hostile environment nearly always requires forms of cooperation (a 'land group' as Pearse, 1975, would say).[21] A harsh (and/or complex) ecological environment is better faced through cooperation (Schejtman, 1980), which often materializes in peasant-managed irrigation systems and/or in socially regulated exchange patterns. Adverse politico-economic circumstances equally require adapted forms of cooperation: in this way, mutual arrangements can function as the 'safety belt' (Tepicht, 1973) of the peasantry. Improving co-production likewise triggers many forms of cooperation, from the exchange of potato seedlings in the Andes (Brush et al, 1981) to the 'study clubs' of Dutch farmers (Leeuwis, 1993). In fact the world provides a bewildering variety of institutions that order and regulate cooperation within peasant agriculture. These range from the *comunidades campesinas* of Bolivia, Peru, Ecuador and parts of Chile, the former *mir* of the Russian countryside, the *zanjeras* of the Philippines and the *baldíos* of northern Portugal, to the newly emerging *territorial co-operatives* in The Netherlands and the *Landschaftspflegeverbände* in Germany. It is important to note that in all of these organizational expressions of the institutionalized need for cooperation there is almost always a well cared for balance between the individual and the collective. Cooperation does not imply, within peasant realities, a suppression of the first part of the implied equation. Instead, it is through cooperation that individual interests and prospects are defended. Let me also underline that through cooperation the struggle for autonomy extends beyond the level of the individual farm unit. Autonomy is often constructed at higher levels of aggregation, as was the case when the first co-operatives were formed at the end of the 19th century in Europe. It also occurred in many impressive episodes of peasant struggle in Latin America during the 20th century. And, if I am not mistaken, it is occurring anew at the beginning of the 21st century again – literally before our very eyes. I will return to discuss one of these episodes in Chapter 7.

Synthesis

Taken together, the elements discussed comprise, I believe, a comprehensive definition of the peasant condition. It is a definition that supersedes the limitations of previous representations of the peasantry. It also allows for a clear demarcation (at least at a conceptual level) between the peasant and other conditions. In particular, it permits what has so far been lacking – that is, a comparative analysis of the peasantry that is not limited to *a priori* delineations that locate the peasantry in the past and/or in the periphery, while neglecting its presence at the heart of the current global system. Furthermore, the definition makes it possible to slot elements such as peasant struggles, agency and culture into the analysis. Above all it allows, as demonstrated later, for an analysis of the *dynamics* of the peasantry and rural and agrarian development processes.

On commonalities, differentiation and change

As stressed many times already, the peasant mode of farming cannot be understood in isolation from the social context in which it emerges and within which it is continuously reproduced. The concept of peasant condition refers precisely to the axis between the peasantry and its social context. The presence of dependency relations and the implied insecurity, marginality and lack of prospects make the peasant mode of farming a necessary *institution*. It is an institution that offers at least some autonomy and possibility for progress. Like every institution it can materialize in a wide array of contrasting expressions that vary from the poor and humiliated *patasucias* in Colombia to the apparently prosperous *boeren* of The Netherlands. Whatever the immediate differences, all these expressions are linked through one and the same substantial rationality, whose progress follows the lines specified in Figure 2.2 where enlarging both autonomy and self-control over resources are decisive. In order to achieve this, labour is central. This constitutes the core of the peasantry. It places labour centre stage, linking it with self-controlled and partly self-shaped resources and with the notion of getting ahead. The specificity of this core becomes clear when compared with other modes of crafting ways forward. Getting ahead is understood within the peasant condition to be the result of one's own labour.

The peasant condition represents a flow through time. It is, at least potentially, a dynamic process that may unfold, depending upon the social formation in which it is embedded, in different directions, with different rhythms and through distinct mechanisms. The process may likewise be blocked: then stagnation or regression occurs, again through a variety of specific time- and place-bound forms. Being essentially a process, it becomes – from an analytical point of view – also possible to discuss the peasant condition in terms of *de-peasantization* and *repeasantization*. The latter concept implies that the peasant grammar is further articulated, in a more coherent and far-reaching way, while it materializes, *in practice*, in stronger, more convincing and more self-sustaining constellations. De-peasantization refers to the opposite tendency: to a weakening, erosion or even disappearance of peasant practices and associated rationality. Both de- and repeasantization can be introduced from the outside or triggered from within. Examples of both are discussed in later chapters.

The struggle for autonomy and the associated construction of a resource base is evidently not limited to situations in which the peasantry has to constitute itself for the first time (i.e. to situations of emigration, land settlement, land invasion and/or the extension of the agricultural frontier). Once a resource base has been constructed, it has to be defended – precisely because the peasant mode of farming continuously articulates with a threatening environment. It is not difficult to lose a farm once it has been built. 'Keeping the name on the farm' (Kimball and Arensberg, 1965) is far from easy (see also Haan, 1993). This applies to the developing world as much as it does to developed countries. It also applies to small as well as large farms. Failure is potentially everywhere. Thus, maintaining – that is, *actively reconstructing* – autonomy becomes a central and universal feature of the peasantry. Continuity is by no means assured: it has to be repeatedly *created* and *recreated*. There is no security whatsoever offered by others, and former successes are no guarantee for the future.

In a poetic and convincing account of Friesian emigrants who tried to escape the poverty and hopelessness at home by settling elsewhere to realize their dream of becoming farmers, Hylke Speerstra (1999) evokes the expression frequently used by the men and women involved – s*konken under it gat krije* – which roughly translates as 'to get feet under your bottom': in other words, to get off your backside and onto your feet, a telling metaphor, indeed, for 'agency'. The expression refers to the tough 'bodily' struggle to build a self-owned and self-controlled resource base. It also conveys the fact that such a resource base gives agency: it renders the possibility of moving forward. Using the same metaphor, it can be argued that, once obtained, it is crucial to *maintain* 'feet under your bottom'. Autonomy and a resource base can easily be dissipated as these same accounts of the emigrants graphically show. It is not difficult to break a leg, especially in a hostile world.

Differentiation and degrees of peasantness

It is important to note that there is no clear-cut demarcation to distinguish in a definitive black-or-white way the peasant from the agricultural entrepreneur, nor are there any clearly cut frontier lines that separate the peasantry from the non-agricultural population. In ideal-typical terms, there are clear and fundamental differences; but in real-life situations there are – alongside clear empirical expressions of these ideal types – extended grey zones that link such expressions and at the same time demonstrate the gradual nature of these linkages. In these grey zones one encounters degrees of peasantness that are far from being theoretically irrelevant. Indeed, they characterize arenas in which, over time, important fluctuations occur with respect to de-peasantization and repeasantization. Hence, it is important to capture, empirically and theoretically, the significance of these changing shades of grey and the associated and sometimes chameleon-like changes that occur (see Laurent and Remy, 1998, for a rich description and theoretical elaboration of the associated rural diversity). This requires systematic longitudinal studies.

Figure 2.4 identifies three important interfaces, each being an arena in which decisive flows are likely to take place.

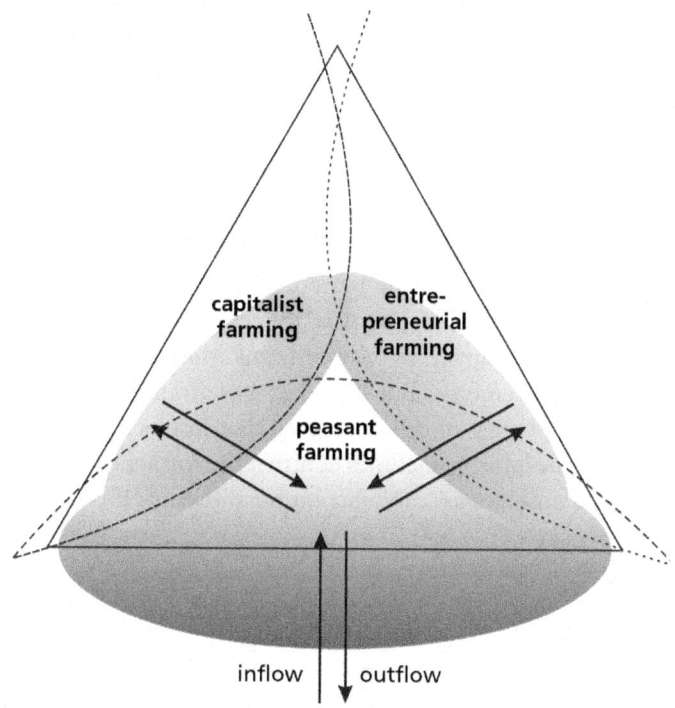

Figure 2.4 *Border zones, degrees and movements*
Source: Original material for this book

At the first interface – the grey zone that both links and distinguishes the peasant and non-peasant – two opposing flows might be identified. Alongside an outflow (e.g. processes of de-agrarianization as described by Bryceson and Jamal, 1997, and Hebinck and Averbeke, 2007), there might also be an inflow. The inflow is made up of non-peasants trying to become peasants. In a study of small farms in The Netherlands, Bock and Rooij (2000) found that a considerable number of these small farms were actually the result of non-farming people (e.g. teachers, policemen, truck drivers and carpenters) investing in the establishment of a farm in order to become peasants. In Chapter 3, I provide a Latin American example. One might even argue that the emergence of urban agriculture in many parts of the world signals the emergence of new numbers of (part-time) peasants and a simultaneous spatial shift of the peasantry from the countryside towards the big metropolises of the world (Veenhuizen, 2006). A second important arena is located at the intersection of entrepreneurial and peasant types of farming (Llambi, 1988). Peasants may constitute themselves as entrepreneurs (e.g. by entering into a widening web of commodity relations as illustrated for Zimbabwe by Terry Ranger, 1985, who talks of 'self commoditization'); but the opposite trajectory is also possible and, in this case, one might talk of a process of repeasantization. Both processes will pass through many in-between situations, thus enlarging the many shades of grey that together characterize this intersection.

Third, there is the complex border zone between corporate (or capitalist) farming and peasant farming, which was once the exclusive focus of studies that centred on 'dualism' in agriculture (Boeke, 1947; Benedictis and Cosentino, 1979). Capitalist farms have disappeared and been reconstituted throughout history, especially during periods of prolonged agrarian crises (as the 1880 and 1930 crises) when it was the capitalist farms that crashed (Zanden, 1985). Some continued as peasant farms while others provided space for new peasant units to emerge. But the other way around is equally possible: following an internal process of differentiation that results in 'poor', 'medium' and 'rich peasants', the latter sometimes contract the former as wage labourers, thus constituting themselves as 'capitalist farmers' (Lenin, 1964).

Dissimilarities

An important aspect of the definition of the peasant condition as presented is that it allows us to assess – in a theoretically grounded way – the many dissimilarities that exist worldwide among the peasantry. All of the elements that have been used to define the peasantry allow for variability – and highlight contrasting empirical situations. Processes of agricultural production reveal differing degrees of co-production: in particular time–space locations agricultural production becomes decidedly artificial (i.e. based on artificial growth factors), while in others it is based mainly, if not exclusively, on ecological capital (on living nature). The resource base might be extensive or limited; it might be controlled by those directly involved or subjected to external prescription and control. Linkages with the markets will likewise vary. The same applies to the concept of survival, where the level and social definition of it will differ considerably from one place to another and from one epoch to another. Reducing dependency will occur everywhere, although the form it takes will differ. In the Philippines, it may emerge through the construction of a *zanjera*. And in The Netherlands, it might emerge as selling part of the land (and/or the quota) to accelerate repayment of loans in order to reduce financial burdens in the next cycle, which, in turn, allows for a less intensive, low external-input type of farming that might render a higher income than in the previous situation. And so on and so forth.

In short, along all variables entailed in this comprehensive definition, significant differences are possible. This implies, in the first place, that at the empirical level, every time- and space-bound expression of the peasantry will represent *specificity*: particular features that reflect the society in which it is embedded and the history upon which it is built. The second implication is that by actively *moving* along one, or several, or all indicated variables the peasantry can constitute itself as being *more (or less) peasant like than was previously the case.*[22]

Peasants of the centre

Worldwide, peasants face dependency and deprivation and the implied danger of further marginalization. Albeit at a different level, European farmers face – as much as do African, Asian and Latin American peasants – the threats implied by the squeeze on agriculture (i.e. stagnating output prices and increasing costs). They likewise suffer from a range of old and new dependency relations, among them the newly emerging regulatory schemes that prescribe the most miniscule details of the labour and production process. As Mendras (1976, p212) argued during the 1970s: 'Today, industrial society is increasingly opposing and condemning the peasantry, because [industrial society] cannot tolerate that people rebel against its rationality.'

I have already referred to the huge, if not abysmal, differences in average income levels realized in the agricultural sectors of different countries such as Peru, Italy and The Netherlands. However, high average levels do not exclude the existence of deprivation. According to a study examining poverty in agriculture, some 40 per cent of Dutch farming families derive less than a minimum income from farming. And even when additional incomes from pluriactivity are added, more than 20 per cent remain under the legally defined minimum (Hoog and Vinkers, 2000). In Italy similar data apply (MPAF, 2003). Poverty, and especially the threat of it, is everywhere. It is not limited to developing countries; it is, as a socially (and legally) defined condition, equally present in the centre.

Both in the developing world and in the West, farming *articulates* with markets. Frank Ellis (1993, p4) tried to specify a theoretically significant line of demarcation between European and developing world agriculture in terms of the markets by arguing that 'peasants are only *partially* integrated into *incomplete* markets', whereas 'their nearest relation, the commercial farm ... is wholly integrated into fully working markets'. This reasoning is helpful for detecting the commonalities between the North and the South, for one can argue that 'competitive, non-distorted markets' not only do not exist in developing world countries, but neither do they in Europe (or in Australia, New Zealand, the US, Canada, South Africa and parts of Brazil, for that matter). In the developed world, agricultural and food markets are not governed by an 'invisible hand' that emerges out of the encounter of anonymous supply and demand forces, but (if not especially) by political interventions and regulations, as well as by the strategic operations of agribusiness groups (see, for Latin America, Guzman-Flores, 1995; for Europe, Benvenuti, 1982; see also McMichael, 1994). And the more that policy interventions are reduced (in the aftermath of the WTO negotiations), the more powerful the ordering of the main food empires will become. This is one argument. The second concerns the issue of 'full' or 'partial' integration into markets. Yes, developing world peasants are mostly only *partially* integrated within markets – *but so are European farmers*. When degrees of commoditization are carefully scrutinized and compared (see Table 2.1), it turns out that developing world peasants are probably even more 'fully integrated' than their European counterparts and that this high degree of 'integration' (or high degree of market dependency) is their main *problem*. Thus, perhaps European peasants are far more peasant than many farmers in the developing world and this explains why they are somewhat better off.

Table 2.1 crosses, as it were, land frontiers. It compares different indicators of market dependency for The Netherlands, Italy and Peru. The data[23] show that European agriculture is, generally speaking, less entrapped in dependency relations and, consequently, less commoditized than Peruvian agriculture. If we take Peru as being indicative of peripheral agricultural systems on a world scale, one may conclude that, in general, agricultural systems of the periphery are more dependent, more commoditized, more based on a 'complete circulation of commodities' than systems of the centre.

Table 2.1 *Different degrees of market dependency in The Netherlands, Italy and Peru (1983) (all figures in percentage)*[24]

Dependency upon the market for the following	The Netherlands: dairy farming	Emilia Romagna, Italy: plains, dairy farming	Emilia Romagna, Italy: mountains, dairy farming	Campania: mixed farming	Peru: coast, co-operative farming	Peru: mountains, potato production, peasants
Labour	6.6	9.1	0.1	13	100	25
Land	NA	28.7	20.2	8	100	21
Short-term loans	1.9	4.6	1.9	12.1	65	27
Medium- and long-term loans	17.8	13.5	5.8	11.1	50	0
Machine services	20.5	30.7	10	14	70	60
Genetic material	13.7	7.2	7.6	8	65	43
Main inputs	NA	43.8	37.8	26.3	85	35
Composite index	NA	26	15	NA	NA	NA

Note: NA = not available.

Source: Ploeg (1990a, p275)

None of this implies, of course, that *all* European farmers should be perceived as peasants: on the contrary. The heterogeneity detected at world level is repeated within Europe – and within every country. Even within relatively small areas, characterized by one and the same set of ecological, economic and institutional conditions, there will be considerable heterogeneity, as illustrated in Table 2.2.

Table 2.2 *The variability in interrelations between dairy farms and the markets (The Netherlands, 1990; n = 300)*

	Average	Standard deviation	Minimum value	Maximum value
Capital market				
Debts per farm (in Dutch guilders)	817,200	603,600	77,270	3,989,000
Debts per labour unit (in Dutch guilders)	462,500	282,500	33,600	1,662,000
Debts per 1000kg of milk (in Dutch guilders)	1540	900	140	6690
Labour market				
Salaried labour as percentage of total labour	10%	16%	0%	70%
Machine services per hectare (in Dutch guilders)	371	243	12	1410
Input markets				
Industrial feed per 1000kg of milk (in Dutch guilders)	104	24	45	166
Total expenses for feed and fodder per cow (in Dutch guilders)	900	249	217	1833
Total feed and fodder expenses per 1000kg of milk (in Dutch guilders)	133	34	43	255
Bought cattle per annum	10,860	22,900	0	197,300
Synthetical index				
Total monetary costs as percentage of gross value of production (GVP)	48%	8%	33%	75%
Total monetary costs + 7% interest over debts as percentage of GVP	60%	10%	35%	95%

Source: Original material using data from Landbouw Economisch Instituut (LEI); see also Ploeg et al (1996, p37)

Table 2.2 concerns the interrelations that have been created between Dutch dairy farms and the most important markets for production factors and non-factor inputs. It shows that dependency upon, for example, the capital market varies significantly. Whereas in some farms total debts per cow (assuming a yearly production per cow of 8000 litres) reached a level of 5100 Dutch guilders (now some 2320 Euros; calculated as mean minus standard deviation, M – s) per cow, in neighbouring farms this could be four times higher: 19,520 Dutch guilders (equal to 8845 Euros; M + s). Due to interest payments such differences in dependency are associated with considerable differences in income level (other conditions being equal), which implies that farmers need to structure the process of production in a completely different way.

What Table 2.2 shows, in a synthetic way, is that a *part* of Dutch dairy farming is based on a relatively autonomous, self-controlled flow of resources, while another part is highly dependent upon external markets. The former refers to farmers who, at least in this respect, structure their relations with markets in a peasant-like way, while the latter clearly refers to an entrepreneurial ordering of the same relations.

Harriss (1982, p22) is right in asserting that 'the process of commoditization ... or the linking up of rural household producers with capitalist production in various ways ... is perhaps the dominant process of change in contemporary agrarian societies'. At the same time, however, processes of commoditization are far from being uni-linear: they unfold in different directions (Marsden, 1991) and can go forwards as well as backwards, are contested, actively accelerated and/or slowed down. Processes of commoditization encompass many arenas (see Long et al, 1986) in which different actors, having different interests and prospects, take different positions. Sometimes they align, at other times they engage in tough and enduring struggles. Consequently, processes of commoditization and their outcomes are highly differentiated both within and between countries.

From peasant condition to the peasant mode of farming

The comprehensive definition of the peasantry discussed here is not only multidimensional, it is also multilevel. It concerns the location of the peasantry in society as a whole by stressing the struggle for autonomy in order to counter dependency, deprivation and marginalization. Simultaneously, the *peasant condition* flows into and embraces a specification of the *peasant mode of farming*. The two concepts are located at different levels; but I strongly believe that the one cannot be understood without the other. The specific location of the peasantry within wider society has important implications for the way in which peasants farm – that is, the peasant condition translates into a distinctive ordering of the agricultural processes of production and reproduction.

The first important feature is that the peasant way of farming is geared to *the production and growth of as much value added as possible*. This focus on the creation and enlargement of value added evidently mirrors the peasant condition: hostile environments are dealt with by generating independent production of income by using basically, though not exclusively, self-created and self-managed resources. The focus on the production of value added clearly distinguishes the peasant mode of production from other modes. The entrepreneurial mode is oriented as much towards the *takeover* of the resources of others as it is towards the production of value added with the resources available (see Chapter 5 for a further discussion of this strategic difference). The capitalist mode of production centres upon the production of *profits* (surplus value) even if this implies a reduction in the total value added. And Empire, the new mode of ordering, produces nothing itself – it is basically oriented towards *draining* off the value added produced by others.

Second, within the peasant mode of farming the resource base available per unit of production and consumption is nearly always *limited* (Janvry, 2000, pp9–11). Although relative well-being might be achieved, the notion of plenitude is definitely at odds with the life worlds of peasants, especially since the threat of losing parts of the resource base is always present. This is not only due to its origins, but also to the intergenerational reproduction that mostly implies a distribution between several children and, consequently, a reduction of the resources available per unit of production. Usurpation of land by others, theft of water, and exclusion and major hindrances in the access to important services will have similar effects. An expansion of the resource base through the establishment of substantial and enduring dependency relations with markets for factors of production is avoided – it runs counter to the struggle for autonomy and would imply high transaction costs. The (relative) scarcity of available resources implies that technical efficiency (Yotopoulos, 1974) and disembodied technical change (Salter, 1966) become central: in the peasant mode of production as much output as possible has to be realized with a given amount of resources[25] and without deterioration in their quality.[26]

Third, concerning the quantitative composition of the resource base, labour will be relatively abundant,[27] while labour objects (land, animals, etc.) will be relatively scarce. In combination with the previous characteristics, this implies that peasant production tends to be intensive (i.e. the production per object of labour will be relatively high) and that the development trajectory will be shaped as an ongoing process of labour-based intensification.

Fourth, the resource base is not separated into opposed and contradictory elements (such as labour and capital, or manual and mental labour). The available social and material resources represent an organic unity and are possessed and controlled by those directly involved in the labour process. The rules governing interrelations between the actors involved (and defining their relations with the implied resources) are typically derived from local cultural repertoires and gender relations, while the Chayanovian type of internal balances (e.g. that between drudgery and satisfaction) also play an important role (Djurfeldt, 1996).

A fifth characteristic, which follows on from the above, is the centrality of labour: levels of intensity and further development critically depend upon the quantity and quality of labour. Associated with this is the importance of labour investments (terraces, irrigation systems, buildings, improved and carefully selected cattle, etc.), the nature of applied technologies ('skill oriented' as opposed to 'mechanical'; see Bray, 1986) and novelty production (Wiskerke and Ploeg, 2004) or peasant innovativeness (Osti, 1991).

A sixth characteristic regards the specificity of the relations established between the peasant unit of production and the markets. As outlined in Figure 2.5, the process of production structured as a peasant mode of farming is typically grounded upon (and embraces) a relatively autonomous, historically guaranteed reproduction. As Schejtman (1980, p128) puts it: 'peasant production is only partly commoditized'. Each and every cycle of production is built upon resources produced and reproduced during previous cycles. Thus, they enter the process of production as use values: as labour objects and instruments that are used to produce commodities and to reproduce the unit of production. Such a pattern contrasts strongly with market-dependent reproduction (as summarized in Figure 2.6), where all resources must be mobilized on the corresponding markets, after which they enter the process of production as commodities. Thus, commodity relations penetrate into the heart of the labour and production process (Paz, 2004, 2006b). Figure 2.6 thus refers to the entrepreneurial mode of farming.

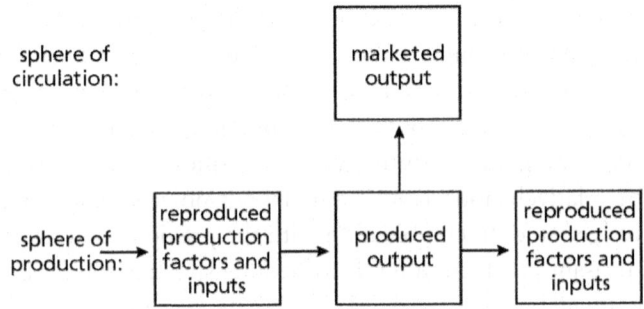

Figure 2.5 *The relatively autonomous, historically guaranteed scheme of reproduction*

Source: Ploeg (1990a, p14)

From a neo-classical point of view, the differences between situations of 'self-provisioning' (Figure 2.5) and those characterized by high market dependency (Figure 2.6) will be irrelevant. Instead, perceived from a neo-institutional perspective, they occur as clear examples of the basic dilemma: 'make or buy' (Saccomandi, 1990, 1998; Ventura, 2001).

Together the discussed characteristics define the distinctive nature of the peasant mode of farming, which is basically oriented towards the search for and the subsequent *creation of value added and productive employment*. In the

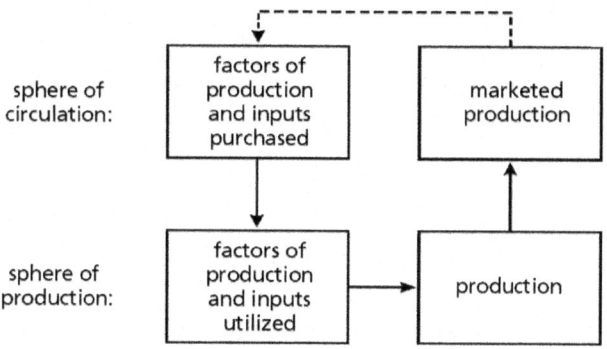

Figure 2.6 *Market-dependent reproduction*
Source: Ploeg (1990a, p17)

capitalist and entrepreneurial modes of farming, profits and levels of income can be increased through reducing labour input, and thus both develop through the outflow of labour. Due to its location in the peasant condition, as well as because of the nature of the *family* farm (Schejtman, 1980), this does not occur with the same intensity in the peasant mode of production (and if it does it might easily translate into regression). Emancipation (successfully facing a hostile environment) coincides here with the enlargement of total value added per unit of production. This occurs through a slow but persistent growth of the resource base (i.e. through the active creation of additional and/or improved resources), or through an improvement of technical efficiency. Mostly, however, the two are combined and interwoven and thus obtain an autonomous moment of self-reinforcement.

Labour-driven intensification

Analytically, intensification implies a steady, but ongoing increase in the production per object of labour – that is, production per hectare of land, and/or per animal (and/or per tree) is increased. Technically speaking, such yield increases are due to an increased use of production factors and inputs per object of labour, or to improved technical efficiency. The key to increased yields is the amount and quality of labour. Through labour investments (e.g. levelling and building irrigation systems) and the often time-consuming improvement of resources (creating more productive animals through selection, obtaining better plant varieties through plant breeding), both the resource base and the process of production are improved. More yields result in higher earnings which, in turn, compensate for the increased labour input.

The development of peasant agriculture typically occurs as labour-driven intensification. Theoretical expressions of labour-driven processes of agricultural intensification are to be found (at the macro level) in the work of Esther Boserup (1970), who stresses demographic growth as the driving force of agricultural growth. Another expression – directed at the micro level – is contained in the work of Chayanov (1966), who shows how the demographic cycle within every farming family (basically the ratio of mouths to feed and hands to do the work) governs a considerable part of the dynamics of the productive units in agriculture.

Farmers' freedom (understood according to Slicher van Bath, 1978, as a double set of relations) is an indispensable ingredient for labour-driven intensification. An intriguing aspect of both rural and development studies, however, is that – even when history entails an impressive testimony of labour-driven intensification – this peasant trajectory of development has hardly been elaborated upon theoretically. Labour-driven intensification has met, at the level of theory, with three series of mystifications. The first assumes 'a technical ceiling' beyond which peasant farming definitely cannot go (see Schultz, 1964, but also Bernstein, 1977 and 1986, who hypothesized an intrinsic backwardness of peasant agriculture). Such a 'ceiling' is, according to prevailing theories, inherent to given resources: poor land, poor instruments, meagre cattle, non-improved varieties, deficient irrigation systems and insufficient knowledge implying that nothing but poor, backward and stagnant agriculture is possible. Peasants make the best of the available resources (in this respect they are seen as highly efficient); but since their resources are 'poor', peasants themselves are (according to this approach) also poor and thus unable to drive development. Parallel to this kind of 'technical ceiling' reasoning is a socio-economic approach that limits peasant production to the level of subsistence. It argues that once immediate and perhaps enlarged needs are satisfied, there is no drive towards further development. Since peasant agriculture does not aim for profit maximization, it is a hindrance to growth and accumulation.

A second series of mystifications centre on the 'law of diminishing returns' as formulated by neo-classical economics. But this 'law' has already for several decades been rejected in theoretical agronomy. Whenever diminishing returns emerge this is seen as a temporary exception, which after correction will make way again for constant or even increasing returns (Wit, 1992). In rural and development studies, however, the ghost of diminishing returns continues to bewitch research and theories. A special case of this 'law' is the theory of 'agrarian involution' (Geertz, 1963; Warman, 1976). The continuing absorption of labour results, in the end, is nothing but counterproductive arrangements that are assumed to govern the redistribution of poverty. It is amusing that Lenin (1961) criticized such a position a long time ago. Nonetheless, the 'Leninists' of today (see, for example, Sender and Johnston, 2004) continue to reject the alternative entailed in labour-driven intensification.

The third set relates to a wide range of empirical examples of stagnation that are taken as reflecting the inherent backwardness of peasant farming. The problem is that no thorough enquiry is made into the specific causes of such stagnation, nor are the available counter-indications systematically taken into account. Thus, the misery entailed in practice is turned into the poverty of theory. In 1850, employment in Dutch agriculture amounted to 300,000 full-time units. In 1956 it had risen to 650,000 (see also Bieleman, 1992, who presents somewhat different data; nonetheless, Bieleman's data confirm a considerable increase in absolute numbers of the economically active population in the agricultural sector of The Netherlands). During the same period, Dutch agriculture witnessed enormous development, a boom that made it one of the world's strongholds in agriculture. Labour-driven intensification was the key to this prosperous development, which was accompanied not only by a general disappearance of capitalist farms, but equally by specific periods of repeasantization, as Dutch historians describe them.

Many similar episodes could be described here, along with a range of cases that seemingly point to the opposite. Evidently, labour-driven intensification might be blocked – but for specific reasons, not because this is intrinsic to it. In Latin America many peasants used to, and still do, describe the situation in which they live as one of *tierra sin brazos y brazos sin tierra:* land without hands to work it and a labour force without land. In such a situation, it is difficult if not impossible to develop or even to maintain a resource base (yet even then yields in the peasant sector might be superior to that of large-scale enterprises). It may also be the case that all the benefits of enlarged production could be appropriated by others, and in such a circumstance enlarging would be ludicrous – as was the case with the Italian *mezzadri* before the land reform. Here, playing the 'backward peasant' indeed became a major line of defence.[28] It is also possible for peasant culture to be erased – as occurred in many parts of apartheid South Africa. However, in all such cases the resulting stagnation tells us more about dramas occurring than about any *intrinsic* backwardness.

Recent developments in the European Union equally offer interesting counter-indications to the 'laws' suggesting that agricultural development equals, per definition, the reduction of labour input and employment. Organic farming, the many emerging expressions of multifunctionality and the evolution of farming economically (a farming style that aims at realizing low monetary costs) all imply an increase in labour input – both at the level of the units of production and sector as a whole. Simultaneously, the same development tendencies generate the extra added value needed to cover the increased input of labour. Parts of European agriculture are – at this very moment – going through a process of labour-driven intensification. More and, in particular, new products and services are produced with the same set of resources. The key to this new mode of intensification is again the quantity and, especially, the quality of labour.

In view of the high levels of marginality and unemployment existing in Africa, Asia and Latin America, labour-driven intensification emerges as a strategic, if not unavoidable, development trajectory (Ploeg, 1997a). Martinez-Alier (2002, p146) refers in this respect to:

> ... the large question, which is still outside the political and economic agenda. ... What is the agronomic advice that should be given, not only to Peru or Mexico, but even more in India or in China: should they preserve their peasantries or should they get rid of their peasantries in the process of modernization, development and urbanization?

A similar reasoning has recently been developed by McMichael (2007). The practicalities of programmes aimed at bringing about labour-driven intensification have been outlined in Figueroa (1986) and Pollin et al (2007). Griffin et al (2002) discuss labour-driven intensification in general terms as an alternative path to development.[29]

When discussing peasant-driven processes of agrarian and rural development, special attention must be given to the importance of reciprocity. Reciprocity (see Sabourin, 2006, for an up-to-date discussion) implies that resources can be mobilized, whatever the market structure. This means that resources are 'liberated' from non-use in order to expand production and spur development. The classical situation by means of which this might be explicated is that of the typical Andean small farmer with excess labour and another who has a pair of oxen (a *yunta*) that is not needed full time on the farm. Thus, two important resources, labour and animal traction, remain partly unused. Normally this is resolved through relations of reciprocity. As European farmers (who often engage in similar relations) would put it, this is arranged as a 'closed wallet operation' – that is, the exchange is socially regulated according to rules such as, for example, 'one day of oxen equals three man days of labour' (the ratio will differ according to local scarcity relations).

Through reciprocity markets can be avoided. Assume that both farmers face a shortage of money. If they had had to operate through the labour and animal traction markets, then neither would have been able to mobilize the missing resource. In such a situation, reciprocity, indeed, 'liberates' resources. Yet, even when money is not a constraint, reciprocity is highly advantageous when compared to the market alternative, especially since reciprocity functions as a mechanism to sustain quality. The work must be done well, just as the land must be ploughed with sufficient depth and precision – if not, a detrimental rupture in the mutual exchanges might occur. Through the market such qualities are far more difficult to achieve and to sustain. Reciprocity cuts out the opportunism that is intrinsic to the functioning of the market (Saccomandi, 1991, 1998).

Multilevel distantiation and its relevance in the 'modern' world

As argued, the peasant mode of farming represents an institutionalized distantiation of farming from markets, especially but not only on the input side. Box 2.1 briefly summarizes the main mechanisms through which distantiation occurs. Distantiation has been patterned in practice in manifold ways, and institutionalized within vested routines and a range of cultural repertoires that stress the virtues of autonomy, freedom, work and progress obtained through the co-production of man and nature. Distantiation has not been there since Genesis – it is the outcome of a complex historical process through which the peasantry has constituted (and reconstituted) itself.[30] This process has moved forward, probably even regardless of the intentions of the actors involved, through many painful lessons that had to be learned and learned again.

A major problem in any discussion of distantiation is that the accountability techniques currently used to represent the economic situation of farm enterprises (and/or the economic situation of the agricultural sector as a whole) do not allow for a clear reflection of the differentiated effects of distantiation and integration. As a matter of fact, these potentially highly relevant differences are *obscured* by dominant accountability approaches and the neo-classical concepts upon which they are grounded. Neo-classical theory assumes that it does not matter whether cows are bred on the farm or bought on the market. The same applies, for example, to hay (a famous issue in agrarian history) and capital. The entrepreneur should calculate as though all these resources were mobilized in corresponding markets. Their particular social history does not matter. It is irrelevant. The only justified parameter is their price as defined by the market. The main consequence of this approach is that the relative advantages obtained through distantiation are eliminated from the representation (and theoretical understanding) of farming.

Most Dutch farms are able to operate – and operate well – precisely because their functioning is distantiated from the immediacies of the different markets. If all the resources used on the farm had to function as *capital* (i.e. generate at least the average level of profitability) and all labour was to be remunerated as *wage labour*, then nearly all Dutch farms, as well as the Dutch agricultural sector as a whole, would go broke. For outsiders, this seems to represent, at first sight, a paradoxical if not perverse situation: it seems that

> **Box 2.1** *Mechanisms of distantiation*
>
> 1. The required (additional) resources are preferably produced and reproduced through the labour process instead of being mobilized through the markets. This applies both to factors of production and to non-factor inputs.
> 2. The technical lifespan of bought and self-produced artefacts is made dominant over economic lifespan. The lifespan of such items is extended through careful use and, in the case of technical artefacts, through adequate maintenance.
> 3. Wherever possible, reciprocity and socially regulated exchange are preferred over market transactions for obtaining (or mobilizing) lacking resources.
> 4. In cases where 1 and 3 are impossible, the required transactions are preferably built upon (i.e. financed with) savings from farming or from pluriactivity – that is, the required resources are obtained as commodities but then *converted* into non-commodities. Thus, recurrence to credit is avoided. The bought items – once commodities – enter the process of production as resources that are no longer to be valorized. Their value resides in the circumstance that they will, from now on, improve the labour and production process in the farming unit.[31]
> 5. In several instances use might be made of so-called *family capital*. Thus, financial resources are obtained that might be used according to rules that differ considerably from those reigning in the capital market.
> 6. Through the creation, further unfolding and widespread application of *novelties* it might become possible, in specific cases, to go beyond available technologies and obtain more production with the same set of resources, thus avoiding an additional recourse to markets for factors of production and inputs.
> 7. The intergenerational transfer of farm units also implies a distantiation from markets. In order to match their own rules and needs, families only partially follow the relations and prices that reign in the commodity circuits.
> 8. In some instances, specific solutions are found that ameliorate the effects of market-dependency in those cases where it cannot be avoided. This is the case with the pooling of machinery by a larger group of farmers.
> 9. By regrounding agriculture on available ecological capital, and simultaneously enlarging the latter, dependency upon a range of artificial growth factors (and associated markets) might be reduced.
> 10. By re-patterning relations with market agencies on the output side of farm units, considerable distantiation might be created. This will translate into higher degrees of autonomy and higher levels of value added.

farmers are, indeed, obliged to exploit themselves since they have to accept inadequate levels of remuneration in as far as their own labour and capital are concerned.

In European agriculture, generally, and in Dutch agriculture, specifically, there is, indeed, *poverty* – but such poverty does not validate the theoretical claim of *self-exploitation*. The point is (and I will discuss this at some length because it underpins and strengthens my argument on the centrality of peasant agriculture in Europe) that the resources normally summarized as capital (land, animals, buildings, machines, etc.) are not, or only partly, mobilized on the capital market. *Hence, they do not function as capital within the farm.* They do not have to render levels of profit comparable to those realized through investments elsewhere. This does not imply that Dutch farmers are, as the saying goes, 'thieves of their own wallet' (or that they are inefficient). The point is that *other* processes of conversion are central and *other* benefits matter.

Within most farms in The Netherlands, the value of the available resources is that they make it possible to generate an income (they allow for survival) and enable one in the long run to create a 'beautiful farm' (a powerful metaphor that is used throughout Europe and strongly rooted in cultural repertoires and associated historical processes). The available resources, and especially the land, which represents a highly elevated value should it be sold, do not function necessarily as capital in the classical sense of the word. If they were to do so, they would flow out of agriculture. Their value is that they allow for farming and that they might be converted, in the longer run, into a pension for the senior generation and a comfortable starting position for the younger generation that takes over. In this sense, it is the case that 'fathers work for their sons' (Berry, 1985). We are looking here at a socially regulated and institutionally grounded process of conversion – a conversion that is very different from the conversion of capital into profits subsequently reinvested as capital in order to realize more profits. But being different does not make it an inadequate or meaningless process of conversion. On the contrary, it enables farming to continue both in the short and long run.

Thirty per cent of all agricultural land in The Netherlands is used through tenancy arrangements. The tenancy relation is an important institution, reflecting among other things the former struggles of tenants and leasehold farmers. The Tenancy Law, which is frequently renegotiated and adapted, establishes the maximum rent at 2 per cent of the agrarian value of the land (considerably lower than its commercial value). If this were not the case – if, in other words, the land had to render an average profit of, say, 4 per cent – then this would imply an additional cost, for all tenants together, of some 325 million Euros per year. Due to tenancy as an institutionalized arrangement, this amount now stays within the agricultural sector.[32]

The land directly owned by farmers' families is also subjected to important institutional arrangements. Probably one of the most important of these is the custom to pass the land from the previous to the next generation for a price that is below market value. Such a transfer implies an agreement (in The

Netherlands often referred to as a 'hole-and-corner' arrangement), not only between the generations but also between the young farmer and his (or her) brothers and sisters. Due to such an agreement, the parents obtain their pension, while the new generation can farm far 'cheaper' than would be the case without it. If an intergenerational change is assumed to occur once every 30 years, it follows that Dutch farmers are 'saving' every year – in this way – 660 million Euros. If the effects of both mechanisms (tenancy and socially regulated intergenerational change) are taken together one might conclude that through this institutionally grounded distantiation from the land market, Dutch agriculture saves at least 1 billion Euros per year (compared with a total agrarian income of roughly 3 billion Euros).

Similar reasoning can be applied to labour, to capital goods other than land and to important input flows. Time and again important institutional arrangements are encountered that mediate the immediate functioning of the corresponding markets in such a way that financial costs (and the transaction costs) are considerably lowered. If this were not the case, farming would become very difficult, if not impossible.

The peasant mode of farming represents an institutionalized distantiation of farming from markets. This distantiation resides partly in the strategies operated at the level of single units of production. Distantiation is equally rooted in a wide range of institutional arrangements such as tenancy, family capital, co-operatives, agrarian policy, etc. All of these institutions (through which the peasant mode of farming is linked with wider society) govern processes of conversion in ways that differ significantly from those that emerge when the market directly governs them.

Given its strategic importance, it is ironic that many scholars (and politicians) do their utmost to ignore this strategically created and institutionally grounded distantiation when they claim that, for example, Dutch farmers are (and, consequently, should behave as) agricultural entrepreneurs who definitely operate in ways that strongly contrast with the *modus operandi* (or the choreography) of the peasantry. It is just another indication of science misunderstanding the peasantry – of 'awkward' science – especially when it stems from agricultural universities. It is, however, blatantly *irresponsible*, as currently is the case with neo-liberals, to advocate the deliberate deconstruction of institutional arrangements such as the *comunidades campesinas* in Peru, Bolivia, Ecuador and parts of Chile; communal landownership in large parts of Africa and Asia (Platteau, 1992); peasant-managed irrigation systems all over the world; and collective ownership and free exchangeability of genetic material (Commissione Internazionale, 2006a). A 'free market'-inspired destruction of these arrangements that protect the peasantries in developing countries would not only strongly elevate the numbers of 'wasted lives', but also seriously threaten worldwide food security.

3
Catacaos: Repeasantization in Latin America[1]

Introduction

I came to know the peasant community of San Juan Bautista de Catacaos for the first time more than 30 years ago. The community, located in the lower valley of the River Piura in the north of Peru, is one of the largest communities in the country. At the beginning of the 1970s it numbered some 50,000 *comuneros*, of whom about 2000 were wage labourers (*estables*) employed by one of large cotton-producing *haciendas*. At the beginning of the 1970s, these enterprises, which had controlled 10,000 hectares (ha) of irrigated land plus a lot of barren land, were converted into state-controlled co-operatives. Alongside this more or less stable part of the workforce there were some 4400 *pequeños propetarios*, small farmers possessing small plots of community land. In addition, there were thousands of *campesinos sin tierra* (landless peasants) involved in harvesting cotton in Bajo Piura, and in transplanting and harvesting rice in the valleys of Alto Piura, Chira, Lambayeque and Santa, and sometimes even farther away. Their constant movement from one work place to another gave them the collective name of *golondrinas* (migratory swallows). While *estables* and *pequeños*[2] were poor, the life of *golondrinas* was miserable and insecure. However, regardless of such differences, in everyday language all of the above categories were grouped into one entity as *campesinos pobres* (poor peasants).

In 2004, more than 30 years later, I revisited Catacaos. During my first period there I had worked for more than a year and a half as an affiliate of FEDECAP (Federación Departamental Campesina de Piura), the Regional Peasant Federation of Piura. But this time I was only able to stay for a brief four-week visit. Luckily, several of the leaders of the community and of the peasant movement of the previous decades were still around. Thus, I was able relatively quickly to obtain an impression of the current state of affairs. In the following analysis, I focus on two specific elements: land and yields. Having control over land (through whatever mechanism) is always a strategic feature in the peasant struggle for autonomy and progress; and changing man–land ratios inform us, directly and indirectly, about the nature and rhythm of rural development. The following analysis then explores how yield levels relate to the social relations of production, how they reflect the quantity and quality of peasant labour, and how increased yields translate into increased well-being. Yields thus become a measure and metaphor that link the past, present and future.[3] They relate equally to the micro level of fields, to meso levels as defined by regional economies, and to macro levels that concern food availability at national level. In other words, yields express the role of peasants (as invisible as they are)[4] in both history and society.

Repeasantization

Among the many differences between then and now is the enormous increase in the *absolute* number of peasants.[5] In this respect, Catacaos represents an outstanding example of repeasantization both qualitatively and quantitatively. The latter aspect refers to the change in the number of peasants. However, becoming a peasant should not be understood as constituting a single step. It is, instead, an ongoing and often sharply fluctuating *flow through time*. Consequently, the peasant condition varies by degrees. Dependency upon markets, market agencies and extra-economic coercion, the relative autonomy that can be obtained, the magnitude of, and control over, resources, and the levels of productivity that are created are all relevant in this respect. In synthesis: once peasants are constituted as such, further repeasantization may occur. At this point we are looking at the qualitative dimension relating to the degree to which agriculture is structured according to the peasant mode of farming. Evidently, the quantitative and qualitative dimensions of repeasantization processes might be combined (as is notably the case in Catacaos). However, it is equally possible that repeasantization expresses itself on only one of the two – or that contradictory tendencies occur.

Throughout history there have been many episodes of repeasantization. Alongside the historical references, there are also several contemporary, although highly differentiated, processes of repeasantization.[6] A systematic enquiry into current expressions of this phenomenon is of the utmost importance: first, because repeasantization represents, from a theoretical point of view, a crucial borderline case. In neo-classical and development economics and in nearly all Marxist approaches, a re-emergence of the peasantry is thought to be impossible and certainly undesirable and, were it to occur, it would necessarily represent a regression (Bernstein, 2007b). A second reason is its relevance as a politically and economically appropriate way out of underdevelopment in many developing world countries (see, for example, Figueroa, 1986, for a discussion of this in relation to Peru). And a third reason for seriously studying repeasantization resides in the new food empires that are currently emerging all over Latin America. As much as these new empires tend to destroy the peasantry, they provoke and create new forms of repeasantization.

In practice, repeasantization will always occur through a range of interconnected, often contrasting and sometimes new processes. In Catacaos, repeasantization emerged out of:

- the change of the former *haciendas* into co-operatives, after which the land was parcelled into individual peasant units;
- a massive appropriation of land and water by landless peasants (mainly *golondrinas*), a process in which the formation, development and, finally, the parcelling of communal units of production (UCPs) played a major role;
- a strong rise in the number of individual plot holders, basically associated with a shift of the agricultural frontier into the harsh semi-deserts (*bosque seco*) that surround the community;
- a reallocation of *pueblos jovenes*, or slums, where they are no longer confined to the cities – in this case, Piura, Trujillo, Chiclayo and Lima – but are now literally emerging 'within the fields' (i.e. within rural communities); at the same time, being involved in agriculture becomes an important feature of life in these new slums;[7]
- finally, especially from the 1990s onwards, a further change within the peasantry, which moves away from a high degree of market integration characteristic of the 1970s and 1980s, thus strengthening the peasant nature of the rural economy.

The outcome of the first three processes was an enormous growth in the number of peasants. In this context, repeasantization implied the almost complete disappearance of salaried workers (*peones*) linked to the many very large *haciendas* and, later, the co-operatives.

Although repeasantization has not been absent in adjacent and comparable areas, Catacaos highlights the processes involved. Table 3.1 compares the number of farms 'managed by the producer' in 1972 (Second Census) vis-à-vis 'natural persons' (i.e. *personas naturales* as opposed to formal enterprises) owning and working the land in 1995 (Third Census). Although these data do not describe the total number of people engaged in agriculture (especially since the *estables* – that is, the permanent labour force of the big plantations – are excluded from the first row and included in the second)[8], they clearly bring out the relative differences in the magnitude of repeasantization.

56 The New Peasantries

Table 3.1 *Rates of repeasantization*

Location	Number of peasants in 1972	Number of peasants in 1995	1995 figures as a percentage of 1972 figures
Catacaos	4396	13,030	300
Chulucanas	3308	7065	214
Morropon	527	1271	240
Buenos Aires	480	1532	306
Elsewhere[9]	11,772	19,132	163

Source: Ploeg (2006d, p409)

Apart from the district of Buenos Aires, the community of Catacaos has the highest rate of repeasantization. Buenos Aires is, literally, the exception that confirms the rule. Together with Catacaos, it was the main focus of peasant struggle during the 1970s. In 1973 the permanent and temporary workers occupied the large Buenos Aires plantation of the Rospigliosi family, with the goal of creating the co-operative *Luchadores del 2 de Enero*. This co-operative distinguished itself by engaging in a long, fierce struggle aimed at augmenting productive employment (see Ploeg, 1990a, Chapter 4, for a full account). In the end, this resulted – as in Catacaos – in a relatively high degree of repeasantization.

A typical characteristic of repeasantization in Catacaos was that it resulted in a relatively equal distribution of land among the peasantry. Figure 3.1 compares the distribution of land (according to the 1995 Census) in the

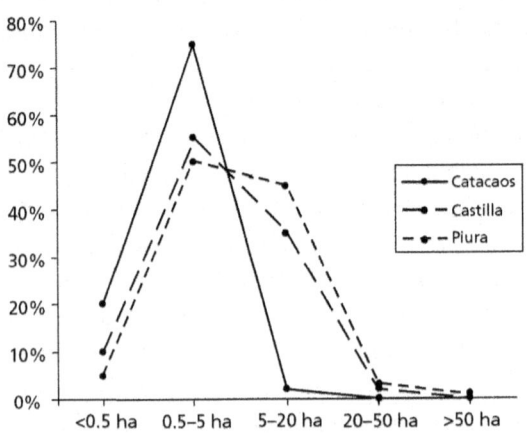

Figure 3.1 *Land distribution in Catacaos, Castilla and Piura (1995)*
Source: Ploeg (2006d, p410)

community of Catacaos with that of the neighbouring districts of Castilla and Piura. A large majority of peasants in Catacaos possess a plot between 0.5ha and 5ha, while in Piura and Castilla, more than 30 per cent control plots of beyond 20ha in extent. That is to say, in Catacaos, 86 per cent of the land belongs to 75 per cent of the peasants. This strongly contrasts, for example, with the district of Piura, where 56 per cent of the peasantry controls 37 per cent of the land, while another 36 per cent controls 60 per cent.

Presenting more exact data on the number of peasants within the community of Catacaos is far from easy. This is not only due to the imperfections of available statistical information, but also to the constantly changing empirical situation and the fluidity and overlap of the usual statistical categories. Table 3.2 summarizes and integrates the information available. Apart from statistical data, it also draws on community archives and information derived from interviews with leaders of the peasant movement. What is relatively clear is the development of the irrigated area. Although this changes from year to year – due to the whims of water availability and distribution – the irrigated area has expanded from some 30,600ha at the beginning of the 1970s (part of which was very poorly irrigated) to 45,500ha in the 1990s. This was made possible by the implementation of the Chira-Piura Project, which diverted water from the Chira River to the Bajo Piura Valley. Although the enlargement of the irrigated area does not automatically translate into an increase in the number of people working the land, in Catacaos the final outcome was repeasantization.[10] Therefore, we have to turn our attention to data on the agricultural labour force and its internal composition.

Table 3.2 considers, in the first place, the salaried labour force that was initially employed by the large cotton-producing enterprises (first *haciendas*, later state-controlled co-operatives). The number of regular workers remained stable at about 2000 during the transition from *hacienda* to co-operative, and with the collapse of the co-operative, this same number was translated into 2000 new peasant units of production. The *haciendas* comprised some 10,000ha of irrigated land, as well as 20,000ha of uncultivated land. Second, there were the production units of the *pequeños*: according to the census, there were, at the beginning of the 1970s, some 4300 peasant units of production in Catacaos. Formally, they possessed 20,600ha. However, full irrigation was hardly possible since there were constant water shortages. The number of peasant units increased to around 6700 by the middle of the 1990s, and in the first decade of the 21st century, this number further increased. Third, there was a series of community-led land invasions into the barren or unused lands of the co-operatives: a process that characterized the 1970s and 1980s. A central vehicle for this process was the creation of communal units of production (UCPs) mainly made up of former *golondrinas*. By the end of the 1980s, there were some 150 such units consisting of a total of 4500 members who together culti-

Table 3.2 The development of agricultural employment in Catacaos

Time axis:	Beginning of the 1970s	Midst of the 1980s	Midst of the 1990s		Current situation	
Sub-processes:	units			average man/land ratio	employment	effective man/land ratio
1. From *haciendas* via co-operatives to peasant units	10 large & 90 medium sized *haciendas*	9 co-operatives on 10,000ha	Parcellation of co-operatives			
	10,000 cultivated ha & 20,000ha of barren land					
	2000 *estables*	2000 members of cooperatives →	2000 peasant units on 10,000ha	1:5	4000	1:2.5
2. From landless peasants via UCPs towards new peasant units	0 UCPs	150 UCPs with 5000ha	Parcellation of UCPs			
	0ha	4500 workers engaged →	4500 peasant units with 6750ha	1:1.5	4500	1:1.5
	struggle for land in and by the community					
3. The increasing number of small peasants	4300 peasant units	creation of some 2400 new units with 8150ha →	6700 peasant units and 28,750ha	1:4.3	13,400	1:2.2
	20,600ha (not always irrigated)					
Total	4400 units with 30,600ha	→	13,200 units with 45,500ha		21,900	1:2.1
		Plus some 15,000 *golondrinas*, shepherds, etc.				

Source: Ploeg (2006d, p412)

vated 6750ha of land. Then, during the early 1990s, these UCPs were also transformed into individual peasant units of production.

Together, these three strongly interconnected processes brought major changes. The initial number of 4300 peasant units increased to 13,200 units. Their number tripled, while the cultivated area as a whole grew by some 50 per cent and irrigated land from 30,600ha to 45,500ha.

Table 3.2 requires a few additional comments. As indicated in the bottom line, there is also a considerable number of landless peasants (*campesinos sin tierra*) in the community. These comprise herdsmen and shepherds who use the seemingly endless semi-desert, or they are people with very small plots (*microfundistas*) or *golondrinas*.

In the last columns of Table 3.2 I have tried to convert the data into some measure of *labour force*. Normally, in small peasant units of production there will be two people working (e.g. man and wife, or father and son), who may also be partially involved in other activities. The same applies to the somewhat larger plots of the former *estables*. But here the larger acreage (on average 5ha per unit) implies, especially under current conditions, that 1ha or 2ha will be rented out to another *comunero*[11] or will lay fallow. Thus, in the final column we can calculate by category the effective average man–land ratio. In synthesis, the table shows that the number of people actively engaged in peasant production is far beyond that of the early 1970s. It is also far beyond the (potential) effects of the enlarged irrigated area. As will be explained in the next section, the effective man–land ratios are also strikingly different from those that underpinned the land reform. In short, what Table 3.2 indicates is *a massive and far-reaching process of repeasantization*.

Mechanisms of repeasantization

Over the centuries, Catacaos has become well known for its fierce peasant struggles (see Cruz Villegas, 1982; Ploeg, 1977, 2006d; García-Sayan, 1982; Revesz, 1989). For instance, in the 1960s, all *haciendas* were occupied by *comuneros* and subsequently de-occupied ('liberated', as some said) through interventions by the military and the riot police. The first small co-operatives were created in order to relieve the tense situation. In 1969 a radical and nationwide land reform was declared and quickly implemented by the then military government. The large *haciendas* were transformed into production co-operatives, within which the man–land ratio had to conform to the guidelines laid down by the so-called Iowa Mission. This North American think tank translated (for each specific ecosystem and starting from production, as it was structured in the then prevailing *haciendas*)[12] the expected price–cost relations, new technologies and a 'savings quote' into the amount of land that could be 'given' to a head of a family. As in the rest of the country, this resulted in a process of marginalization since only some 10 per cent of the economically active agricultural population could be integrated within these new co-operatives. The remaining 90 per cent was denied any (further) access to the land. In Catacaos, the calculated man–land ratio was 1:5.

The community of Catacaos countered the state-controlled land reform process with a massive and well-elaborated response. They decided to fight for all-round full employment through the creation of *unidades comunales de producción* (communal units of production) formed by the occupation of barren lands and characterized by a completely different man–land ratio: initially 1:2, but later 1:1.5, compared with the 1:5 that prevailed in the new state-controlled co-operatives. For those observers not familiar with the peculiarities of agricultural production, the effective increase of the man–land ratio (creating more employment on the same disposable surface area) might be a somewhat surprising, if not irrational, sharing of poverty. However, if the higher labour input per unit of land translates into higher (i.e. more intensive) levels of production, and if, at the same time, costs are reduced (substituting, for instance, costly inputs by labour), then the net result is quite likely to be the opposite: high employment levels spur an intensification of production, while high yields effectively remunerate the increased labour input.

In 1972, the community of Catacaos formed the first 16 units for communal production. These units became the starting point for an impressive process of labour-driven intensification. By 1974 their number had risen to 38, in which some 650 people were actively cultivating a total of 1215ha. With a man–land ratio very different from that of the state-controlled co-operatives (less than 1:2), however, they managed to pay those working the same wage. Key to their performance was the relatively high yields and low cost levels. Two years later, the number of UCPs had risen to 65, composed of 1320 members cultivating 2306ha and making an average man–land ratio of 1:1.7. San Pablo Sur, a UCP that had 60 workers in 1974, had expanded to 200 members, cultivating 300ha (thus entailing, in 1976, a man–land ratio of 1:1.5). Then, by the end of the 1980s, there were 150 UCPs with 4500 workers operating a total of 6750 cultivated hectares. Extremely important at that time was the coalition that was forged between members of the co-operatives and the UCPs since it was possible through this coalition for the community to enter into the spheres of commercialization and processing. Thus, high employment and yield levels, especially in the UCPs, were consolidated.

The creation of *unidades comunales de producción* was the outcome of devoted, persistent and massive peasant struggles that were to last for many decades, but which reached their peak during the 1970s and 1980s. Elsewhere I have described these struggles at length (Ploeg, 1977, 2006d). Here I limit myself to a brief discussion of the *principios de lucha* (the shared values that guide communal actions) that crystallized during the long series of socio-political struggles endured by the community and which, in the early 1970s, became focused on resisting the state-imposed land reform, simultaneously prompting the creation of the communal production units (see Box 3.1). I consider these 'shared values' (or 'principles of peasant struggle') as a beautiful and powerful synthesis of a 'moral economy' (Scott 1976). They synthesize history and skilfully translate it into guidelines for the present in order to create a better future.[13]

Box 3.1 *The shared values of the peasant community of Catacaos*

The communal values of the peasant community of Catacaos comprise the following:

- a united, indestructible and autonomous community;
- a community that is governed through the democratic intervention of all her members;
- a community in which all members are equal in rights and duties;
- a community that recognizes labour as the only source of wealth;
- a community that does not allow the exploitation of her resources and production by foreign elements;
- a community that struggles to secure for all her members the satisfaction of basic needs for housing, health, food, education and employment;
- a community that actively works for both the immediate and the future needs of her youth;
- a community that engages in solidarity with the entire labour class of our country in order to strive for the integral transformation of the country.

The shared values also reflect several dimensions of the peasant condition and mode of farming discussed in the previous chapter, while simultaneously translating them to a higher level of aggregation – that is, the peasant community as collective actor (Long, 2001).

From the middle of the 1990s onwards, the community of Catacaos suffered a serious decline. The state-imposed division of co-operatives and UCPs into small plots, the change from communal landownership and individual possession towards generalized individual ownership (also dictated by the Fujimori regime), the sharp repression that followed the armed struggle of *Sendero Luminoso*, and, finally, a leadership of the community that became, during the late 1990s, strongly biased towards individual enrichment all resulted in a crackdown of communal activities, at least temporarily. What remained, however, was the elevated man–land ratio as created through the preceding episodes of peasant struggle.

The effects of repeasantization: Intensification of production

In situations such as the community of Catacaos, yield levels are basically the outcome of two contradictory but combined processes. On the one hand, there is the struggle for emancipation: working the land as well as possible in order to obtain the best yields and, therefore, the highest levels of added value. This is the only way to proceed in a situation of overall marginalization. When nobody cares for you, you have to care for yourself by obtaining the best possible results out of the co-production of man and nature. More specifically, repeasantization must result in an intensification of agricultural production; if not, a sharing of poverty (or involution) will, indeed, be the result. On the other hand, specific social relations of production are needed to allow for such a process of peasant-driven intensification. Central here is the availability of the required means of production. These may result from previous cycles of production, in which case the required means are historically guaranteed (as illustrated in Figure 2.5). The opposite situation might also occur, in which case the required means have to be mobilized through the markets (see Figure 2.6).

Apart from these *interrelations* between markets and the units of production, the *conjuncture* has also to be slotted into the analysis. This conjuncture might range from one in which market and price relations are favourable for farmers (availability of credit, well-functioning markets, a fair balance between costs and benefits, no one-sided distribution of risks, etc.) to the opposite situation in which the market conjuncture is unfavourable for peasants and farmers. Thus, several combinations emerge. Evidently, a high market dependency combined with a negative conjuncture is the worst scenario. Peasant farms need 'space', as Halamska (2004, p249) convincingly argues in a discussion of the Polish peasantry. Such 'space' is 'not a given once and for all. It [is] in constant flux, mutable and [can] be either reduced or expanded'. A favourable conjuncture offers 'space', just as a deteriorating conjuncture implies its reduction. Following this line of reasoning, it can be argued that *reduction* of space is characteristic of the Peruvian countryside[14] and, consequently, of Catacaos as well, especially from the mid 1990s onwards. The *Banco Agrario* was disbanded and peasants have had to address themselves to private banking circuits in order to obtain required loans. Apart from the exclusion imposed on large categories of people by these circuits and the high transaction costs involved (Fort et al, 2001), the widespread fear of losing their land refrains most peasants from using them – hence, the tragic ironies of families in urgent need of production (for consumption and for cash), on the one hand, and fields lying bare, on the other. Then there are the highly elevated costs of production, natural risks (*El Niño*) and market-induced risks that further reduce the space available for peasants. This applies especially to those peasants who are more market dependent.

Taking these two contradictory tendencies together – that is, the need and willingness of the peasantry to make progress through improved production and the near total lack of conditions that would allow them to do so – it is remarkable that yields in Catacaos are somewhat higher than in adjacent and comparable areas in the department of Piura. Table 3.3, based on data from the regional office of the Ministry of Agriculture, shows that yield levels differ considerably from year to year, due mainly to climate. For instance, in 1999 the aftermath of the last *El Niño* was still clearly felt. Nevertheless, in 2000, when more than 10,000ha were planted (and harvested) in the community of Catacaos, the average yield was 1.84 tonnes per hectare, a bit more than 10 *cargas* per hectare. This was as much as 69 per cent above the average yields in the neighbouring districts. In 2001, the difference was far less accentuated. However, in all three years, production per hectare was higher in Catacaos than in surrounding areas.

Table 3.3 *Cotton yields in the community of Catacaos compared with neighbouring districts (Piura, Castilla, Las Lomas and Tambogrande)*

Year	Catacaos (tonnes per hectare)	Neighbouring districts (tonnes per hectare)	Percentage difference
1999	1.71	1.64	4
2000	1.84	1.09	69
2001	1.57	1.39	13

Source: Ploeg (2006d, p418)

The relatively high yields (which can also be found for other crops, such as maize and beans) realized in Catacaos are a direct effect of the process of repeasantization that results in more intensive farming and higher yields. However, as discussed in the next section, these *relatively* high yields could have been much higher – that is, the highly adverse social relations of production in which the peasantry is currently engaged exclude, for the moment, any further intensification.[15]

Spurred intensification

Over time, the communal production units have constituted the *loci* where further intensification of production was wrought. Already by 1976 San Pablo Sur realized a production of 12 *cargas* (some 2000kg per hectare) that represented a further leap forward from the 1973/1974 level (10 *cargas*), and it was also among the highest levels of production in the whole Bajo Piura Valley. Yet, intensification of agricultural production, spurred and driven forward by the quantity and quality of labour, continued: in 1987 and 1988, yields as high as 25 to 28 *cargas* per hectare were reached in San Pablo Sur. Such yields were made possible by new improved seed varieties, which the UCPs obtained from a specialized plant breeding institution (Fundeal). 'However', as Jorge Vilches Sandobal, one of the agronomists working with the UCPs, explained, 'if you overlook that experience as a whole you can rightly conclude that the improved varieties account for only 20 per cent of such results. The remaining 80 per cent is due basically to organization.' This concept, 'organization', synthesizes several elements:

- In the first place there was meticulous planning of the agrarian calendar, so that all relevant activities (preparation of the land, irrigation, sowing, weeding, cultivation, fertilization and disease control) became well coordinated over time and adapted to the growing cycle of the plants. In combination with the communal organization of work, which makes sure that there is always enough labour available[16] to undertake the right job at the right moment in the right way, this mode of preparation guaranteed good yields.
- Second, the embedding of the UCPs in the overall structure of the then powerful and well-administered community allowed for a proper and timely delivery of water, credit, inputs and additional machinery services. The community of Catacaos negotiated the required credits directly – for all UCPs – with the *Banco Agrario*. It also made sure, through negotiations with the Water Board, that water would be delivered at the right time and in sufficient quantities, and it obtained directly from the manufacturers the needed fertilizer. Thus, a smooth-functioning socio-technical network gave support to the process of production within the UCPs. This allowed not only for high yield levels, but it also reduced the variable costs and associated transaction costs.
- Third, the magnitude of the UCPs (at that time 200 people were working in San Pablo Sur) allowed for a certain internal division of labour. With regard to high yield levels, it is important here to mention *el plaguero*. One of the members of the UCP received extensive training outside the community in the detection and control of plant diseases and plagues, and therefore, during the season, would constantly oversee the fields in order to be able to intervene quickly and appropriately if there was any sign of infection. Equally important was the fact that two or three UCPs could, whenever needed, jointly contract an agricultural engineer to assist them on a daily base in the planning and implementation of the work.
- And finally, the very structure of the UCPs (equal payment for all and a redistribution of profits at the end of the season) meant that those who

took part in production had an interest in obtaining good yields. And apart from the implied financial stimuli, 'people were extremely proud to see such results coming out of their own work', as Jorge Vilches Sandobal explained to me.

Apart from the implied technicalities, this description underlines another basic point. Socio-political struggles not only occur through demonstrations, land occupations and road blocks, they *also take place in the fields,* at first sight in a somewhat hidden form. This often entails a long and painstaking fight to increase the control over the process of production, to improve it, to mould it in ways that correspond with one's own interests and prospects, and to obtain better results from it. This latter struggle is as important as the former. One might even argue that the more visible struggles are, essentially, a condition for engaging more successfully in the hidden struggles in the fields.

With the division of the UCPs into single individual plots and with the demise of the community as a powerful and well-administered whole, this finely tuned socio-technical network collapsed and with it cotton yields went down again to levels of some 10 *cargas* per hectare.

New modalities of repeasantization

The current situation in Catacaos (and in Peru, as a whole) is characterized by two decisive elements: an impoverished peasantry and extremely adverse market conditions. The latter do not allow for any strong engagement in the different input markets. Credit can barely be obtained unless high risk, high interest rates and high transaction costs are accepted. Thus, expensive industrial inputs (improved seeds, fertilizer, pesticides and the associated technical assistance) cannot be obtained. And output markets (for cotton, maize, rice, etc.) are characterized by considerable price fluctuations, which make any engagement in the capital and input markets even more risky and unlikely. Therefore, it is difficult to realize margins, at the end of the productive cycle, which are sufficient to cover household expenses for the cycle to come. Thus, a complete paralysis of the rural economy could be the outcome. What occurs, though, is that peasants develop, of necessity, new responses to face this extremely difficult situation. Together, these new responses might be understood as so many steps in a further process of repeasantization. I will discuss here four such responses, which, in practice, are often combined – albeit in different ways.

First, production for self-consumption again becomes central. Whereas cotton cultivation was important during the 1970s and 1980s, it is now beans and maize. Consequently, family consumption is organized far less through involvement in different market flows (producing cotton in order to convert it into money, to be converted into the required food products). The beans and maize go directly from the fields to the households: both production and consumption are now distanced, more than before, from the markets. Only the *excess* is marketed, just as it is only *lacking* ingredients that are bought (if funds are available).

This translates into another gradual shift. Peasants of Catacaos are now engaged in a multiplicity of activities, far more than 20 or 30 years ago. Since farming renders little and sometimes no money at all, peasant families engage in other activities to earn required cash flows. It struck me, when talking with informants, that most had two, three or even four 'jobs': every opportunity that might yield some money is used.[17] Thus, multiple job holdings or multi-occupation peasant households are constituted.[18]

Second, farming as a productive activity is now organized in such a way that monetary costs are minimized. External inputs are replaced, where possible, by locally available means. Thus, a low external-input agriculture emerges (Reijntjes et al, 1992), often characterized by highly different techniques and practices. This change is accompanied by a new cultural repertoire used by the farmers to understand and order their own practices and the relations in which these are embedded. Figure 3.2 contains an example of such a grammar for farming economically, a mode of reasoning actively used by peasants in Catacaos during the process of agricultural production. *Asistir bien a la planta* (taking good care of the plants) is the central notion that translates into knowledge on ploughing, sowing, cultivation, irrigation, weeding, pest control, their interrelations and their organization through time. All of the associated tasks and subtasks (that together compose the labour process) are aimed at obtaining high yields (i.e. *una linda producción*). Through curiosity (observing and trying to understand small differences in yield) and small, often nearly invisible, experiments (Badstue, 2006), the interrelation between the two (between *asistir bien a la planta* and the *linda producción*) is improved – it is precisely at this point that novelty production in agriculture becomes strategic.

A fine or beautiful production is, in turn, converted into what is mostly termed *la utilidad para la casa* (utility for the household economy). Part of production is consumed by the family itself (and by the animals), part is stored as seed for the next cycle, and the remaining part will be sold. The higher the yields, the better will be the utility for the house.

A high level of production depends upon three crucial conditions that together allow the farmer to 'take good care of the plants'. These are, apart from the availability of land, water and good climatic conditions, the availability of *medios* (i.e. the means to do the job). Such means could have been produced in the previous cycle: manure, compost, good seed, a healthy horse, tools and some savings. They could also be lacking, in which case *medios*

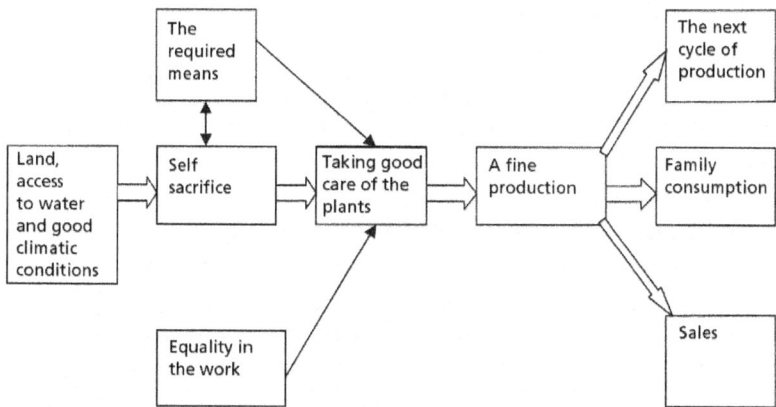

Figure 3.2 *The symbolic organization of the agricultural process of production in Catacaos*

Source: Ploeg (2006d, p422)

become identical to working capital and the possibility of obtaining credit. There is also the need to dedicate serious attention to the process of production. This is referred to in Catacaos as *sacrificio*: self-sacrifice. The work is often hard and demanding. It frequently implies a certain substitution between labour and means. If the means are lacking for whatever reason, they might be replaced, to a degree, by more, and often heavy, labour. Compared to fertilizer application, the collection, storing, (re-)working, transport and application of manure represent drudgery, indeed. But this is often necessary. And, finally, there must be, according to the discussed grammar, *igualdad en el trabajo* (equality on the 'work floor'), but also in the redistribution for final home use: only in this way will all those involved really dedicate themselves to the job. This is why the use of family labour (and engagement in reciprocity relations) is much preferred to wage labour relations; apart from this, the latter would imply increased monetary costs and these are, in the context of 'farming economically', to be avoided where possible.

A third element to be mentioned here is that low external-input agriculture (or farming economically) does not necessarily represent stagnation and/or regression. On the contrary, a new kind of dynamic is emerging in which *novelty production* is paramount. Novelties or peasant innovations might be defined here as a particular combination of new practices and new insights through which peasants raise the technical efficiency of their process of production. Their particularity resides in the circumstance that the *working* of these new practices is, as yet, neither understood nor recognized by agrarian sciences. One amazing and highly illustrative example of such a novelty that I encountered in the fields of Catacaos was the 'rebuilding' of small quantities of fertilizer into a new 'liquid fertilizer' (*abono foliar*), which is not put on the soil, but distributed over the leaves of the plants:

> *We put 12 or 15 eggs into the tank [the so-called* cacorro*: the manually operated tank placed on the back and normally used for spraying insecticides and pesticides], mix them with water, 1kg of* nitrato de potasio *and another kilogram of* nitrato monamonico, *and, if we have it, some 10cm of liquid fertilizer. This mix is then sprayed over the plants. It is applied for the first time eight days after the plants emerge. We follow this with three more applications but only with eggs. In this way we save money. Of course, we do this to survive; if we didn't, we would be in a really bad situation. And it is actually giving good results. We have made comparisons in the fields. If you treat plants this way, with eggs, they will be more beautiful, stronger and the fruits will be larger. We have tried it with maize, with cotton, beans, whatever you want. It strengthens the plants, they are better fed, and well-fed plants will have no plagues and diseases.*
>
> *I know, people laugh at us: they say that eggs are for human consumption and not for throwing on the plants; but we are happy. We learned it from an old man. We are always searching, always experimenting, and, as I told you, it is also out of need that we do so. If you buy 1kg of liquid fertilizer it will cost you 60 [Peruvian] soles. Now, our way of making it ourselves is far cheaper. A dozen eggs will cost you 2 soles and 50 cents. Hence, we cut some wood, sell it, and use the money to buy the eggs. Thus, we get it almost for free.*

This is just one example. There are several others, many of them consisting of the reintroduction of ancient practices (Stuiver, 2006, discusses this as 'retro-innovation'). The point is that technical efficiency of production is raised (i.e. the same amount of resources results in higher yields or the same yields are obtained from a reduced input of resources). Substitution of external (and expensive) inputs by labour is often strategic to novel practices. Thus, through peasant innovativeness, the *utility* of production expands.

Fourth is diversification within farming. Today, people sow and plant far more varieties than 30 years ago. The range of food products has been enlarged considerably. Alongside this change from 'cash crops' to 'food', the number of cattle has also increased enormously, not only for milk and meat (and notably for traction),[19] but also because herds function as savings and insurance.

Taken together, these four elements infer a strengthening of the peasant economy under current conditions of hardship, poverty and exclusion.

Together they indicate how autonomy is wrought under conditions of extreme dependency – indeed, wrought through the creation, reproduction and developing of an autonomous self-controlled set of resources, both social and natural, that are combined, used and further developed in order to make a living, or at least part of it. Thus, the rural economy is increasingly structured as a peasant economy: autonomy is augmented in order to reduce dependency upon input markets, production is diversified to reduce dependency upon output markets, and at the level of the household a multiplicity of income-generating activities is undertaken. Equally, local knowledge is developed resulting in novelties that further spur both yields and autonomy.

These new responses or lines of defence also reveal a range of weak points, and it is at such points that new forms of cooperation are likely to emerge. New cooperative forms of micro-credit can be highly useful in situations where there are hardly any resources available at household level. Knowledge-sharing through informal study groups could spur novelty production and disseminate the most promising outcomes. The formation of strong producer groups can strengthen peasants' position vis-à-vis the Water Board and other public entities. These, then, are the new organizational solutions that could, at this moment and within the current context, strengthen the cooperative side of the peasantry of Catacaos.

Meanwhile: The rise of Empire

Alongside current peasant economies, new forms and new spaces of production are emerging that relate in a completely different way to the local and regional environment. Let me begin by discussing the building of a food empire on top of the existing peasant economy by using an example from the nearby Chira Valley. The Chira Valley has quite a different landscape from that of Bajo Piura. In the Bajo Piura Valley the land is barely above river level (in fact, protective dikes are necessary), whereas in the Chira Valley the river passes through a deep gulley. Consequently, Bajo Piura can be irrigated by gravitation, whereas in the Chira Valley large pumps are needed. Apart from this there is a second important difference. The Chira River always has an abundant flow of water. The Piura is highly capricious: sometimes it contains too much water; mostly, however, it has far too little.

Along the Chira River, pumping equipment is a standard asset of nearly all farming units, including the small ones. Petrol available in local stations, water that runs abundantly, the peasants' own land, labour and craftmanship, the specific knowledge of agricultural engineers, and inputs such as seeds, fertilizers and pesticides available in local shops all form the required ingredients for an efficient rice-producing network. There is just one ingredient lacking: the availability of working capital and/or access to the banking circuit for the credit required. The peasant units of production, with few exceptions, have insufficient working capital (in savings or animals that can be sold). This is due, among other things, to the frequent occurrence of climatic disturbances caused by the phenomenon of *El Niño* – disturbances that cover a three-year period (first very high temperatures, then extreme rainfall, followed in the third year by extremely low temperatures). These natural disasters heavily impoverished peasants of the area and led to indebtedness with the banks, which refuse to lend further credit. Periods with product prices below production costs, as frequently occurs due to globalization and liberalization, have the same effect. Thus, it is impossible for peasants at this moment to obtain and interlink all of the required resources.

The solution to emerge from this standstill is that rich entrepreneurs (mostly with no agricultural background and sometimes of foreign origin) have rented the land and pumping stations. They establish large blocks, one of which I came to know well – a block of 540ha. These new agrarian magnates engage local peasants (for 10 Peruvian soles a day) to work this land, they hire agronomists (for a net salary of 640 soles per month) and they buy the required inputs. The total spending thus amounts to more or less US$1100 per hectare. For 540ha, this amounts to some US$600,000. It is important here to note that no fixed investments are involved. The physical infrastructure (land plus improvements, as well as the pumping stations) is simply rented on a yearly basis – and can, therefore, also be quite simply abandoned. Hence, all of the features of a hit-and-run industry are present.

With the help of agronomists it is possible to produce some 11 tonnes of rice (*arroz cascara*) per hectare. At current prices (1.40 Peruvian soles per kilo), this amounts to a GVP per hectare of US$3500 and therefore to a total GVP of some US$1.75 million.[20] The total profit for this 540ha block, therefore, will be more than US$1 million.

This highlights one of the main paradoxes of developing world agriculture. There is potential for considerable wealth; but local peasants and farmers are unable to access it. It is beyond their reach because of the poverty (and, consequently, the lack of means) in which they are entrapped. Thus, only through the construction of a new food empire, centred around the availability of capital,[21] can rice production be effectively organized and the corresponding wealth garnered. However, due to the structure of this particular food empire, the generated wealth is accumulating in the hands of one (external) agent. There is no 'trickling down', as would have been the case in a loosely structured network of interconnected peasants with working capital or access to it. Thus, the mirror side of the wealth concentrated in the hands of an alien entrepreneur is the 'grapes of wrath' – that is, Empire here equals a truly *parasitic* network (Feder, 1971). Food empires of this nature and structure are *vampires*.

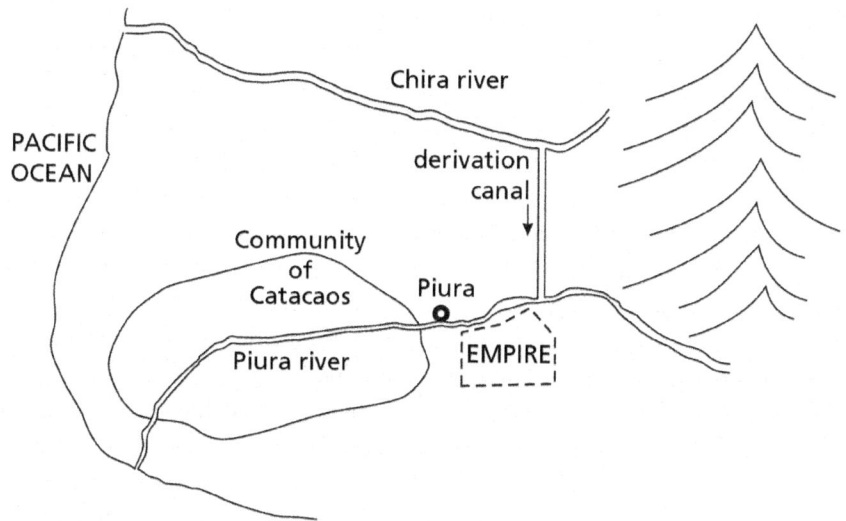

Figure 3.3 *'Clever geography'*

Source: Original material for this book

They digest, as it were, local resources until they are exhausted,[22] and transport the wealth obtained towards other places. Thus, a 'resource curse' is being created: the richness entailed in the available natural resources translates into poverty of people and land (Ross, 1999; Sachs and Warner, 2001; Melhum et al, 2006; Zhang et al, 2007). Empire explains this, at first sight, highly enigmatic nature of the resource curse.

The agent controlling this food empire was able to buy with the profits from just one year 1000ha in the community of Castilla in Bajo Piura, adding the acreage to previously obtained land, so that a new 2670ha enterprise could be formed for the production of organic bananas. This is how Empire expands. It hardly creates any *additional* wealth; it simply taps into locally produced wealth in order to concentrate and reuse it according to its own logic.

The type of food empire that has emerged in the Chira Valley does not function in a similar fashion in Catacaos. The high degree of repeasantization excludes, at least so far, the rise of a similar constellation (although incipient forms are widely detectable). The way in which Empire is emerging and manifests itself in Catacaos is not so much *on top* of the existing peasant economy, but rather *alongside* it. Nonetheless, it also 'drains' the community. In other words, it behaves as much as a vampire as it does in the Chira Valley. Control over the water is crucial in this case – as it always has been in the long and conflictive history of Catacaos (Revesz, 1989).

At first sight here, Empire is located in a 'non-place': in the semi-desert behind the small village of Chapairas in the Medio Piura Valley. Of course, the location is a very clever one (see map in Figure 3.3) because it is located very close to where the derivation canal that links the Chira to the Piura River enters the Piura. Here, the Piura is filled with a large volume of water. Empire taps off this water before it passes the *boca tomas* that feed the irrigation canals which run to Bajo Piura, where the community of Catacaos is located. Empire, in synthesis, is built on a very clever use of geography. This is also reflected in another fact. During the establishment of the enterprise (the first 400ha) and then during expansion to the current 1500ha, the land could be bought from the Regional Ministry of Agriculture for very little because it was semi-desert.[23] Empire here basically concerns, as in the Chira case, a (re)assembling of resources already available. These are combined with a few resources that have, so far, been lacking. The (re)assembly involves available desert land; irrigation water; high-powered electricity; barbed wire and arms; machinery; processing plants; seeds; cheap and abundant labour; professional expertise; access to, and knowledge of, international trading channels; cooling containers; the facilities of the Paita harbour; support of the authorities; organizational capacities; capital and access to credit; fertilizers (including organic fertilizers); drip-irrigation technology; and so on (Figure 3.4 gives a schematic overview of these, initially de-linked, resources). Out of all these elements a specific sociotechnical network is assembled (see Figure 3.5). The creation of new interlinkages is, in this respect, strategic. As such, the semi-desert is nearly useless (unless you want to combine it with the hardship of being a herdsman

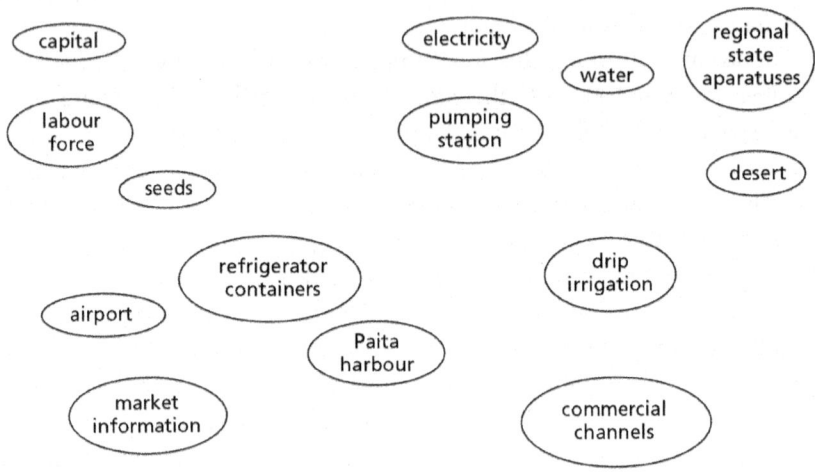

Figure 3.4 *Available but de-linked resources*

Source: Ploeg (2006d, p429)

or poor peasant). It only becomes valuable (extremely valuable) if you are able to link it to other elements, such as water. That is precisely what happened. Along the already discussed section of the River Piura (between the entrance of the derivation canal and the *boca tomas*) a huge pumping station was constructed, a high-powered line was installed to feed this station and a canal running to the acquired semi-desert lands was established. Typically, this canal is not made with concrete (as is normally the case), but just vested with a special type of plastic: ready, as it were, for the next move – no so-called *sunk costs* will bind Empire to this specific place.

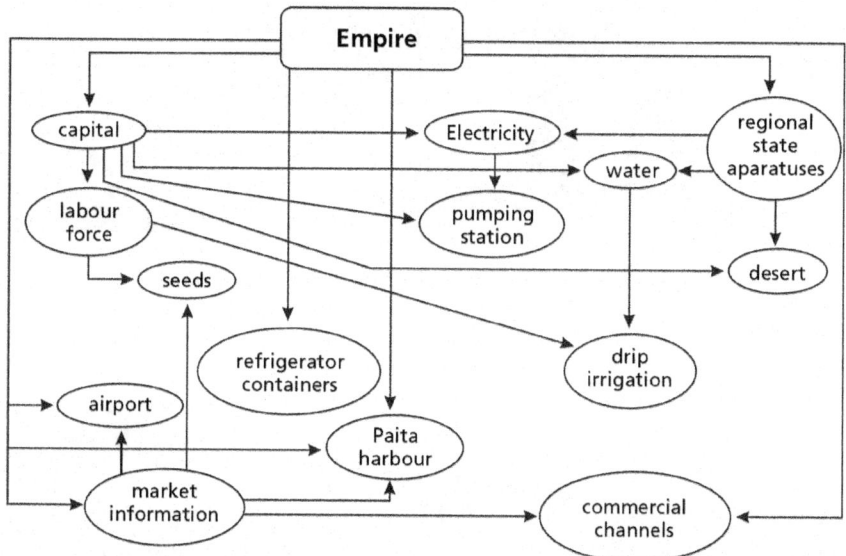

Figure 3.5 *Modelling the world according to Empire*
Source: Ploeg (2006d, p430)

Visiting this expression of Empire for the first time hurt my eyes. Empire represents here in too many respects a complete denial of recent Peruvian history. Some 30 years after the radical land reform that began in 1969, Empire appears to represent the re-emergence of the *latifundio*, the large-scale plantation controlled by foreign capital.[24] As they say in Catacaos: 'the *gamonales* are back'.[25] Empire is equally vested with tens of kilometres of barbed wire (the *culebra gigante*, or 'giant snake' in the novels of Manuel Scorza, 1974: a snake that slowly garrottes the local communities) (see Figure 3.6), with armed guards (see Figure 3.7) that are continuously patrolling and huge machinery (see Figure 3.8) preparing the next cycle of expansion. Within Empire there are artificial lakes (see Figure 3.9) that sharply contrast with water scarcity in the surrounding peasant communities (see Figure 3.10).

Figure 3.6 *Representations of Empire: Barbed wire*

Source: Photograph by Jan Douwe van der Ploeg

Figure 3.7 *Representations of Empire: Armed guards*

Source: Photograph by Jan Douwe van der Ploeg

Figure 3.8 *Representations of Empire: Machinery*
Source: Photograph by Jan Douwe van der Ploeg

Figure 3.9 *Representations of Empire: Artificial lakes*
Source: Photograph by Jan Douwe van der Ploeg

76 *The New Peasantries*

Figure 3.10 *Representations of Empire: Water scarcity*
Source: Photograph by Jan Douwe van der Ploeg

Empire is dedicated here to the production of high-quality peppers, paprika, organic bananas, organic sugar, rice, onions and consumption grapes, using drip-irrigation technology. All of these crops are highly labour intensive (especially during the harvesting).[26] At times, some 1500 people are employed. When the harvest is ready on one side, they restart on the other. Empire also produces *cangrejo de rio* (river crab) in its artificial lakes. Most products are processed in an internal processing plant. There is daily transportation (in Maersk cooling containers) to the harbour of Paita. There are other expressions, other 'tentacles' of Empire that specialize in, for example, the cultivation of asparagus. This is the case in the valleys near Trujillo. The asparagus is transported by aircraft and offered as fresh produce on European markets for the incredibly low price of 1 Euro for 0.5 kilos (see Figure 3.11). In addition, they are bottled and frozen. Frozen asparagus goes mainly to Poland where it is broken up and integrated within frozen pizzas for the Western European convenience food market.[27]

So far, I have accentuated the material elements of the assembled sociotechnical network: the pumping station, plastic-lined irrigation canals, semi-desert, processing plants, cooling containers, etc. It goes without saying that the same network also contains a range of specific social elements, such as the centralized planning-and-command structure; favourable export regulations (more generally speaking, a 'free trade regime'); political support from central government that favours the agro-export economy and from the regional ministry for the cheap sale of land; and power companies that deliver electricity

Figure 3.11 *Fresh Peruvian asparagus sold on European markets*
Source: Photograph by Jan Douwe van der Ploeg

at a low price and the Water Board that likewise supplies water. In addition, the network includes a cheap and abundant labour force (which goes back not only to national legislation, but also to the demolition of trade unions, peasant communities and other popular organizations during the 1990s). Without these specific institutional arrangements the assembled network would be impossible.

Empire is a mode of patterning (see again Figure 3.5) – a specific way of assembling material and institutional resources into a network, the structural characteristics of which imply hierarchy, continuous conquest, submission and exclusion (these are aspects that are also identified and elaborated by Colás, 2007). It constitutes a complex techno-institutional network that is not designed to coordinate ongoing activities and processes; instead, it *imposes its own order* – even if such an imposition is highly disruptive when perceived from angles other than the imperial one. Capital plays a confusing role in such a network: on the one hand, it is indispensable; on the other, it is of only secondary relevance. It is crucial to obtain and assemble the resources (both material and institutional) in a particular way. It is because of the availability of a certain level of capital that the high-power line, water, land, seeds and drip technology can be mobilized and combined. And it is the logic of capital that defines the specific modalities of both mobilization and assemblage. Resources are not moulded, developed and combined in order to spur development – their only rationale concerns capital accumulation. At the same time, though, it is also the case that hardly any capital is brought from outside into the local situation. Capital is basically mobilized on the national capital market through the promise that the cash flow generated by means of the new enterprise will render considerable profit and security. The local situation and the resources and potentials entailed in it are used as *collateral*. Hence, capital is just a part of Empire – and definitely not its core.

The first impression of Empire, here in Chapairas, is that it has been made out of nothing, which gives the strong impression that Empire brings 'development'. Previously, the semi-desert produced barely anything at all; today it is green and contributes to exports. The contrast with the neighbouring peasant community is, at first sight, overwhelming. The latter is arid and hardly productive, while Empire is blossoming and growing. It is the spade and the donkey *versus* heavy tractors, processing plants and computer technology that provide up-to-date market opportunities in Europe and the US. On closer inspection, however, one cannot neglect the aspects of *substitution* and *exclusion*. In the Chira case, the main feature was direct expropriation; but in this case it is, above all, substitution. Production areas such as Bajo Piura, of which the community of Catacaos represents the largest part, are simply being substituted by new spaces of production – seemingly arising out of the blue – that become the major producers and exporters. Literally, a new order has been put in place.

Evidently, such a substitution does not mean that the two spaces, Empire and Catacaos, are disconnected – on the contrary. Pivotal in this interdependency is the control over water. The water used in Empire is no longer available to Catacaos – that is, substitution also brings *draining* – even in the literal sense of the word. The community faces increasing water shortage, and it is likely that the cultivation of rice will soon be forbidden (it consumes too much water). Water is also very limited or insufficient for second harvesting. Engineers directly relate this to the diminished capacity of the Poechos reservoir, even though the water flows to Empire play a decisive role. Indeed, it has been argued that drip irrigation[28] is highly efficient compared to the flooding techniques practised in Bajo Piura. This might be true; but it is equally the case that during field visits I noticed considerable leakages in Empire. Water efficiency in itself is no target whatsoever. What counts is profitability, which in the end drains water, labour and development opportunities. That is how 'the future is stolen' – yet again.

In the cases discussed so far, Empire *contributes* nothing. Empire just links (or re-links) *already available* resources. Empire is nothing but a network that (re)assembles already available resources into a specific pattern – a pattern that allows for control and for extraction. In order to establish such a pattern, alternative modes of patterning (such as that entailed in peasant farming in the Chira Valley and in the community of Catacaos) are to be expropriated or substituted.

Earlier I referred to the clever geography of Empire. In this respect, it is telling that along the diversion canal that runs from the River Piura to the River Chira there are now some ten enterprises that are more or less identical to the example described above. Among them is one belonging to the army. Initially, this canal was meant to strengthen agriculture in Bajo Piura (Catacaos included), yet is increasingly being taken over by large enterprises.

Thus, the peasant economy is simultaneously being subordinated and destroyed by Empire, as much as the latter is fed by the cheap labour provided by the peasant economy and by the shift of resources and development opportunities from the peasant economy to Empire. This typical combination represents a sharp rupture with the past – not only with the decades in which the peasant economy was dominant, but also compared with times prior to the land reform of 1969. In other words, Empire is not just a tragic return of large landholdings. Empire is structurally different from the cotton-producing *haciendas* that once dominated the Bajo Piura area in three fundamental ways. The first concerns the co-existence – as uneasy and unequal as it was – that linked *hacienda* and peasantry. *Latifundia* and a myriad of *minifundia* co-existed alongside each other: a specific and unequal division of labour and land tied the two together. The peasant units provided cheap temporary labour needed by the big enterprises during harvesting periods and at other peaks in the labour cycle. And, in their turn, peasants earned much needed cash when working for the *haciendas*. In addition, peasants produced a range of food crops required by the households of permanent *hacienda* workers. Typical of this co-existence was the fact that peasants could also produce the same crops (in Catacaos, cotton was especially important) produced by the large enterprises – they only had to sell it at unfavourable prices to the large landowners who controlled the cotton mills. Today, such a co-existence – as unequal as it was – is increasingly missing. Evidently, peasants are no longer needed. They are, as Bauman (2004) puts it, increasingly doomed to live 'wasted lives'. At best, Empire needs their resources, land and water. It also needs what remains once the peasantry is destroyed – namely, the cheap labour of people with no alternative. Co-existence in the form of delivering (despite unequal terms) the same products is likewise excluded today. It would require far too high a transaction cost to collect the small and non-uniform product flows from peasants and add them to the large and homogeneous flows that result from Empire. A second difference concerns the prospect of discontinuity compared to the relative stability associated with the former *haciendas*. The latter realized huge investments, not only in units of agricultural production but also, and increasingly, in units for processing and trading. The associated 'sunk costs' translated into a certain 'rootedness' (that could only be broken by radical land reform). In the case of Empire, such investments are basically lacking (and anyway only marginally relevant). Empire is a hit-and-run phenomenon. As soon as conditions for production and trafficking are better in some other place, Empire will move its 'roots', leaving behind only ecological destruction and a generalized impoverishment.

In the third place, there is, I think, an important difference in so far as spatial patterns are concerned. The typical *hacienda* once represented a kind of complement (albeit highly unequal) at the global level. The *haciendas* in South America produced raw materials needed for the industries in Europe: cotton for the textile industry, soya for intensive husbandry and so forth. In turn, European agriculture specialized in high-value products such as vegetables, meat, etc. Thus, the dynamics of the *hacienda* system was relatively indifferent to European farmers. Empire, on the contrary, produces a twofold set of negative effects. By producing a range of vegetables in Peru, it not only negatively affects Peruvian peasants, it also marginalizes and, in the end, destroys a wide range of vegetable growers in Europe. Asparagus is a case in point.[29]

The peasant community and Empire

Are peasant communities such as Catacaos able to operate as a countervailing power vis-à-vis Empire as they did in the past under the *hacienda* system?

At first sight this question cannot be answered but in a negative sense. At the end of 2004, the community of Catacaos was not even a shadow of the vibrant, strong and creative peasant community of the 1968 to 1995 period. Through a range of interlocking processes and factors, both internal and external (such as increasing poverty, strong repression, changing legislation, internal conflicts, corruption, etc.), the community seems to have disintegrated markedly. A dramatic expression of this was not only the disappearance of the co-operatives, but, in particular, the disintegration of the communal units for production – once the pride and joy of the community. True, co-operatives had been eliminated almost everywhere in the country; but the disintegration of the UCPs into individual holdings was a tragedy. They had represented not only a peasant response to the land reform as designed by the state, but the UCPs were – with their highly intensified production – understood to symbolize the peasant road to the future.

Similarly, the impressive range of services the community previously supplied to its *comuneros* (e.g. health centres, dental services, machinery, drinking water and information) has now almost completely disappeared, while the capacity of the communal leaders to intervene, within and outside the community, has evaporated. Beyond this, I had the impression that there was far more poverty now compared to the situation of 30 years ago. But probably even more serious was the lack of hope and expectation of a better life and the willingness to fight for it. Where once hope governed, there now exists despair and wrath. However, in 2004, there was laughter as well, and where there are jokes, irony and laughter, rebellion and protest cannot be far away.

Throughout the 1970s and 1980s the community of Catacaos was unified in terms of political *and* economic power. Through this unity important reforms were possible and important concessions could be obtained from both the state and the capital groups that controlled the processing and commercialization of cotton. Technically speaking, the financial transfer of 3 per cent of GVP from the co-operatives and UCPs to the community – which was used to create and finance all kinds of social services – was a strategic mechanism: it was built upon, and at the same time it strengthened, the unity of economic and political power, and it supported, both materially and symbolically, the 'unity of all poor people of the countryside'. It was precisely here that the state's induced demolition of the co-operatives and UCPs played a crucial role in the demise of the community. It destroyed the symbiosis of political and economic power embodied in the community,[30] and with it the decisive role of the community in the spheres of marketing and processing, and in the provision of social services. The backbone of the community was 'effectively broken' – to paraphrase a once well-known slogan that accompanied the state-promoted land reform. The same went for other features: possessing individual property titles, it was easy for *comuneros* to ask themselves why they still needed the community.[31]

Nevertheless, I disagree with those who argue that with the demolition of the co-operatives and the UCPs, the peasantry of Catacaos lost its potential for struggle and reform. Neither do I believe that the community will fade away, that only a memory will remain. Throughout history, peasant communities have frequently been 'reconstituted'. This point of view is beautifully illuminated in Diez Hurtado's (1998) study of processes of community formation in the Sierra of Piura covering the period of 1700 to 2000.

The peasant mode of farming always entails a balance of communal versus individual interests. The precise nature of such a balance depends, of course, upon its allocation in time and space. Cooperation, in whatever form, is always a strategic and indispensable *institution* within peasant societies, especially when faced with a hostile environment. Cooperation represents a much needed, though not always effective, line of defence. However, as a fundamental institution, cooperation does not imply that its organizational *form* will always remain the same.[32] In fact, there are many different forms, of which some may be more apt, more adequate and more efficient than others – depending upon the situation. And, with a change in conditions, forms that were initially highly effective might be considered inadequate, whether through internal degradation or through changes in the context.

Given the current conjuncture, it would be ludicrous to dream of the recreation of co-operatives and UCPs. Even if the previous years represent, in collective memory, the 'epoch of the powerful community' (as many *comuneros* in Catacaos argue), the once available legal framework no longer exists. Nor is there any willingness among peasants, in the current situation of misery, to lose whatever piece of control they have over their individual plot, their own labour and the few other resources they dispose of – the more so because distrust is now highly generalized after the previous episodes.

The co-operatives and especially the UCPs have been, without doubt, very important as a mechanism of the transition that ended *gamonalismo* – that is, the social formation characterized by fierce control by large landowners over the local and regional economy. It is equally the case that the impressive repeasantization in Catacaos would have been impossible without the UCPs. Facing a hostile ecosystem such as the semi-desert and fighting the Water Board and the *Banco Agrario* to obtain water and credit would have been impossible if undertaken by individuals alone. In that context, the UCP as a specific *form* of cooperation was indispensable.

However, the organizational expressions of cooperation should not be discussed outside of their specific contexts – they should not be separated from, nor 'exported' from, their time and space (which is exactly what some political parties from the left and several non-governmental organizations, especially those with Roman Catholic inspirations, are doing). The central question now is how to design and mould *new* organizational modalities of cooperation (and self-defence) that are in line with the current situation and that correspond to the current needs of the community as an institution. Behind the *organizational* disorder of the moment, the community of Catacaos seems, nevertheless, stronger than ever. There are three structural reasons for believing so. To begin with, the community is, as an institution, omnipresent 'in the fields' precisely because of the massive process of repeasantization that I described earlier. When the historical context is taken into account – a state-controlled process of land reform aimed at severe de-peasantization – one has to conclude that Catacaos, with its high degree of repeasantization, represents a significant rupture in what seems a general tendency. Due to the communal struggles of thousands – sometimes tens of thousands – of deprived people aimed at constituting themselves as peasants, the community is now materially present in the fields as an elevated man–land ratio, and as more than 20,000 people (see Table 3.2) who try to craft the land in an effective way that allows for both survival and future prospects. This material presence in the fields (an 'indestructible' fact), which is equal to and evokes the strength of the community as an institution, translates into new claims for conditions that allow for working the land in an improved and meaningful way. Given the correlation of socio-political forces, these claims cannot be solely individual ones; they have to emerge, sooner or later, as communal claims. And it is no surprise that such claims will focus, in particular, on access to and use of water.

A second reason for believing that the community of Catacaos is and will remain a strong institution – which could, at appropriate moments, translate into a new organization for action – relates to the question of collective memory. Due to the structure of the *caserios* (the hamlets of the community), the type of story-telling and the ardent political struggles during and after the land reform process, the community has reached a widely shared and clear formulation of its own specific trajectory and the values that are central to it (see Box 3.1). This is perceived in nearly every aspect of everyday life; as local people say: '*carajo*, we really need *la comunidad*'. Collective memory is associated with a notion of superiority. Just as the UCPs were able to realize levels of employment (man–land ratios) far higher than those of the *haciendas* and the state-controlled co-operatives, as well as yields that went far beyond those of other modes of farming, the community as a whole will be able to function – *again* – as the ordering principle that combines available resources with the capacities and the needs of the peasant community in a way that is superior to that of Empire. Figure 3.12 tries to outline, albeit roughly, the essence of this peasant mode of organization that pivots upon the unifying axis of the concept of community. It shows that locally available resources, such as land, labour and water, combined with the historically developed mode of intensive land-use practices and their concomitant modes of organization, might, in the end, be further developed to meet and overcome the social, economic and political needs of the peasants of Catacaos. In this respect, Figure 3.12 reflects the constellation that was *de facto* emerging during the late 1970s and early 1980s. As opposed to the centrality of Empire, the community has to take the lead. In Catacaos, collective memory focuses on the notion that 'we can do better than others'. This notion, that triggered the fierce struggles of the land reform period, now applies to the ordering of regional society and economy as a whole. Every hectare being irrigated in neighbouring Empire implies the loss of 3ha of productive and producing land in Catacaos. It implies the loss of prospects and dignity, as well as the loss of productive employment. Therefore, sooner or later, the claim for new and better modes of ordering will emerge. And this means that the community will enter again the immediacies of political struggle and power.

The third structural element underlying the potential strength of the community as an institution – and also as an organization – resides in the bitter need to build countervailing power. The old *Agrarfrage* (Kautsky, 1970) or, more in line with the case presented, the *problema de la tierra* (Mariátegui, 1925) has in no way been resolved. It is more pressing, widespread, complicated and urgent than ever. The ranks of those who have to live a 'wasted life' are growing fast, while collective memory stresses how unjust this is – a degradation that is actively provoked by the presence of Empire.

Hence, Catacaos, like many other peasant populations around the world, faces new and radical contradictions between community and Empire. It is especially bitter in the case of Catacaos because these contradictions centre on water and its unjust usurpation by others. In fact, Catacaos is currently suffer-

84 *The New Peasantries*

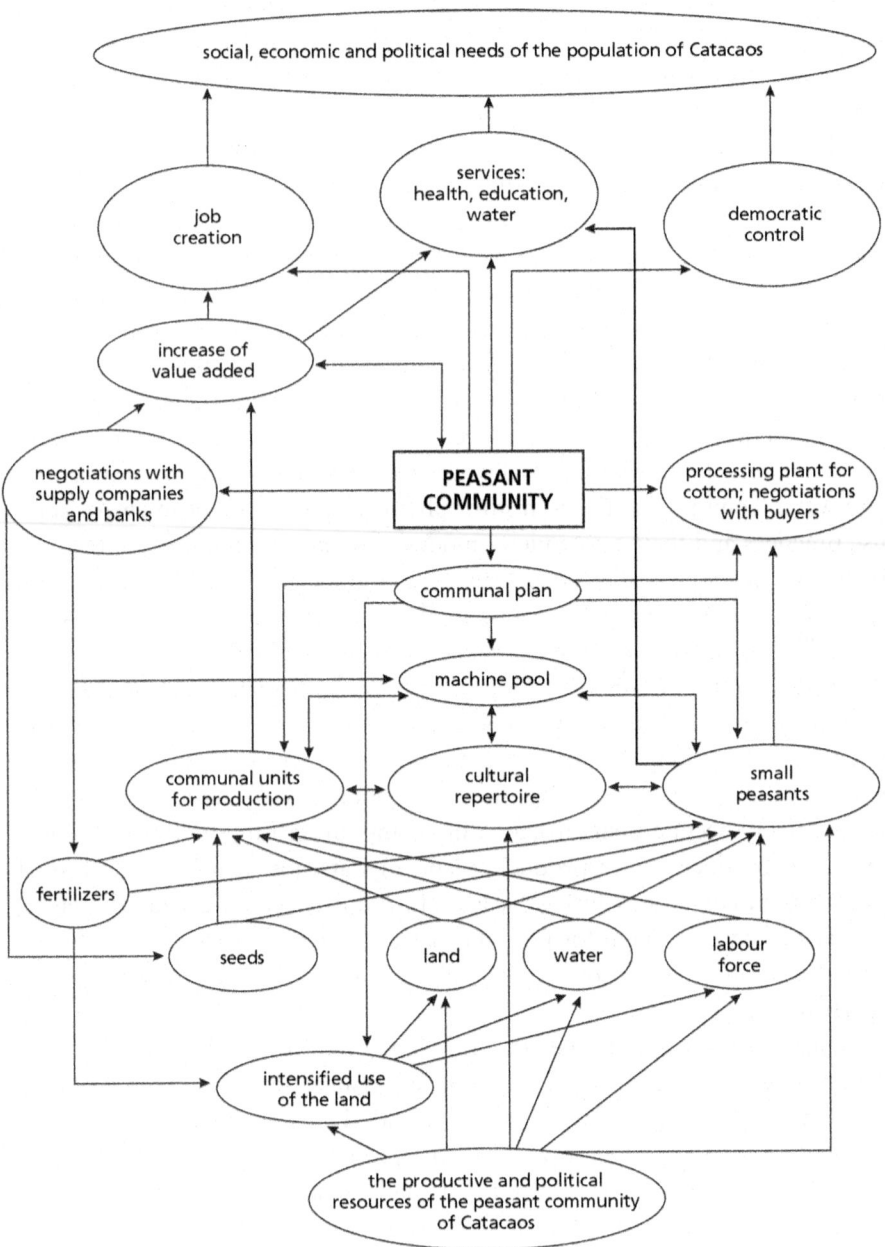

Figure 3.12 *An alternative patterning*

Source: Ploeg (2006d, p433)

ing its third wave of water appropriation. The first two have been redressed (see Ploeg, 1977, 2006d). The current wave, provoked by Empire, will likewise be redressed. For the three reasons I have already indicated, the community of Catacaos will reconstitute itself around the claim for water and access to resources and markets that are currently denied it. And, due to the nature of Empire as a permanent threat to the peasantry, the resulting struggles will probably be more radical and far-reaching than ever before.

4
Parmalat: A European Example of a Food Empire[1]

Introduction

Chapter 3 provided a discussion of two Latin American expressions of Empire. I now wish to explore further the nature and dynamics of newly emerging food empires, focusing, in particular, on the *patterns* they create, which increasingly order significant segments of the world's food industries. Central to the analysis here is the case of Parmalat – the Italian multinational that crashed at the end of 2003 when it became clear that more than 14 billion Euros had been 'lost'. I will also make occasional reference to other cases, such as that of Ahold, the world supermarket chain that almost collapsed at the beginning of the same year.

Parmalat is exemplary of Empire:[2] first, because it was never other than a specific pattern that interlinked already existing productive and distributive activities, while at the same time submitting them to centralized control (to a new *cupola*) and to new ordering principles; and, second, because Parmalat managed to create, within this globalizing pattern, new linkages between places of poverty and places of wealth. Using a new assemblage of already existing technologies it could link, for example, poor production areas from the East to the prosperous consumer markets in Western Europe. *Latte fresco blu* is the icon for this part of the story. Linked to these first and second aspects is a third that lends weight to the exemplary nature of the case – that is, Parmalat never represented or created any additional value whatsoever: it simply centralized the value already produced by others, while simultaneously destroying other wells of value and systematically degrading the notion of value as such.

Intrinsic to Empire is a far-reaching redefinition of the notion of value: a redefinition strongly associated with an unprecedented extension of the process of commoditization. This actively spurred extension is also inherent to Empire as an ordering principle (Alexander et al, 2004). Extension proceeds as the ongoing and often drastic conversion of non-commoditized domains of social and natural life into new commodity spheres. *Access* to particular domains, for instance, is itself increasingly turned into a commodity.[3] Thus, new commodities and commodity circuits emerge that define and carry new values, while original values are simultaneously redefined and frequently subordinated to those newly defined. New levels (constituted by new commodity definitions and circuits) are put over existing ones. They are not simply an extended expression of existing levels. They introduce, instead, new rationales, thus reshuffling and reorienting the already existing definitions and circuits. If, for example, milk was, in a classical pre-Empire situation, a commodity (whose exchange value depended upon freshness, taste, health effects, etc.), then the dairy factory was, above all, the place where this commodity was being processed. Although this dairy was to be managed as an enterprise (implying well-timed depreciations and acceptable levels of profit), its value was, in the first place, its capacity to process 'raw milk' into a range of commodities and, second, its capacity to reproduce itself as enterprise,[4] if possible on an enlarged scale.

In the new Empire-like situation, however, the processing unit itself becomes the important, if not decisive, commodity. It might be sold, mortgaged, bought and/or leased – not just once and not only in exceptional circumstances. Within an Empire-like framework, it is permanently 'for sale'. Consequently, its primary use value is no longer that it allows for the processing of milk and the associated production of profits (or, in the co-operative model, the associated production of reasonable off-farm prices and long-term security). Its value is, above all, that it is an (exchangeable) asset in a global enterprise that aims at large and growing market quotas, which in their turn allow, for example, the possibility of attracting additional flows of capital, high share prices and/or the opportunity of further expansion. This redefinition (that superimposes global criteria upon local ones) impacts strongly upon the 'original' commodity flows. Collecting milk from primary producers, its processing and the subsequent distribution and selling of dairy products are *only* relevant in as far as they contribute to the newly created (or induced) Empire-like type of dynamics (e.g. through elevated levels of cash flow; see Smit, 2004). Hence, the value of these activities and, consequently, the value of their associated artefacts is redefined. This often goes together with an overall restructuring of the initial activities: they are to be reassembled according to new patterns that allow for control and appropriation.

The mechanics of global expansion

The Parmalat group grew, in a relatively short period, from a small trading company in Collecchio in the province of Parma in northern Italy to one of the biggest multinationals in the dairy industry – becoming what we now call *a global player*. In 2000 it realized total yearly sales of US$6.5 billion and was

29th in the world's top 50 food groups (Lang and Heasman, 2004, p154). In 2003, total sales amounted to 7.6 billion Euros and the group as a whole embraced 260 different commercial societies, with 139 establishments in 30 countries. It counted 36,000 dependents. In Italy it ranked number eight on the list of the largest industrial groups (Franzini, 2004, p12). It expanded at a phenomenal rate. In 1999 it bought the Lactona group, the Union Gardarense in Argentina (for 65 million Euros), a distribution chain in Nicaragua (for 20 million Euros), and, in the US, the Farmland Dairies (for 125 million Euros). Its total cash flow in 1999 equalled 458 million Euros; but 612 million Euros were invested in acquisitions. Its debts also rose in that same year by 500 million Euros, bringing the total debt level to 2.2 billion Euros.

This pattern was repeated in the following years. In five years, from 1998 to 2002, Parmalat invested 2.1 billion Euros in 'shopping enterprises', as Franzini (2004) calls the ongoing takeovers. In the mean time, one of the participating banks, J. P. Morgan, announced (in April 2002) that Parmalat's net profits would more than double to 349 million Euros in 2005, and that bonds would rise from 3.7 to 4.4 Euros (Franzini, 2004, pp93–101).

Mortgaging is one of the main vehicles for this accelerated expansion. Assume that there is a first enterprise (number 1 in Figure 4.1). This enterprise is mortgaged in order to buy enterprise 2, which is subsequently mortgaged to obtain number 3, and so on. This type of expansion is organized in such a way as to obtain increased market shares. As I argue later, an increased market quota can represent a 'value' on its own. This, in itself, rather simple scheme (that can be encountered everywhere) is often modified in the following way: having obtained enterprise 2, the loan (or mortgage on enterprise 1) is converted into shares to be commercialized through the stock exchange. This occurred for the first time in 1989 as a response to Parmalat's first big financial crisis. This shifts the risk from the participating bank towards the stockholders. The emission is, in part, legitimized by the expected ('prognosticated') increase in market quota.

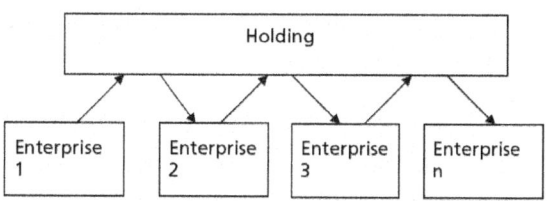

Figure 4.1 *The mechanics of expansion through mortgaging*

Source: Original material for this book

This pattern of expansion was developed into highly complicated financial schemes that involved Parmalat-related establishments in The Netherlands, the Antilles, Luxemburg, Ireland, the Cayman Islands and the US. In the final years before the crash, *financial engineering* increasingly included innovations (which have since become generalized), such as acquiring new enterprises with loans to be repaid by the bought enterprises themselves, which then restructure in order to squeeze out maximum profitability. Another interesting element is that immediately after acquiring a new enterprise, the buildings are sold on lease and then leased back. Thus, even in the short run, considerable additional cash flow can be generated.

The expansion-through-mortgaging strategy and other forms of financial engineering may, of course, come with some specific dangers, especially when driven by the need to obtain an increased market share. Enterprise 2 or 3 or n may be bought at too high a price (to prevent competitors from increasing *their* market share), or possess some form of 'skeleton' in the cupboard (as was, for example, the case with the takeover of the Cirio company).[5]

Nevertheless, the consequence of this pattern of growing through mortgaging is that debts grow proportional to the expansion of the enterprise. The enterprise as a whole is not growing due to *extra value* produced in the first unit(s), but simply through mortgaging them. Hence, debts and magnitude grow hand in hand. Debts might even grow quicker than the magnitude of the concern as a whole, as would be the case if bought units were 'overpriced' or if 'skeletons' emerged. Within this game, *institutionalized trust* is strategic. As long as banks and shareholders trust that the expanding enterprise will produce profits, expansion can be continued. Thus, for the expanding enterprise, positive scores on indicators such as profitability, market share, future prospects, etc. are decisive: they feed and maintain the required trust.[6]

If, on the other hand, such positive scores should be lacking, then the constructed whole will come tumbling down: its 'value' (i.e. the prospect to remain stable and profitable in the longer run) will fade. Banks will require enhanced repayment (or will raise interest rates due to the risky nature of the project as a whole) and stock values will decrease. Thus, a crash will emerge. And that is exactly what happened on 22 December 2003: the bonds and obligations lost 66 per cent of their value in one day. When trust no longer exists (in the form of continued lending and/or stable stock value) to counterbalance debts, only debts remain.

Trust is essential in building food empires – especially trust related to expected financial performance. Expectations are central, and a considerable part of corporate functioning is thus geared towards meeting these raised expectations. Hence, data on growing market shares, and on expected and realized levels of turnover and profitability become strategic. From this two consequences emerge. The first is that a process of ongoing, if not accelerated, growth and expansion is institutionalized within the firm. Accelerated expansion becomes a material and inbuilt need. The second consequence is that a specific form of *opportunism* becomes tempting – that is, the units that compose the firm are put under considerable pressure to construct the most optimistic (indeed, the most *tempting*) data possible. This was what occurred on a large scale in the Ahold holding and which nearly brought it to melt down (Smit, 2004). It happened in exactly the same way in Parmalat, where reported turnover was 25 per cent above actual turnover.

There is a clear interrelation between the two aspects. If accelerated growth implies specific risks (and puts levels of profitability under pressure, rather than raising them), *further* expansion and new *promises* appear to be the most evident way forward to compensate for the relatively disappointing performance realized. Thus, a *bubble economy* is constituted within which promised and subsequently expected performances dominate over current results. Here we encounter the other side of the equation: the future is mortgaged as much as enterprises 1 to n (in Figure 4.1). Future performance is turned into the main justification for current practice, which amounts to a complete turn around of the interrelations between past, present and future (Ploeg, 2003a). Within this new organization of time, trust is no longer historically grounded – it becomes, instead, *future dependent*. It also applies that the proposed performances need to be *made* real – if not, a crash could be immanent, for if the expected (and announced) performances do not materialize, the bubble will burst: the virtual image of the enterprise as it was thought to be no longer convinces anybody.

This, in synthesis, is what occurred with Parmalat. It was, in the end, as simple as that. In public, the *Parmacrac* or crash was discussed mainly in relation to fraud. There had, indeed, been fraud, nepotism and illegal payments to political parties. However, these do not explain the crash. The burst of a bubble is to be explained, in the first place, by the bubble itself.[7] The real mystery is that the nature of the bubble remained unnoticed for so many years. Even at the height of the crash, many people would or could not believe that they where witnessing the breakdown of a giant.[8]

The expansion of the Parmalat group was in no way an exception. During the same period, the Ahold group from The Netherlands (built upon the initial Dutch Albert Heijn supermarket chain) used the same mechanism to grow into another global player (for a thorough analysis, see Smit, 2004). The names of Enron, Worldcom, etc. are enough to recall that this type of expansion occurs everywhere in the modern globalizing economy, just as it had already occurred in Italy with, for example, the Ferruzi concern of Raul Gardini. All of these specific expressions of Empire appeared to be giants – but they were, in fact, giants with feet of clay. Their very expansion created the relations that made them collapse.

The cases indicated are clear expressions of the rise, expansion and, in the end, the vulnerability of Empire (see Ploeg and Frouws, 1999, for a theoretical discussion). The emergence of Empire is related to a range of interacting conditions. A very important one among these is the enormous availability at world level of freely moving capital in search of the highest levels of profitability. It allows for the mortgaging type of expansion described in Figure 4.1. Second, the increasing push to liberalize markets (including food markets) and the availability of technologies that allow distance in time and space to be bridged (especially important for food markets) also permit another crucial characteristic of Empire – that is, the possibility of linking and controlling different spaces. Third, there is the new 'managerial revolution' that introduces new all-encompassing 'planning-and-control cycles' based on a widespread use of information and communication technologies (ICTs) to secure the making of future profits (of whatever kind) to compensate for the historically created debts. It also allows for a centralization of value at the superior level (that is, within the 'centre'). The second and third conditions, in their turn, support the first: increased market shares and centralized power support the mobilization of additional capital.[9]

Empire is a set of more or less interconnected networks, each of which is oriented towards the planning and control of large segments of society. Central to Empire is that it increasingly shapes and reshapes the concrete practices within these segments. Through the access mechanisms it controls, it increasingly becomes impossible to reproduce practices (and the units directly involved) *outside* of Empire. Everything becomes subjected to it – that is, the logic introduced by Empire penetrates and reigns nearly everywhere.

Food empires such as Parmalat are the outcome of this much wider transition in and through which Empire (as a generalized reassemblage of the world) is being constituted. Empire is not only *superimposed* upon the specific domain of food production and consumption. Through food empires, the practices of production, processing and consumption of food are becoming drastically *remoulded*, just as the organization of food flows over the globe. Empire is not just another way of delivering food to the table; it deeply transforms food itself – the way in which it is produced and the way in which it is consumed. Food empires reshape considerable parts of life itself, just as they induce their own new sciences and technologies to re-engineer life.

The rise of large organizations (especially in economic life) is normally justified by reference to their efficiency, which is basically grounded in the fact that they are able to operate with far lower transaction costs than alternative (and smaller) units. The point here, however, is that low transaction costs do not characterize Empire (the opposite might very well be the case). For Empire does not arise out of competition; it is increasingly being created and enlarged through the mere exclusion of other, alternative, forms for organizing life. This was already clear in the description of Empire as manifest in the Chira Valley in northern Peru. It was through the *exclusion* of rice producers from the credit markets that Empire could emerge there. The same applies to fruit and vegetable production in the Piura Valley. It was basically through the *expropriation* of water (e.g. from the community of Catacaos) and the associated exclusion of peasants that the expression of Empire encountered there could be created. Again, in the case of Parmalat, we meet similar instances of actively organized exclusion and expropriation.

Parmalat as a three-tiered network

Figure 4.2 outlines some of the main elements (or resources) that constituted Parmalat as a socio-technical network. Probably the most striking element is that none of these resources is unique to Parmalat. In this respect even the name itself is a telling metaphor. Literally, Parmalat means *milk from Parma*. In reality, very little milk from the Parma province was ever transformed or marketed by Parmalat.[10] And the more the industry grew, the more this tiny part decreased. Nor did Parmalat itself develop any technology. Its main technology, Tetrapak, was hired from Sweden. The same goes for the other elements indicated in Figure 4.2. They were all resources that figured (and figure) in other patterns as well – that is, they were operated (and in many places are still operated) through other resource combinations (through different patterns). Parmalat did not *add* anything to the already available resources, it just represented *control* and *access*. Having access to banks, the stock exchange and to circuits of political power, it could continuously enlarge its control over growing commodity flows, enlarged consumer groups, more and more processing units, etc.

Networks such as Parmalat are increasingly conceptualized as being composed of three levels. The first level concerns *physical infrastructure*. In the Parmalat case it consists of the 'entry' or 'intake points' through which the milk passes from dairy farms (wherever located) into the Parmalat structure: the transport facilities and associated logistic capabilities (e.g. the processing plants, the supply lines that pass the processed products to supermarkets, shops and wholesalers).

The second level concerns the actual *movements* of milk and derived products through this infrastructure. Just as people and freight are moved through a railway system (and cars through a web of highways), milk is moved into, through and out of this network. These movements have a price (as have tickets and highway tolls). As I will indicate later, both the milk-producing and milk-delivering farms and the consumers pay a considerable price (other than the price of milk) for using this network. Society as a whole and farmers who do

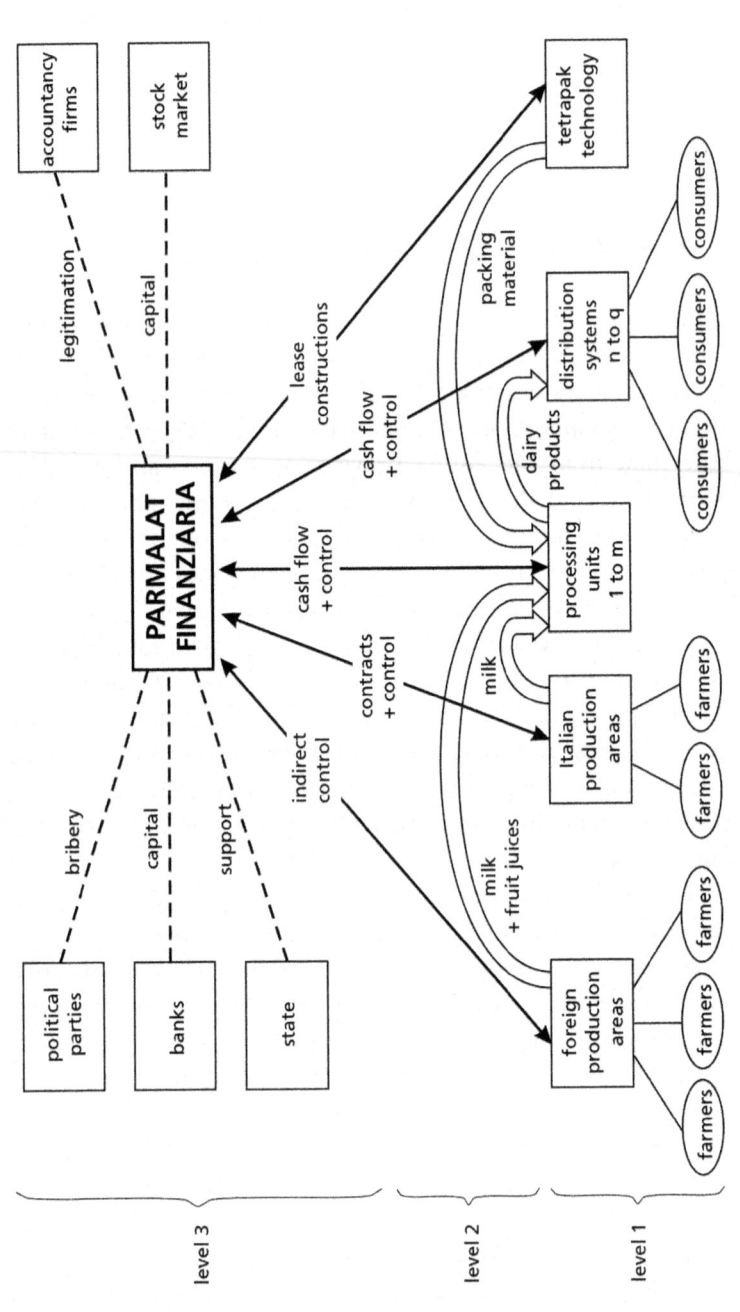

Figure 4.2 *Parmalat as a socio-technical network*

Source: Original material for this book

not deliver to Parmalat *also* pay a price. The benefits associated with these payments are accumulated at a third level, which is properly the level of Empire. From this third level, the first and the second are governed and controlled, expanded or, if need be, contracted, while the values produced through the second-level *movements* (movements made possible by the first level) are *accumulated* at the third level. In Italy, this third level is publicly known as the 'parallel Parmalat' – something existing alongside the 'real' Parmalat (i.e. the first and second level). This 'something' was composed of a jungle of dozens of enterprises distributed all over the world (The Netherlands being one of the strategic nodes) through which complex financial games were orchestrated. Parmalat itself often referred to this third level as the '*Parmalat Finanziaria*'. Major accountancy firms such as Deloitte & Touche and Grant Thornton assisted Parmalat in designing and constructing this 'parallel enterprise'.

Although the third level squeezes profits out of the first and second (even to a degree that turns level-one enterprises into units that are no longer profitable), it could be argued that profits are not the main and foremost objective. What really matters at the third level are new values, such as market share, expected shareholder value, rate of expansion and expected rise of profitability. Profits are important in as far as they sustain these new values (or commodities, as Alexander et al, 2004, would call them), which in their turn attract additional flows of capital (through stock emission or as loans from the banking circuits). This implies that the function of level one, within this framework, is changing in a significant way. Level one is no longer meant primarily to capture milk and channel it (after processing) to consumers, or to deliver dairy products (and fruit juices or whatever) towards consumers. Level one now functions mainly to deliver the new commodities needed at level three.[11]

Level three does *not* represent value,[12] neither does it generate any additional value (understood as social wealth). The financial values managed at level three, and which attribute it with the image of power, are *derived* from levels one and two (and from an outer periphery composed of farmers, consumers, foreign countries, etc.). Level three is '*matrix*': it is nourished by underlying and often somewhat '*subterranean*' activities and entities of levels one and two. The 'value' of level three is that it organizes *conquest*: the takeover and subsequent domination of increased parts of the social and natural world.

The different levels may unfold through seemingly different trajectories. Parmalat is a case in point. As a whole, Parmalat represented after the crash an enormous debt and also industrial and logistic machinery that, due to this debt, could no longer function as an enterprise to collect, process and distribute dairy products. It was broke. However, through the Bondi intervention that followed the crash (Bondi was nominated by the Italian government to intervene in Parmalat to save what could be saved), Parmalat was split, with almost surgical precision, into two parts: the remains of level three (an enormous debt to be negotiated with the participating and partially responsible banks and others), and the initial level-one and level-two infrastructures, processes and movements that (now liberated from level three) could function again as a valid and economically sound set of enterprises.

Did Parmalat ever produce value?

Within and through the three-tiered network described above, Parmalat was able to accumulate considerable wealth, especially through the sometimes near monopolistic position it obtained through its expansion. It was a wealth not so much produced within the centre of this network, but rather wealth that was moved from its periphery towards this centre. I will discuss, in this connection, five flows (see also Figure 4.3).

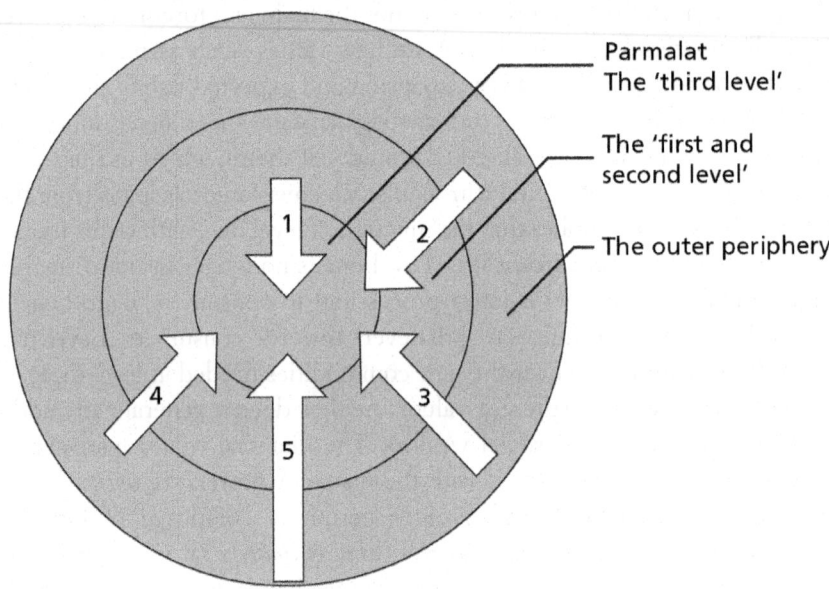

Figure 4.3 *Value flows*

Source: Original material for this book

A first flow of value to the centre of the network evidently stemmed from level one: the productive and logistic units directly controlled by Parmalat. The cash flow generated by these units was directly centralized within *Parmalat Finanziaria*. As a consequence, these units could no longer be reproduced outside of Parmalat. The required working capital was derived from the centre, just as the centre controlled the different outlets. From the centre perspective, the production of milk (or whatever product) was of only secondary importance. The primary concern was cash flow (which, in the end, creates the 'power' of the centre). The levels underlying Empire (the units that carry forward its productive and logistic practices) are only relevant in as far as they produce and enlarge a cash flow that meets central criteria (see Smit, 2004, for an identical description of the mechanics of Ahold).

A second flow originated out of the typical relationship between Parmalat and the dairy farmers delivering 'consumption milk' to it (in Italy, a distinction is made between *latte alimentare* (milk for consumption) and milk for cheese production; the two have to meet highly distinctive criteria at the farm level). Throughout Europe, farmers delivering milk to the dairy industry (especially when it is milk for consumption) are paid 14 days or, at maximum, one month after delivery. Within the Parmalat consortium, however, payment to farmers was from the beginning fixed at 180 days. Over the years this period became extended to 250 days. Given the short cycle between delivery and consumption, this implied that Parmalat, on a permanent basis, disposed of an extra capital of some 400 million Euros. This construction represents an amazing entanglement of local and global levels. The long extended payment period would not have been accepted by dairy farmers anywhere else in Europe. But in the broad surroundings of Parma (the provinces of Fidenza, Parma, Reggio Emilia, Modena and parts of Bologna), this was different. Dairy farmers here have produced from ancient times so-called 'cheese milk', which is processed into Parmesan cheese. This cheese is stocked for at least 18 months before it can be sold. In combination with the co-operative structure of the processing plants, this usually meant that farmers in the area were only paid after 450 days. Thus, the offer from Parmalat of 180 days was not unattractive. Once dairy farmers accepted this offer and changed to the production of consumption milk, they could hardly return to the former production of cheese milk. They became 'entrapped', among other things, because it incurred changing to a different breed of cattle, different feed and fodder, different storage techniques, etc. Thus, Parmalat created a dependency that allowed them to slowly extend the payment period. This increased the financial capital available for Parmalat. One might argue that this was, above all, *virtual capital*, since although it seemed to be part of Parmalat, it actually belonged to others.

Apart from the postponement of payments, which highlights the unequal power relations that existed, there is another telling expression of the relation between Parmalat and its milk producers. Compare the production of consumption milk, for example, with the production and processing of 'cheese milk' (see Table 4.1). In the same area, many dairy farmers deliver their milk to the small co-operative *caseifici* plants, in which milk is converted into Parmesan cheese. When calculated properly (14.30kg of milk are needed for 1kg of

Parmesan cheese), the consumer pays 0.99 Euros for 1 litre of this cheese milk, of which the dairy farmer receives 51 per cent (i.e. 0.50 Euros per kilogram of milk). This is in sharp contrast[13] to dairy farmers related to Parmalat, who received only 25 per cent of the average consumer price (i.e. 0.33 Euros).[14]

Table 4.1 *Contrasting value chains (Euros/kg, January 2004)*

	Consumption milk		Cheese milk for Parmesan cheese	
Milk production	0.33	25%	0.50	51%
Processing and storage	0.70	52%	0.24	24%
Storage and distribution	0.30	23%	0.25	25%
Consumption	1.33	100%	0.99	100%

Source: Ploeg et al (2004a, p23)

Cheese milk, especially cheese milk required for the production of Parmesan cheese, has to conform to exceptionally high-quality criteria – hence, the obvious price differential. But it is difficult to understand why the processing and storage of consumption milk, which is far simpler and shorter, should require a share that is both absolutely and relatively far higher than that of cheese milk. If the same *relative* distribution were applied, the farmers' price for consumption milk would be far higher.

Unequal power relations are also reflected in the fact that there is a small price differential within the realm of consumption milk. Some producers of consumption milk in Italy, such as Granarolo, pay, on average, 1 Eurocent extra per litre of milk. If this difference alone is introduced into the calculations, then it amounts to a total transfer, over 25 years, of 212.5 million Euros.

The flow of value from dairy farmers towards the 'centre' of the system also passes through other channels. It is estimated that some 1 million tonnes of *black milk* (i.e. milk illegally produced outside the quota system and associated controls) circulate each year in Italy. In the case of black milk, farmers get paid just 40 to 50 per cent of the official milk price. This implies again a considerable transfer of value. In addition, there are indications that at least some milk is 'constructed', using milk powder and butter oil. Although this is forbidden (except for some yoghurts and cheeses), control over it is, especially in Italy, of dubious quality.

A third flow of value runs from consumers towards the 'centre'. It is related to the excessive difference – in Italy – between the prices paid by consumers and the prices paid to the farmers. Consumer prices in Italy are among the highest in Europe (Menghi, 2002). In *absolute* terms, the price paid to farmers is also high; but in *relative* terms (i.e. compared to the consumer price), it is low (see Table 4.2) – just 35 per cent of the price paid by consumers

– while in Germany, Belgium and The Netherlands this is between 50 and 60 per cent. France and Denmark have a level comparable to that of Italy. The UK shows the lowest level (23 per cent) since the demise of the Milk Board dramatically changed relations between producers and consumers.

Table 4.2 *Prices for farmers in relation to prices paid by consumers*

Product	Netherlands	Belgium	Germany	Switzerland	Italy	France	Denmark	UK
UHT milk (1 litre) for long-term conservation (VAT excluded) (Euros)	0.51	0.52	0.59	0.85	1.02	0.96	0.96	1.12
Off-farm price (Euros)	0.31	0.31	0.30	0.34	0.36	0.31	0.33	0.26
Off-farm price as percentage of consumer price	61	60	51	40	35	32	34	23

Source: Unalat (2002, p25)

If Italian consumers had paid 0.89 (the European average) instead of 1.02 Eurocents for UHT milk and 1.30 Euros for each litre of milk, they would have saved between them some 470 million Euros per year. However, due to the mechanics of food empires, this amount went the other way. It was shifted from the 'outer periphery' of the network towards the centre of Parmalat.

A fourth value flow relates to the massive import of milk, mainly from Germany, but also from France and later from Poland.[15] In these countries milk can be obtained at prices far lower than in Italy. During 2002, Parmalat processed 850 million kilograms of consumption milk produced in Italy and 380 million kilograms from imports. Taking into account the price difference (on average, 0.06 Euros per kilogram compared to Germany, and 0.17 Euros per kilogram compared to the Polish price level) and deducting 0.01 Euros per kilogram for transport and logistical costs, this implies that through *outsourcing*, Parmalat obtained at least an additional value flow of 19 million Euros per year.

A fifth value flow relates to Parmalat's capacity to tap the savings of civilians, mainly in Italy but also internationally, through share issues and the banking circuits. The accountancy firms that were supposed to judge the solidity of Parmalat, the institutional 'watchdog' of the stock markets (especially the Milan one) and, indirectly, the Central Bank of Italy all played a role in the production of the required trust. In retrospect, one can only wonder about their responsibility and the way in which this was translated (or not) into factual checks. Anyway, with the crash, private shareholders suffered losses of billions of Euros. In one day (22 December 2003), 7.9 billion Euros were reduced to 2.8 billion Euros: 5.1 billion Euros just evaporated.

A new mode of 'harvesting'

When analysing the *Parmacrac* it is crucial to distinguish between the different levels. Level one featured the plants, the networks and logistical procedures. They ran and channelled to level two: the movement of milk, dairy products and other commodities. And beyond them at the third level was Parmalat as a continuously expanding global player. One of the major surprises after the crash was that the first level continued to produce, and it actually augmented its total output[16] and turned out to be profitable again. But as discussed earlier, this was due to the Bondi intervention, which almost surgically separated the first and second levels from the third level. It was this third level, with its huge debts (debts which Bondi tried to negotiate with international banks, accountancy firms and others that had played a decisive role in the expansion and crash of Parmalat), that played a decisive role in the expansion and demise of Parmalat.

This distinction is important for making the point that this third level (i.e. Parmalat as food empire) never produced any added value in contrast to the already existing first-level units that were gradually taken over and submitted to the global network. Empire produces no value. It simply takes over (expropriates) and accumulates value produced at the lower levels and at the periphery of the system.

'Harvesting' was not limited to direct interrelations within and between the first, second and third levels. Summarizing the value flows discussed above, it might be argued that Parmalat obtained value especially due to the unequal exchange patterns that could be created and sustained due to its oligopoly nature. Parmalat, in this respect, represented above all *extra-economic power*. Consequently, farmers received a relatively lower price (see Table 4.1), and consumers paid a relatively higher price (see Table 4.2); Parmalat was able to achieve this because farmers and consumers became increasingly dependent upon Parmalat as a network, just as farmers elsewhere (e.g. in parts of Poland). Through all kinds of 'non-economic coercion' applied in the associated domains, Parmalat, as a third-level corporation, could obtain and centralize considerable wealth – not because it produced this wealth but because it could reap it from those who became dependent upon it. In this, there is a remarkable difference from 'classical capitalism': an industrial enterprise embedded in relations of competition (i.e. in a 'free market' context) could never have established price levels (as illustrated in Tables 4.1 and 4.2) that differed so much from the ones established by competitors. In *Latte Vivo*, an Italian analysis of the functioning of Parmalat, colleagues make a detailed analysis which shows that, over a 25-year period, Parmalat probably accumulated – by paying lower prices to farmers, confronting consumers with higher prices, postponing milk payments and outsourcing – the enormous sum of some 12 billion Euros.

This is a worrying resemblance to the expressions of Empire discussed in Chapter 3. Neither in the Peruvian or Italian cases did Empire add any substantial set of resources, new technologies and/or superior organizational capacities to the constellations discussed. Empire adds nothing; it simply combines and recombines already available resources. It can do so because it has at its disposal, through a combination of political and economic power, multilevel entrances, throughput facilities and delivery systems that are closed to others. Thus, extra-economic coercion is created. Empire – wherever located – combines resources into new socio-technical networks that establish at their frontiers new kinds of dependency patterns. The same point has been made in an analysis of the Chilean tomato empire: it functions 'according to the principles of patron–client organization – i.e. centralized authority, dyadic command structures, personalistic incentive systems and no room for grassroots organizations' (Peppelenbos, 2005, p11).[17] This 'non-economic coercion' is further nourished through the strong intertwining of food empires and the state, both in Italy and in Peru. Lang and Heasman (2004, p127) point to the same phenomenon when they assert that 'corporate power is now so great within and between national borders that it is redefining what is meant by a "market"'. They likewise observe that 'corporate policy is becoming more fully engaged in public policy to further its own interests, thus raising questions about accountability' (Lang and Heasman, 2004, p127). Thus, another intriguing element occurs. While the different food empires are internally constituted as a kind of internal market, the 'outside' markets, in their turn, are subjected to hierarchal and unequal power relations and 'non-economic interventions'. Markets and the state, once separated through clearly defined lines of demarcation, now co-penetrate each other. State intervention represents 'regulation for, rather than against, the market' (Burawoy, 2007, p7). I will further elaborate this tendency in Chapter 9.

The last resort: Fresh blue milk

At the end of the 20th century it became quite clear that – at the third level – enormous debts were threatening the very existence of Parmalat. At the end of 1998 the overall debts amounted to 2.1 billion Euros, *which was more than the net value of Parmalat as a whole* (Franzini, 2004, p61).[18] Harvesting the already available fields was increasingly insufficient to turn the tide. Thus, a new project was constituted and set in motion. The key term is *latte fresco blu*: fresh blue milk. This project is exemplary of Empire and its associated *conquests*, and was meant to be the ultimate rescue operation. *Latte fresco blu* was meant to greatly increase the required levels of profitability needed to sustain the urgently required levels of trust of financial markets and private stockholders. It aimed to re-establish trust by inflating profitability data, as Ahold had tried to do. The difference, though, is that the *latte fresco blu* project not only affected the outcomes of (manipulated) accountancy procedures, but potentially entailed a considerable risk to food quality, public health and the very continuity of dairy farming in Italy.[19] *Latte fresco blu* represents the most drastic disconnection of time and place as far as the production and consumption of milk is concerned. To put it bluntly, through the *latte fresco blu* operation it became possible to collect low-quality milk at low prices in Poland, for example (and later probably in the Ukraine), in order to reconvert it and offer it three months later on the Italian market as fresh and first-class milk.

Until the beginning of the 2000s, there was a clear division of labour in the Italian diary industry. Parmalat had specialized in UHT milk (long-life conservation milk), while Granarolo and several smaller co-operatives controlled the market for fresh and high-quality milk. Margins are far higher in the latter market segment. Through the *latte fresco blu* project, Parmalat aimed to conquer and take over this segment. In politico-economic terms, it came down to the following: Polish milk was to be acquired on a massive scale (for 0.24 Eurocents per kilogram of milk) and submitted to a new, although not unknown, treatment referred to as micro-filtration (also described as ultra filtration). Through this treatment (to be combined with several other technical interventions which I will describe later) foreign milk could then be offered as fresh blue milk (*latte fresco blu*) on the Italian market for a price, initially, of 1.5 Euros per litre, and later for 1.2 to 1.3 Euros per litre. It was projected that, in the end, total sales on the national market for consumption milk could be augmented through this project to some 1.6 billion Euros per year. Net profits could easily jump, according to the internal forecasts, to 1 billion Euros per year. This would have been sufficient to make the outstanding, albeit it not yet publicly known debts, melt like snow. The banks involved, very aware of these debts, were impressed by this 'revolutionary project' and decided to refinance the debts.

From a technological perspective *latte fresco blu* basically represents a new and extended form of food engineering. The separate techniques were well known and widely applied in different places. What was new was their combination. As summarized in Figure 4.4, the production of *latte fresco blu* basically comes down to the skimming of milk, after which the cream is pasteurized and homogenized, and the skimmed milk is heated and, subsequently, microfiltrated. Thus microbic flora is almost completely eliminated (as a matter of fact, the first versions were presented as 'sterile milk'). In a next step the cream

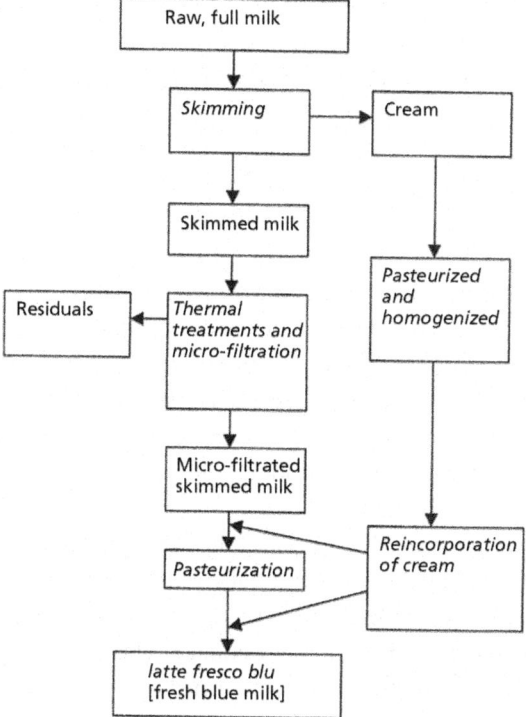

Figure 4.4 *The making of* latte fresco blu

Source: Ploeg et al (2004a, p64)

is again added – that is, the 'milk' is rebuilt. After this, pasteurization is again carried out. The thus constructed milk, the biophysics of which are strongly altered compared to raw milk, could be conserved for a considerable time (according to the law for ten days, whereas fresh milk had a legal durability of three days). Equally, the time between milking and processing could be extended, and what was initially poor hygienic milk corrected for quality.

The main advantage, however, goes beyond the particular techniques. The point is that the technological combination allows for a far-reaching disconnection between the production and consumption of milk. Through its separation into distinct elements and thanks to the different treatments, milk can now literally 'travel' over long distances through time and space. This, in its turn, allows for the creation of a new pattern that is essential to Empire: the direct connection of cheap places of production to rich places of consumption. In doing so new 'non-places' are created. It no longer matters where milk comes from – it might stem from everywhere (literally so, since repeated thermal treatments make any traceability impossible). It likewise implies a loss of identity. Since milks originates from a 'non-place' and since it is no longer what it says and at first sight seems – that it is fresh – its identity is increasingly a 'non-identity'.

Latte fresco blu is not neutral in so far as food quality and safety are concerned. The milk to be processed through the chain described in Figure 4.4 might be of poor initial quality. Through micro-filtration several valuable elements will be lost (at least in part). The process as a whole (especially the repeated thermal treatments) almost completely eliminates the microbic flora, and quality control along the chain becomes increasingly problematic due to the implied distances (the first plant for micro-filtration was located in Berlin, the milk came from Poland and the milk was consumed in Italy). The biological quality of fresh *blue* milk is inferior to fresh milk as we once knew it. This does not imply that it is a dangerous product. However, the risk that something might go wrong is evidently higher.

If we consider *latte fresco blu* as a politico-economic project, then we must emphasize, in the first place, that it potentially represented a major shift in the relations of competitiveness in the Italian market for consumption milk, which is roughly divided into two more or less equal parts: fresh and high-quality milk, and UHT milk. As mentioned, the former segment was controlled by Granarolo and some smaller co-operatives operating at regional level, while the latter segment belonged to Parmalat. Price levels in the first segment were and are considerably higher than in the second. Within this specific arena, the *latte fresco blu* project was considered to be a frontal assault on Granarolo, the co-operatives and associated interests (mainly of Italian dairy farmers). The *latte fresco blu* project evidently aimed at a takeover of the relatively prosperous and profitable segment of the consumption market. *Takeover* is the key word – just as it was the case in expressions of Empire described elsewhere in this book. Had the project succeeded, it would have implied the simultaneous destruction of industries such as Granarolo and the regional co-operatives, while most Italian farmers would have lost their outlet, have gone bankrupt and might possibly have disappeared as dairy farmers. Together with a centralization of value added in Parmalat, there would have been a simultaneous destruction of value-added production elsewhere. The *latte fresco blu* project was not meant to increase value at the level of society as a whole. It was meant to redistribute and centralize value under the aegis of Parmalat – even if this implied a reduction of social wealth at higher levels of aggregation. Thus, it is indeed the case that 'in the name of creating new wealth, humanity, in fact, impoverishes itself' (Korten, 2001, p2).

To launch the *latte fresco blu*, Parmalat invested enormous amounts in publicity and even more in political bribery. The Italian law defined very precisely the conditions under which milk can be presented on the consumption market as *fresh* milk: it must be delivered within a day after milking, be processed within a further day and consumed within three days. Obviously, the proposed project could never meet these criteria. Hence, the law needed to be changed. After considerable pressure which involved several organizations, power groups and key persons, a new legal category was introduced for microfiltrated milk. Completely different conditions were linked to this new category making it possible to introduce *latte fresco blu* on the market – a new product that sharply differed from fresh milk (*latte fresco*), but which, through the additional '*blu*', could be presented under the umbrella of freshness.

However, the crash materialized before the *latte fresco blu* project was fully off the ground. It was introduced on the Italian market in the course of 2002. Parmalat crashed in December 2003. Afterwards (i.e. when Parmalat no longer represented considerable political and economic power), the law was again changed and *latte fresco blu* was forbidden. The consequence is somewhat ironical since, at present, Italy is now the only European country that has a strong legal defence of *real* fresh milk and, simultaneously, an *explicit* prohibition of the technological assemblage and associated patterning that results in *latte fresco blu* and similar artefacts. The European law is extremely vague in this respect, while national legislation, in nearly all European countries, delegates responsibility to the big corporations: 'fresh' is what *they* define as such, and 'milk' is what *they* introduce on the markets.

The distorted development of food production and consumption

According to Moquot (1988), a leading French expert on dairy industries, a high-quality level of dairy products basically depends upon the quality of the milk produced at farm level (and thus on cattle feed, hygiene, milking, selection, care, etc.) and *not* on all kinds of technological remedies for errors made at the level of primary production, nor for those that emerge in the long trajectory of processing and distribution. Equally important, in this respect, is the organization of time, which basically comes down to three (interrelated) time spans: the time between milking and collection; the time between collection and processing; and the time between processing and consumption.

With a well-tuned and controlled process of primary production and a short and equally controlled delivery to consumers, it is possible to produce and consume untreated raw milk (the Italian expression is far more beautiful: *latte vivo* – living milk). Indeed, there is an increasing trend throughout Europe to produce and deliver such milk,[20] which in quality and taste is highly attractive.[21]

The main trend, however, goes in the opposite direction. Milk is increasingly subjected to a complex combination of sophisticated treatments and extended time spans. Again, in themselves, neither are necessarily problematic – although one might wonder, knowing that each treatment implies increased costs, why industry opts for it. The potential dangers, as demonstrated by the Parmalat case, are in the interaction of three processes. Due to competition (but currently also due to the effects that the 'third level' exerts on the 'second') food industries are constantly looking for the cheapest acceptable raw materials. This translates among other things into outsourcing, and the unavoidable opportunism that reigns in every market[22] strengthens the process further, especially by constantly redefining the limits of the 'acceptable'. This first process is then made possible and even accelerated by a second process: the development of new technological mixes that allows for the introduction of distance in both time and space in the trajectory that links production and consumption. Through this second process the new patterns potentially entailed in the first – linking poor areas of production with rich areas of consumption – are made possible. But inclusion of poor areas is often identified with poor primary production conditions, and the long distances in terms of time and space often require the use of additives. Thus, combined with the other two, a third process is created: continuous degradation. At this moment, the worldwide reorganization of agricultural production goes along with massive degradation – which is masked, corrected and reproduced through the engineering and re-engineering of what in the end is presented as food.

In debates following the collapse of Parmalat, the focus was mainly on Parmalat as something exceptional. From the foregoing analysis, however, we see that Parmalat is absolutely no exception. It represents, in all respects, Empire as an ordering principle. It is an example *par excellence* of systematically organized re-patterning, re-engineering and degradation, just as it is exemplary of the main drivers for the active organization of this combination – that is, conquest and opportunism. Neither the re-patterning (milk from Poland) nor the re-engineering (its conversion into *fresco blu*) was quality driven. It was driven by the need to extend, enlarge and improve the profit-making machinery. *Latte fresco blu* was a device for reaping faraway fields of production and nearby fields of consumption.

The non-exceptional nature of food degradation: The rise of 'lookalikes'

In the world created and patterned through food empires, everything loses its identity. Food products are no longer produced in a particular place by particular people at a particular moment, and then passed through more or less known or at least knowable circuits towards consumers. Food is becoming a range of 'non-products', the origin of which no longer matters, just as its voyage through time and space no longer matters. Although several systems for 'tracing and tracking' have been put in place, they are nothing more than a mechanism to delegate responsibility, risk and associated cost in cases of severe food disasters. They refer, in any case, only to abstract origins, which might today be in China, tomorrow in Poland, while yesterday, for example, it was located in Peru. Abstract origins are 'non-origins'. Traceability only demonstrates that origin, understood as a knowable and trustable place, no longer exists – at least within the context of Empire.

However, the processes of (re-)patterning entailed in Empire are not only about places converted into non-places, or given time horizons and clearly defined notions, such as freshness, transformed into directives that basically misinform, or about the drastic reorganization of the networks through which food is moved, transformed, stored and delivered. These same transitions drastically alter food itself – not just the concept of it, but food as material entity. Empire introduces non-food. It makes non-food appear as food in order to distribute it under the etiquette of the latter. Food empires, in other words, increasingly produce 'lookalikes'.

A first level of lookalike products emerges when food production is ever more disconnected from a well cared for primary process of production. When agricultural production, of whatever kind, is not done with care, contaminations emerge (of whatever type), and all kinds of interventions are then required, either directly during production or further along the food chain in order to correct the initial lacunae. The effects of such interventions will affect the end-products. Produced in this way, the food will be different from what it ought to be – and different from what consumers expect it to be. It becomes a 'lookalike'.

Additives are crucial in this respect. The main problem with chemical additives to food is that although there is considerable knowledge regarding direct effects, the effects of their interaction are hardly known. The interaction of two or more in themselves 'innocent' additives might produce within the human body as yet unknown, but potentially far-reaching, effects on human health. Likewise, very little is yet known about the long-term effects of many individual additives.[23] Finally, there is an ongoing and harsh race between the agencies controlling food quality and those running illegal laboratories. The latter produce flows of new forms of growth hormones, for example, which cannot as yet be traced by existing detection techniques.

Italian machine industries are delivering harvesting machines to China for tomatoes. Payment of these deliveries is, as in many cases, built on so-called *tied sales* mechanisms. This implies that the imported goods are paid for in kind, especially since many developing world countries lack hard currency. In this case, the imported technology is paid for in large quantities of tomato concentrate, subsequently sold on the European market to different food processors to be converted into tomato juice and tomato sauce for pasta, etc. According to European law, these latter (i.e. converted) products can be sold as European products. As long as the final conversion has taken place within Europe, the Chinese origin no longer matters. Consumers are hardly aware of such complexities; but for food processors, such mechanisms represent a considerable competitive advantage. The tomato concentrate, however, turned out to be extremely deficient: colour, taste and aroma were all far below standard. The tied sales mechanism evidently encourages such incidents. Be that as it may, the missing qualities were corrected through a range of technological interventions. Only in retrospect did it become clear that one of the applied additives was highly carcinogenic and one of the food processing industries involved, Barilli, had to withdraw from shops almost a whole year's complete production. This has contributed considerably to the financial troubles in which Barilli is embroiled.

During the last years there has been a near avalanche of similar incidents. I refer only to the recent discovery of the poisonous Sudan colorant encountered in many British food products, the rotten Christmas rabbits of Albert Heijn on the Dutch market, the application of methyl bromide in worldwide container transports, the presence of MPA hormones in pork meat, and residuals of pesticides on vegetables. Such detected cases are probably only the tip of the iceberg.

'Lookalike' food of the second kind

The foregoing cases all have in common the fact that they resulted in contaminated food. What was thought to be and what was presented as healthy, tasty and trustable food products turned out to be quite the opposite. But the change from 'real' to 'virtual' food is not limited to cases of contamination. It goes beyond that. A typical example illustrating the change towards food products that in the end are scarcely food products at all, but 'lookalikes' of a superior order, stems from northern Italy. It concerns the making of food from products that were never destined for food or that are no longer appropriate for conversion into food.

For several years a Milan-based enterprise, Agricomex SRL, imported massive quantities of waste products, among them milk that had passed its durability and milk powder to be used for animal feed, etc., in order to transform them into milk. The Italian newspapers quickly coined a new word: '*simil-latte*': not 'milk as such', but something similar to milk – a 'lookalike', indeed. '*Simil-latte*' was sold for years and throughout Italy as UHT, or long-life milk, under a range of well-known brand names. Micro-filtration again was crucial. When skimmed milk is treated with micro-filtration, then not only the bacterial contamination but also some milk sugars and minerals such as calcium, salts and probably even some proteins are taken out. The remaining material (the 'residuals' in Figure 4.4) might be heated for some 30 minutes and/or treated chemically (in order to raise the pH) and then again be subjected to micro-filtration. The thus obtained material is normally used as animal feed. However, in the processing plants of Agricomex SRL (located in Mantova and Brescia), this same material was combined, in huge quantities, with cream, milk that had passed its durability, and water and waste products such as whey in order to construct '*simil-latte*'. After this, ammonia was frequently added in order to reduce the pH again. After the treatment the ammonia evaporated out of the obtained lookalike. Nobody died (to put it crudely), but neither did anybody know that they had been consuming something that was definitely not milk but 'rubbish we would not even feed to our pigs', as the tapped phone calls revealed between some of those arrested (Rassegna Stampa Italiana dal Ministero delle Politiche Agricole e Forestali, 2005).

The described practice continued for several years before it was detected and denounced. It is evidently impossible to assess the dissemination of such practices. The point is that food is increasingly 'engineered' in order to span distances in time and space and to construct profitability. New food engineering technologies emerge with such a velocity and from so many sources that it is scarcely possible to keep pace with adequate laws and control systems. And, finally, harsh competition within the food industry increasingly spurs a grazing of the globe to encounter the cheapest ingredients and the cheapest solutions, thus stimulating the opposite of 'well cared for' primary production. Taking these elements together, they make for a mix that cannot but result in a widening range of lookalikes – detected or not. As long as the (global) market allows for, or favours, opportunism, the situation will remain the same.

It is estimated that Italian food industries lose, every year, 2.8 billion Euros due to the creation of 'lookalikes'. This is more or less 50 per cent of the total turnover of Italian food industries. These lookalikes are produced everywhere, from Brazil, through the US, Canada and Australia, to The Netherlands. Several lookalikes are also produced within Italy itself (*Agrarisch Dagblad*, 2007, p2).

Empire compared with a contrasting mode of patterning: Regressive centralization versus redistributive growth

The Italian food industry as a whole has been capable, over the last years, of slightly improving its position in the world market. This is demonstrated in Figure 4.5, which summarizes the changes in the relative share of the world food market of different countries over the 2000 to 2003 period. A more detailed analysis shows that this improvement is basically associated with wine, dairy products (mainly cheese) and processed meat products – that is, 'the market segments in which there is a particular incidence of high-quality products known as typically Italian, and with a DOP [Denominazione di Origine Protteto] or IGP [Indicazione Geografica Protteta] guarantee' (ISMEA, 2005, p10). Second, it is remarkable that 'prices paid to farmers have undergone a change in trend: from 2000 onwards they have slightly increased, while at the level of the EU-15, off-farm prices have suffered a decline' (ISMEA, 2005, p7). A third feature, which might be relevant as far as the analysis of this constellation and its development tendencies are concerned, is that apart from a few very large Empire-like corporations such as Parmalat, the Italian food industry is basically composed of small- and medium-sized enterprises (often co-operatives). The average number of workers in food processing enterprises in Italy was 6.4 (in 2002), compared to 24.4 in Germany.

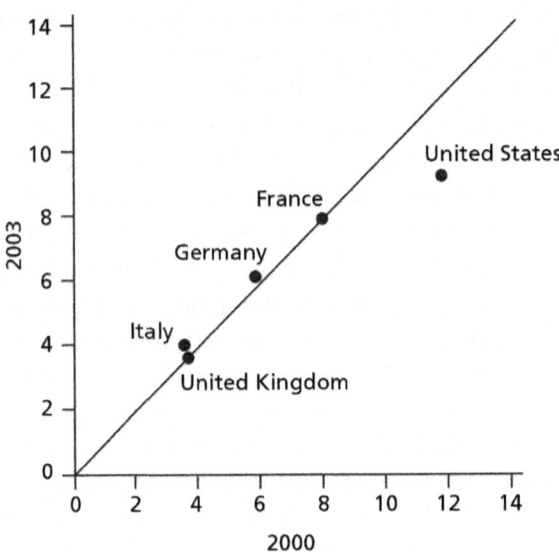

Figure 4.5 *Relative shares in world food market*

Source: ISMEA (2005, p89)

The features indicated are highly interdependent: it could be argued that together they constitute a model of quality-centred redistributive growth. The presence (and the relatively good performance) of such a model is the background against which the Empire type of operation needs to be discussed. It shows, first and foremost, that there are *alternatives* to Empire. Linking to the world market does not necessarily imply the construction of, and subordination to, Empire-like corporations. There are other modes of patterning, such as those entailed in the small and medium quality-oriented enterprises of the Italian food industry (for a convincing case study, see Roest, 2000). Second, the comparison highlights the specificities of the production and distribution of value within a world patterned in an Empire-like way. The latter tends, as the Parmalat case makes clear, towards a centralization of value added. Value added is shifted from primary producers and concentrated within the corporation. This is associated with a *decline* of the value added realized within the primary sector. It is even possible that *overall* value added (or social wealth) will be reduced; then a *regressive centralization* occurs. The 'blue milk' project of Parmalat might be understood as the supreme expression of such a regressive redistribution. While it would have increased the value added to be realized within the Parmalat corporation, it would have simultaneously implied a drastic reduction of value added within the dairy sector as a whole. Empire induces, in the societal segments affected by it, a negative trend in the development of social wealth. That was central to the expressions of Empire encountered in Peru, and it is equally central to Empire as it operates in Europe. This strongly contrasts with the upward tendencies produced by decentralized types of food processing and distribution.

5
Peasants and Entrepreneurs (Parma Revisited)[1]

Introduction

Chapter 2 introduced and discussed in general terms the peasant condition and its associated mode of farming. I now aim to compare the peasant with the entrepreneurial mode of farming. The discussion is illustrated with a case study of dairy farming in the northern Italian area dedicated to the making of Parmesan cheese. I also analyse the historical origins of the entrepreneurial mode of farming. This feeds into a section that focuses on the political economy of entrepreneurial farming, which describes how it currently interacts with Empire and how, indirectly, it introduces the effects of Empire into the fields, the landscape and the regional economy. I also give attention to what I refer to as the entrepreneurial condition. Just as the peasant mode of farming resides in, and is reproduced by, the peasant condition, so the entrepreneurial mode assumes a specific set of conditions from which it emerges and which allows it to further unfold. The difference between these two conditions turns out to be strategic in the epoch of globalization and liberalization.

The multiple contrasts between peasant and entrepreneurial farming

The basic difference between the peasant and the entrepreneurial modes of farming resides in the degree of autonomy that is built into the resource base. Autonomy is also encountered in the relations in which this resource base is embedded, as well as in the way that it is operated, extended and further developed. This many-sided autonomy is constructed along a number of dimensions, which are summarized in Table 5.1. Some of these dimensions directly concern the way in which the process of agricultural production is ordered (see also Barlett, 1984; Salamon, 1985; Strange, 1985); others pertain to higher levels of aggregation. In order to discuss these dimensions further, I focus on one specific time- and place-bound constellation – namely, dairy farming in Emilia Romagna or, more specifically, the production area of

Parmesan cheese. I do so because the peasant and entrepreneurial ways of farming here do not represent a sequence (Gorgoni, 1987); they co-exist alongside each other, thus allowing for comparative analysis.]

Table 5.1 *The main differences between the peasant and entrepreneurial modes of farming*

Peasant mode	Entrepreneurial mode
Building upon and internalizing nature; co-production and co-evolution are central	Disconnecting from nature; 'artificial' modes of farming
Distancing from markets on the input side; differentiation on the output side (low degree of commoditization)	High market dependency; high degree of commoditization
Centrality of craft and skill-oriented technologies	Centrality of entrepreneurship and mechanical technologies
Ongoing intensification based on quantity and quality of labour	Scale enlargement as the dominant trajectory; intensity is a function of technology
Multifunctional	Specialized
Continuity of past, present and future	Ruptures between past, present and future
Increasing social wealth	Containing and redistributing social wealth

Source: Original material for this book

'Artificialization' versus co-production

In the peasant mode of farming, co-production is central, by which I mean the many-sided and continuously evolving interactions between man and living nature (i.e. the process of production in which nature is converted into goods and services for human consumption) (Ploeg, 1997b, p42). The utilization, maintenance and further unfolding of ecological capital are central in co-production (Toledo, 1992). The resources that together compose ecological capital are continuously being transformed and improved through co-production. In this respect they represent objectified and accumulated labour (Bourdieu, 1986, p241). A strong rooting in ecological capital and, consequently, an ordering and unfolding of the agricultural production process as co-production are decisive features of the peasant mode of farming. These features not only translate into the resilience that has characterized peasant farming through the ages; they also underpin the currently emerging attractiveness of peasant farming as far as sustainability is concerned. Co-production requires – and equally results in – a specifically ordered type of knowledge, referred to in the French tradition as *savoir faire paysan* (Lacroix, 1981; Darré, 1985) or *art de la localité* (Mendras, 1967, 1970). Respect and admiration for, and patience with, living nature are mostly an integral part of such knowledge (Kessel, 1990).

Entrepreneurial farming differs from this in several respects. Although 'nature' remains an unavoidable ingredient (it composes the required 'raw material'), development in the entrepreneurial mode is focused on increasingly reducing its presence. 'Nature' is too capricious – it excludes standardization of the labour process and thus becomes a hindrance to accelerated scale increase. It also limits (or retards) increase in productivity. Therefore, the presence of nature within the agricultural production process is reduced and that which remains is increasingly 'rebuilt' through an all-embracing process of 'artificialization' (Altieri, 1990). Several expressions of this process are well known: well-matured manure is replaced by artificial fertilizers; grass, hay and silage by industrial concentrates; care of animals by preventive medicine use; fertile and clean land by artificial substrates; sunlight by artificial light; labour by automation; *savoir faire* by computerization; manual labour in weeding by herbicides; etc. The real extension of this process of artificialization, however, has already moved far beyond this. By using genetic modification and through the creation of aseptic ('hygiene') conditions, a new artificial nature is created that allows for further industrialization.

In the entrepreneurial pattern, the processes of agricultural production are progressively disconnected from nature and the ecosystems in which they are located. These disconnections translate into growing levels of counterproductivity (Ullrich, 1979). The efficiency of nitrogen use in Dutch dairy farming, for instance, declined from 60 per cent during the 1950s to only 16 per cent in the late 1980s (Reijs, 2007); the longevity of dairy cattle and sows fell sharply (Ploeg, 1998; Commandeur, 2003); efficiency in the use of scarce irrigation water declined by some 50 per cent (Dries, 2002); and energy use was multiplied several times, while its efficiency declined (Ventura, 1995). Agriculture thus became an activity that produces large flows of waste.

Entering into market dependency versus striving for autonomy

Partly as an effect of the 'artificialization' of the agricultural process of production, the entrepreneurial mode of farming is characterized by an elevated degree of externalization – that is, many subtasks of a once integral process of production and labour are shifted towards and taken over by external institutes and market agencies. Once this occurs, new dependency relations are created between these institutes and agencies and the farms involved. These dependency relations are of a double nature: they embrace new commodity relations as well as technico-administrative relations through which the farm labour process is prescribed, conditioned and sanctioned (Benvenuti, 1982).

In a multidisciplinary research programme in Emilia Romagna undertaken between 1979 to 1982 (and later replicated; see Benvenuti et al, 1988), a range of situations was identified that ran from relatively autonomous farms (in which the majority of the required resources was produced and reproduced on the farms themselves), to highly market-dependent farms (Ploeg, 1987, 1990a). In the latter, labour, capital, land, knowledge, cows, feed and fodder, as well as machine services, were mainly mobilized through markets. The reproduction of highly commoditized farms is far from being historically guaranteed (unlike those of the relatively autonomous type). Instead, they depend upon the results of future production: their reproduction relies heavily upon future market constellations. Table 5.2 summarizes some of the relevant data. From this we can observe that peasant farms are less integrated within markets on the input side of the farm compared with entrepreneurial farms. In particular, the level of so-called 'intermediate consumption' (or the level of 'variable costs') is far lower. This is evidently due to the centrality of ecological capital.

Table 5.2 *Contrasting degrees of commoditization in Emilia Romagna, 1980*

	Peasant mode	Entrepreneurial mode
Labour (percentage of labour input mobilized through the labour market)	14	35
Machine services (percentage of total machine services delivered by contractors)	23	57
Short-term capital, mainly related to variable costs (percentage of short-term finance that is covered by short-term credits)	0	9
Mid-term finance primarily related to mechanization and cattle (percentage of mid-term finance that is covered by mid-term credits)	8	37
Long-term capital mainly related to land and buildings (percentage of long-term finance that is covered by long-term credits)	3	19
Land (percentage of rented land)	17	32
Feed and fodder (bought feed and fodder as percentage of total cattle alimentation)	24	67
Cows (bought cattle as percentage of total herd)	1	14

Source: Ploeg (2003b, p60)

Entrepreneurship versus craftsmanship

In the way in which peasants practise farming, craftsmanship (the capacity to realize in a sustainable way high and rising productive results per object of labour) is strategic. Local knowledge of the *savoir faire paysan* type is an indispensable ingredient and the artisan character of the labour and production process allows for the development and enrichment of this type of knowledge. The entrepreneurial mode of farming represents a strong contrast in this respect. Here, entrepreneurship becomes the central capacity – that is, the capacity to pattern the labour and production processes according to *market* relations and prospects becomes decisive. Whereas within the framework of craftsmanship internal indicators are normative (e.g. 'What, regarding the behaviour and history of a particular cow, is the ration that best fits her'?), *external* indicators become the main beacons within the framework of entrepreneurship ('Given the relations between the price of milk and the costs of different feed ingredients, what is the best ration?'). Based on these external indicators, day-to-day farm operations are constantly modified, at least in so far as entrepreneurial practices are concerned. Peasants would hesitate or be unwilling to do this: 'By doing so you'll only ruin your cows; they need what best suits them and they need continuity as well.'

Together, craftsmanship, local knowledge and the patterns of communication and exchange that support them constitute the *quality* of labour. We could equally refer to it as the *human capital* entailed in the sector (i.e. the capacity to govern and develop the processes of production in an endogenous way). This human capital resides, above all, on the peasant side of the equation. It is strategic for progress. On the entrepreneurial side, externalization inevitably brings a reduction in the overview of, and insight into, the relevant whole, while the possibility of adapting (to 'remould') the different resources (especially the bought ones) is evidently reduced.

A frequent error is to interpret the centrality of craftsmanship as an expression of non-economic behaviour. This is most definitely a misconception: in the peasant mode of farming the unit of production is related to the markets in a manner that essentially differs from the way in which market relations are patterned in entrepreneurial farming. In the peasant mode (grounded on distantiation and relative autonomy) the market is basically an *outlet* – it is the place where the products are sold, for better or worse. In the entrepreneurial mode, the market is, above all, an *ordering principle*. Due to the high integration within and dependency upon markets, the unit of production has to follow 'the logic of the market' – which is precisely why entrepreneurship instead of craftsmanship becomes the central mechanism for patterning both the social and the natural within and around the farm enterprise. The focus on craftsmanship does not imply that peasants are un-enterprising persons. On the contrary, they are keen to grasp new opportunities. They are as the early birds that get the worm. Peasants are enterprising, inventive, keen and shrewd. But they do not operate in an entrepreneurial way. The underlying logic is different.

In the research programme mentioned, particular attention was paid to the *logic of farming*: the way of perceiving, calculating, planning and ordering the process of production. Two contrasting logics were found: that of the *contadini* (i.e. peasants) and that of the *imprenditori agricoli* (the agricultural entrepreneurs). In the peasant logic (see Figure 5.1), the notion of *produzione* (good yields) has a central position and significance. Within this logic, *produzione* refers to the production per labour object (i.e. per cow or per unit of land). *Produzione* must be high and sustainable, but it is not to be 'forced', as peasants will argue. It should be as high as is possible within a framework defined by *cura* (care). One has to care well for the animals, the plants, the fields – and if work is done with care, then production per labour object will be high. *Cura* equals craftsmanship. It refers to the quality of labour or, in more general terms, to ordering the processes of production and reproduction in such a way as to guarantee that good yields and steady increases are the outcome.

In the worldview of *contadini*, high levels of *produzione* are justified because they produce and sustain incomes (*guadagno*) in the short run; probably even more important, in the long run, they allow for the making of a beautiful farm (*la bell'azienda*). Thus, through *cura* – through one's own work – a promising future may be built.

Cura, in turn, depends upon several conditions. There must be *passione* (passion), *impegno* (dedication: but here it refers also to a high quantity of labour input and to hard work), *professionalità* (knowing the job) and, finally, *autosufficienza*: the farm unit must be as self-sufficient as possible. Hence, through such a logic a relatively autonomous, historically guaranteed process of production and reproduction is created, maintained and, wherever possible, further developed.

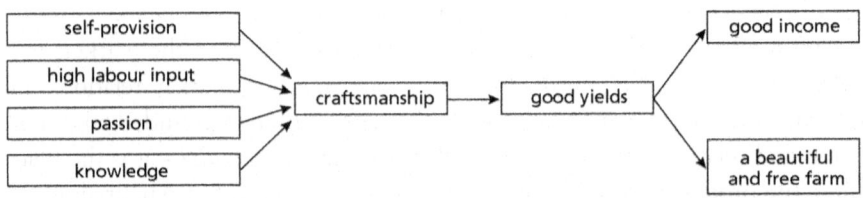

Figure 5.1 *The* contadini *(peasant) logic*

Source: Ploeg (1990a, p59)

The logic of the agricultural entrepreneur (see Figure 5.2) is structured in a different way. External parameters within this mode of ordering are decisive, for reigning market relations and the price–cost ratios entailed in them define what the margin (*il margine*) will be.[2] Similarly, it is the available technologies (and the process of technology development) that define *la scala*: the scale of operations. Together, the margin and the scale define the income (*il reddito*) realized – not only within this logic but also at the material level: the farms of the *imprenditori* are structured in such a way that income is highly dependent upon the scale of farming. This again constitutes a notable contrast. *Contadini* farms are structured in such a way that income is relatively independent of scale. Thus, natural and social worlds are moulded in different, mutually contrasting, ways, each entailing different models for generating income.

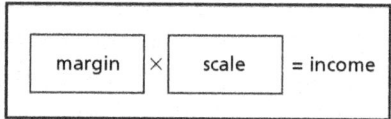

Figure 5.2 *The logic of the* imprenditori agricoli *(agricultural entrepreneurs)*
Source: Ploeg (1990a, p69)

Scale enlargement versus labour-driven intensification

A crucial difference between the two logics regards the focus, if not the gravitational centre, of the farm development process. The *contadini* mainly aim at continuous improvement in yields and, thus, in the value added (VA) per object of labour. In ideal typical terms, they will only enlarge their farm if:

- such an enlargement does not negatively effect the VA per labour object; and
- the enlargement (at least the biggest share of it) can be founded on (i.e. financed with) their *own* available means.

Hence, a step-by-step process of growth is mostly the outcome (Ploeg et al, 1990). For *imprenditori*, the scale of farming is the main lever for further farm development – also because the conditions to enlarge yields are less well developed on their farms. Hence, they frequently opt for a substantial scale enlargement, the more so since under current conditions scale enlargement has a 'self-propelling' character: an increase in scale results in a decrease in the margin per object of labour, which in turn induces the need to further accelerate the growth at farm level. In this respect, French colleagues use a telling expression when concluding that farm development becomes *une fuite en avant* (a race to the bottom) (Eizner, 1985).

Entrepreneurs typically aim to acquire the newest technologies and to restructure their farms in such a way as to 'fit' the new technological models. Taking credit to finance expansion becomes strategic. It is true that the yields may also rise (sometimes even considerably); but this basically depends upon the purchase of technologies and inputs that *embody* particular rises in yields (e.g. Holstein cattle in combination with energy- and protein-rich feed and fodder; high-yielding varieties (HYVs); precision agriculture; etc.).

120 *The New Peasantries*

Our research allowed us to reconstruct, on the basis of farm accountancy records, the different farm development trajectories over the 1970 to 1980 period. Figure 5.3 presents a schematic summary of the main findings. It shows that the development trajectories of the peasants and the entrepreneurs do, indeed, differ markedly (P and E, respectively, in Figure 5.3). While the former mainly increased the intensity of farming, the latter largely increased the scale. The graph also refers to a group of capitalist farms (C) that also operate in the area. The development over time of entrepreneurial farms is quite close to the typical pattern for capitalist farming (see also Raup, 1978).

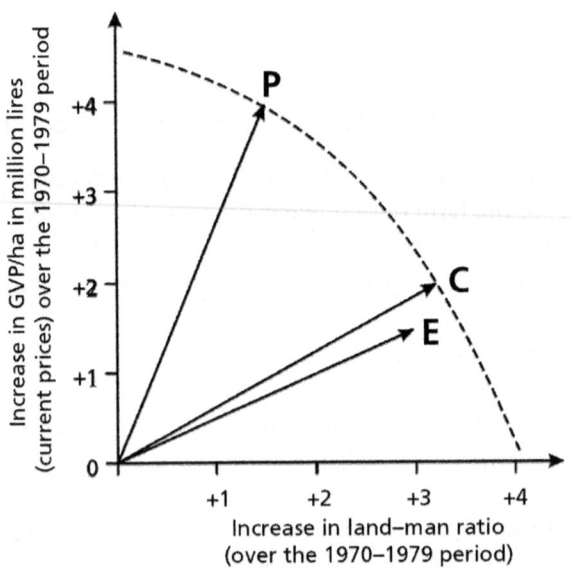

Figure 5.3 *Differential farm development trajectories in Emilia Romagna, 1970–1979*

Source: Ploeg (1990a, p45)

In the international literature, the assumption made is that development trends in agriculture will reflect relative factor prices (Hayami and Ruttan, 1985). Where labour is abundant and cheap, but land (and, more generally, capital) is scarce and expensive, intensification will be the outcome. If, however, relative factor prices are inverted, then scale enlargement will dominate. What we witness here argues against such an assumption. Within one and the same homogeneous situation (relative factor prices are the same for each and everyone, new technologies are accessible to all, etc.) there are, in spite of everything, highly diverging farm development trajectories. This does not imply, of course, that factor prices do not matter: on the contrary. But what equally matters are the *interrelations* between farms and the factor markets. The point is that the entrepreneurial segment will closely follow the logic entailed in the labour, capital and land markets, precisely because (as shown in Table 5.2) it is highly integrated within and dependent upon them (Friedmann, 1980). Within peasant realities these interrelations are patterned differently, partly due to the logic (or strategy) that is used. Thus, a relative autonomy arises: the processes of agricultural production and development are actively *distantiated* from the markets and can, consequently, follow a different route.

Specialization versus multifunctionality

A fifth aspect of the many differences between peasants and entrepreneurs and the contrasting realities that they create relates to the degree of specialization. Inherent to the peasant way of farming is what Tepicht many years ago referred to as its polyvalence. The entrepreneurial mode, instead, results in a strong specialization and, consequently, in encapsulation in complex networks that result from social and spatial divisions of labour and which often reduce the farms involved, in an almost Fordist way, to the repetition of just one simple routine (see, for example, Bonnano et al, 1994; McMichael, 1994). For a long time the *multiple use of resources* entailed in, for example, mixed farms and pluriactivity was perceived by the expert systems of the day as a superior expression of the intrinsic 'backwardness' of the peasant farm. Since the late 1990s, however, a completely different view emerged from the innovative work of researchers such as Saccomandi and his students, who began to apply neo-institutional analysis to the agricultural sector (Saccomandi, 1991; Ventura, 2001). What is theoretically interesting here is that they conceptually link 'economies of scope' with multifunctionality, while specialization is related to 'economies of scale'. This view is increasingly being echoed, albeit in a somewhat modified form, by some of the large expert systems (OECD, 2000). It is also telling that there has been, from the beginning of the 1990s onwards, a clear trend in practice (often referred to as 'rural development') that results in the deliberate creation of new forms of multifunctionality (such as the development of agro-tourism; the agrarian management of nature, landscape and biodiversity; production of energy; the production, transformation and marketing of high-quality products and regional specialities; taking care of disabled people; the retention of scarce water; etc.). Nearly always, it is the peasant farms that function as the point of departure (and as the resource pool) for the creation of such new multifunctional entities (see Broekhuizen et al, 1997; DVL, 1998; Coldiretti, 1999; Stassart and Engelen, 1999; Joannides at al, 2001; SARE, 2001; Scettri, 2001; Ploeg et al, 2002c; Wolleswinkel et al, 2004). Along with these new expressions of polyvalence, new forms of integration, cohesion and mutual understanding are emerging that link agriculture and society in new ways. In other words, a new *cultural capital* is unfolding.

In the first round of empirical enquiries in the Italian research area (1979 to 1983), we saw only minor differences in degrees of specialization. However, the course of events since then has resulted in a panorama full of significant contrasts. Later in this chapter I detail some of them.

The social patterning of time: Continuity versus rupture

One of the most articulate exponents of the neo-classical interpretation of peasant farming, Theodor Schultz (1964), claims that peasant farming represents a standstill that is deeply grounded in history. Peasants cannot go beyond the 'technical ceiling' entailed in the resources with which they work. Thus, they are, it is assumed, bound to the past, and the future cannot be but an endless repetition of the past. On the other hand, 'modern farming' is built upon and represents a chronic disequilibrium – it is constantly on the move towards a new future (see also Heynig, 1982).

The Schultz thesis, as demonstrated in many historical and anthropological studies, is blatantly wrong – both in general terms and in a more applied sense (see, for example, Bieleman, 1992; Wartena, 2006). Yet, this does not imply that there are no differences, in this respect, between peasants and entrepreneurs. In peasant farming the future is constructed through a specific unfolding of the available resources that have been created in the past. Thus, a flow through time is created that unfolds as endogenous development. Entrepreneurial farming, on the other hand, develops far more through the creation of ruptures (Ploeg, 1990c, 2003a).

Enlarging or containing the production of value added

As outlined in Chapter 2, the peasant mode of farming centres essentially on the creation and growth of value added, which – at a higher level of aggregation – translates into the creation and growth of social wealth; thus, in comparison, peasant farming contributes more to the generation of social wealth than entrepreneurial and corporate farming. This is the case both in Europe and in developing world countries.

Table 5.3 is based on a comparative analysis of the dairy farms already discussed in this chapter. From the sociological research carried out it was possible to construct two groups, one in which farmers clearly reasoned and operated according to an entrepreneurial logic and another in which farmers' strategies clearly reflected peasant logic. Subsequently, the farm accountancy data of the farms of each group were analysed and made comparable by translating them into an imaginary block of 1000ha. This was done for 1971 and

1979. Later, in 2000, I revisited all of the involved farms and obtained the most recent farm accountancy data for the year 1999. Thus, Table 5.3 presents a synthesis of the differentiated development patterns of entrepreneurial and peasant agriculture as they co-exist in one and the same homogeneous region.

Table 5.3 *Differentiated growth patterns of production and value added (dairy farming in Parma provinces at current prices, per 1000ha)*

Entrepreneurial agriculture	Peasant agriculture
1971	**1971**
195.5 labour units	168.8 labour units
Gross value of production (GVP): 735 million lire	GVP: 844 million lire (+ 15%)
Gross value added (GVA): 479 million lire	GVA: 638 million lire (+ 33%)
GVA as a percentage of GVP: 65%	GVA/GVP: 76%
1979	**1979**
116.0 labour units	141.7 labour units
GVP: 2845 million lire	GVP: 3872 million lire (+ 36%)
GVA: 1770 million lire	GVA: 2616 million lire (+ 48%)
GVA/GVP: 62%	GVA/GVP: 68%
1999	**1999**
63.5 labour units	85.1 labour units
GVP: 8235 lire	GVP: 12,815 million lire (+ 56%)
GVA: 3956 lire	GVA: 6142 million lire (+ 55%)
GVA/GVP: 48%	GVA/GVP: 48%

Source: Original material for this book

Table 5.3 shows, in the first place, that (apart from 1971)[3] the peasant mode of farming generates more employment than the entrepreneurial mode – which in itself, of course, is not surprising. Second, the table shows that this imaginary block of 1000ha would produce considerably more when tilled in the peasant compared to the entrepreneurial way, and the difference increases over the decades. In 1971, gross value of production (GVP) realized through the peasant approach constituted 15 per cent more than realized through the entrepreneurial mode. In 1979, the difference was 36 per cent and in 1999 it amounted to 56 per cent. This difference was partly due to the deactivation that began to express itself in the entrepreneurial group of farms. This clearly demonstrates that there is no 'intrinsic backwardness' to peasant farming. It also stresses that the frequently articulated view that peasants are unable to feed the world is unsound since it depends upon the 'space' they dispose of.

Third, peasant farming produces the highest total amount of (gross) value added (GVA). This is not only due to the fact that total production is higher, but also because within the peasant mode of farming GVA represents a larger part of total GVP. In 1971, for instance, GVA represented 65 per cent of total GVP in entrepreneurial farming, while in peasant farming it amounted to 76 per cent. In synthesis: if farming is structured according to the peasant mode, not only more production and employment are generated, but *the peasant mode also generates more income*. This applies to the agricultural sector as a whole – *it equally applies to per capita income levels* (at least in this case).

In 1971, income levels per unit of labour force were equal to 2.5 million Italian lire in the entrepreneurial group, and 3.8 million in the peasant group. In 1979, the income levels per unit of labour were 15 and 18 million in current lire, and by 1999 this was 62 million lire and 85 million lire, respectively. Hence, peasant farming (or, more specifically, labour-driven intensification) is not, by definition, identical to the often assumed distribution of poverty and does not necessarily result in involution. Within the mathematical models of neo-classical economics, intensification may run counter to assumed diminishing returns: in real life, peasants are patterning development (as an organized flow of activities through time) in such a way that incomes remain at acceptable levels or are even augmented.

I am, of course, aware that there are many instances in time and space that entail differently structured patterns. As a matter of fact, there are many places where ongoing intensification is blocked (as described in Chapter 3) and where, consequently, 'diminishing returns' emerge. There are, equally, instances where poverty is socially distributed. The point, though, is that such phenomena are not *intrinsic* to peasant farming – they are repeatedly induced into it through interaction with wider society. Depending upon these relations, peasant constellations might seemingly die or show considerable superiority.

Within rural economies the difference between 3956 and 6142 million lire (the latter, thus, 55 per cent more) is far from irrelevant. Translated into the Euros that were later introduced, this is – for an area of 1000ha – a difference of 1 million Euros. Extended to the Parma province as a whole (and assuming a minor difference for the hills and mountain area), this would imply an additional income of some 70 to 80 million Euros per year. Again, at macro-economic level this is hardly relevant (Parma is one of the richest areas of Italy); but for the rural economy of Parma, it is definitely not irrelevant.

In developing countries, however, the potential superiority of the peasant mode of farming could make a considerable difference – that is, it could contribute far more than other modes of farming to the generation of productive employment, income and growth of production (see also Figueroa, 1986; Hanlon, 2004).

Table 5.3 also reveals the Achilles heel of peasant farming as it is currently constituted. Even if total value added remains at a high level, as part of GVP it decreased from 76 per cent in 1971 and 68 per cent in 1979 to only 48 per cent in 1999. Thus, one of the central pillars (and lines of defence) of the peasant mode of farming (i.e. the capacity to generate at given levels of production a value added that is higher than in other modes) is visibly eroding. Due to the high intensity levels strived for (and *de facto* realized), more inputs are needed and, thus, variable costs (per hectare per cow) are rising. Especially when input markets are increasingly dominated by Empire, such dependency can become highly problematic. In Chapter 6, I show how European peasants are currently reversing this trend.

From deviation to modernization: The historical roots of agrarian entrepreneurship

Within all peasant societies there is considerable heterogeneity, not only in terms of socio-economic status, but also in the way in which farming is organized. This is beautifully reflected in the title of Zuiderwijk's thesis (1998) on farming in the north of Cameroon: *Farming Gently, Farming Fast* (see also Steenhuijsen Piters, 1995). The big divide between capitalist agriculture (large scale, extensive) versus peasant farming (small scale, intensive) is repeated – in a miniaturized way – within peasant farming itself. Some peasants will dedicate a lot of work and attention to each object of labour and, thus, obtain high yields, and related to this, they will not work too many labour objects. They are farming 'gently'. In contrast, there are peasants who tend to work far more labour objects, and who therefore do not have to dedicate so much labour (less care) to each single labour object.[4] They rush. They 'farm fast'. Peasants in Guinea Bissau characterize this as *lavrar quente-quente* (work in such a way that you get overheated; see Ploeg 1990b). As a consequence, yields will be somewhat lower and people will also talk about it as 'farming roughly', an expression often used in The Netherlands.

The tension between 'farming fast' and 'farming gently' is characteristic of all peasant societies – at least the ones I know personally and the ones that I have read about. It is potentially an explosive tension. It can introduce, once farming fast is extended, ferocious internal competition (to accumulate the most objects of labour).

The threats entailed in farming fast are controlled and maintained through the commitment to a 'moral economy', which is encountered in peasant communities everywhere. This concept, coined by James Scott (1976), refers to cultural repertoires that specify how to work and how to relate to others, etc. Such moral economies normally put a strong taboo on working fast and roughly, or on working 'beyond one's own forces' and 'putting more hay on your fork than you can carry' since 'pride goes before a fall'. The moral economy thus kept the threatening deviation and the pitfalls of arrogance and megalomania to the proverbial dimensions of the exceptional. The limited exceptions reaffirmed, above all, the rule as entailed in local morals. The modernization of agriculture represents, in this respect, a breathtaking change. Through the modernization project, what was initially taboo became generalized into the quickly expanding entrepreneurial mode of farming.

The deviation known as farming roughly emerges from the 1960s onwards as the dominant trend in all farming systems where modernization is introduced. Central to modernization was (and continues to be) a disproportionate scale increase at farm level and, consequently, a decrease of labour input per object of labour and less care for land, animals and crops. The same disproportionate scale increase (which also implied that existing farm units were 'too small', 'backward' and 'inadequate') rose considerably above the endogenous growth potential of the farms and of the sector itself. Farmers had to enter into dependency relations in order to finance the proposed expansion and change. They also had to reshuffle the relations existing within the peasant community: socially regulated relations of reciprocity had to make way for transactional relations that implied that many resources (of whatever kind) needed to be redefined as commodities.

In many places, modernization started as a cultural offensive (Karel, 2005): as a frontal attack on existing moral economies and associated practices. Fruit trees were uprooted (they 'had become too expensive to be kept') and cattle breeds that represented the pride of many farmers were slaughtered and replaced. Furthermore, the proposed expansion and scale increase at farm level materialized, sooner or later, as the *takeover* of development possibilities of others, who had to be socially redefined as no longer having the right to stay in the agricultural sector. All of these elements evidently ran counter to the then existing cultural norms.[5]

In yet other places (such as Brazil), modernization literally started and proceeded as a kind of war: smallholders and their crops were moved from the land in order to convert the area into, for example, coffee-producing areas in which the former smallholders could now work as labourers – at least some of them; others had to move to the big cities (Cabello Norder, 2004). The military government made sure that all of this occurred without too much resistance.

As a cultural and/or military offensive alone, modernization could never have succeeded. But modernization also embraced:

- massive state intervention in the markets;
- new technologies that allowed for elevated production results; and
- a new division of labour and space that allowed for an externalization of negative effects.

Let me briefly explain each of these elements and their interrelations. First, as part of modernization, a huge technico-administrative apparatus (Benvenuti, 1975b) was developed, first at the national level and later, in Europe, at the supranational level. This apparatus was meant to regulate, homogenize and stabilize off-farm prices and to directly finance a part of the required on-farm investments. Thus, the relevant factor prices that initially were relatively unfavourable for capital were reshuffled: capital was made cheaper through interest subsidies, tax reforms and additional premiums, while labour became more expensive. Over the decades, this has implied huge transfers of capital towards the agricultural sector and agribusiness. Through this, an entrepreneurial condition (see Box 5.1) was created that materially allowed for the creation, maintenance and development of the entrepreneurial mode of farming. Had there not been such state-financed maintenance of prices and the associated assurance that this would be continued, peasants would not have changed their well-proven *modus operandi* for the new entrepreneurial mode.[6]

Second, from the 1960s onwards, new technologies were developed and widely disseminated (often through state extension services) that allowed for the combination of abrupt scale increases with simultaneous increases in the levels of intensity. Such new technologies turned intensification from a mainly labour-driven into a technology-driven process. In this way, scale enlargement could be pushed without provoking stagnating or deteriorating income levels. 'Vanguard farms', '*les grands intensifs*' and '*le aziende di punta*' became new realities and new points of reference – they seemed to testify materially to the assumed superiority of entrepreneurial agriculture.

Third, while nearly every farm unit of production once composed a mosaic of elements, practices and relations that mutually strengthened each other, modernization gave birth to the highly specialized, somewhat monotonous, agricultural enterprise. New patterns for the division of labour and space emerged. Some areas specialized in breeding, others (possibly thousands of kilometres away) in fattening (the transport of calves over long distances being a new phenomenon), while third areas specialized in the production of feed and fodder. In addition, there was a fourth category specializing in the administrative destiny of manure surpluses and a fifth that provided cheap labour. Evidently, such new patterns could not have emerged had the state not accepted or even positively facilitated them. For the core groups of entrepreneurial farms located at the centre of these newly emerging networks, the quick expansion of the division of labour implied considerable cost reductions, while many of the negative effects were shifted to other peripheral locations.

Together with the cultural offensive, these elements explain how and why the initial deviation (farming roughly) could be turned into the general rule. Together they explain the transformation of peasant agriculture into the entrepreneurial mode of farming. As became clear only later, this transformation was, however, only partial. Many different shades of peasant farming remained, and when entrepreneurial farming started to show its immense contradictions, these shades increasingly obtained more profile. It also became clear that modernization and the rise of the entrepreneurial mode of production could not be seen as changes that had occurred once and for all – neither were they eternal or irreversible. This became clear when the required entrepreneurial condition (see Box 5.1) started to erode.

Box 5.1 *The entrepreneurial condition*

To prosper, farms structured according to the entrepreneurial mode of production need a specific politico-economic context that is characterized by, and secures, the following conditions:

- Relatively stable prices and the avoidance of steep fluctuations are imperative since entrepreneurial farms require large investments due to their dimensions, structure and accelerated expansion. These can only be realized when there is long-term stability in off-farm prices – with too much turbulence, planning, investments and further expansion become difficult, if not impossible.
- Price levels must allow for a positive margin between costs and benefits.
- Markets must be ordered in a way that prevents a sharp rise in costs, interest levels, energy prices, etc.
- State intervention is needed to 'rescue' affected farm enterprises when dramatic events interrupt the normal reproduction and growth of entrepreneurial farms (e.g. BSE, foot and mouth disease and climate conditions that cause massive harvest failures).
- The state must secure relatively cheap capital and/or allow for cheap labour.
- The state must create the spatial and institutional conditions that allow for ongoing and accelerated expansion.
- Entrepreneurs need to control farmers' unions and be able to impose their programme on the majority of farmers.
- There must be expert systems that allow for a flow of well-tested innovations that contribute to further scale enlargement and industrialization of the farm labour process.
- There must be an outflow of 'bad' farmers who manage 'economically unviable farms'. This liberates resources badly needed by the expanding entrepreneurs. Whenever hindrances occur to this outflow, the state must remove them.
- Civil society and the state are to grant entrepreneurial farming 'safe havens', where spatial, ecological, social and economic conditions that best fit with such farming may prosper.

The political economy of entrepreneurial farming

Worldwide, agriculture is subjected to an economic squeeze (Owen, 1966) that is increasingly tightened by Empire. Off-farm price levels are held down or reduced through the worldwide re-patterning of the interrelations between food production and consumption. This squeeze drains an enormous amount of social wealth from the countryside that is accumulated in Empire. As a matter of fact, the food industry throughout Europe is the sector that represents the highest growth level of value added. This is illustrated in Figure 5.4 (concerning Italy). It shows that in 2003 the value added of the food industry was 48 per cent higher than in 1980. This growth by far exceeds the levels realized in other economic sectors.

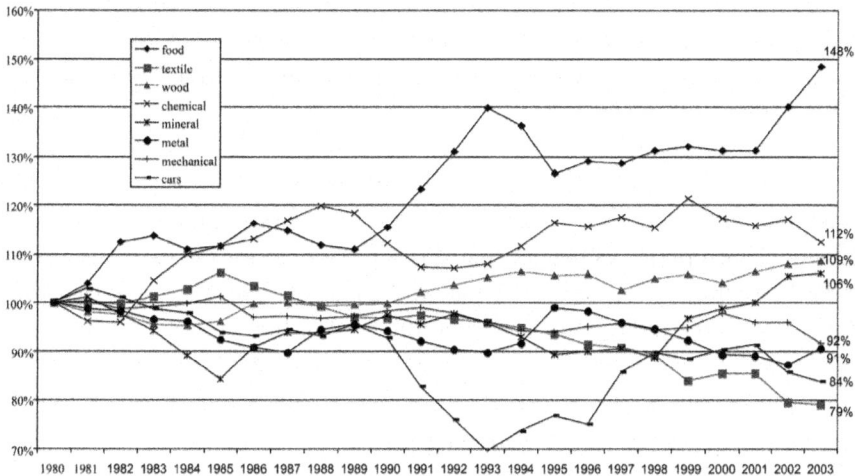

Figure 5.4 *Added value for main industrial branches in Italy (1980 = 100)*
Source: ISMEA (2005, p73)

The squeeze on agriculture confronts farmers with considerable downward pressure on their incomes, as well as an erosion of prospects in the longer term. One of the responses to this has been the extension of entrepreneurial farming at the expense of peasant farming. Under current conditions, growth proceeds through the *takeover* of other farms – or more precisely, through the takeover of the capacity to generate value added. Currently, scale enlargement means that the capacity to generate value added is concentrated in a small and diminishing group of farms. However, this process of takeover and concentration is far from neutral: it translates into a reduction of the *total* value added at sector and regional levels. This is due to two intimately related effects. The first effect is that the expanding farms, in which the production of value added becomes concentrated, are at the same time industrializing agricultural production, thus augmenting cost levels and further narrowing the margins. Second, the transactions needed for the takeover also imply that a considerable amount of value flows out of the sector.

Thus, under current conditions, scale increase augments the cost levels in agriculture. The large and rapidly expanding farm enterprises are characterized by cost levels that are higher than those of smaller slowly growing farms. According to textbooks of farm economics, it *ought* to be different; but then the real economy may be quite different from any straightforward textbook application. As a consequence, agriculture is not subjected to just one but to a double squeeze: alongside an 'external squeeze' there is now an 'internal squeeze' (see Figure 5.5). The overall effect is that social wealth (total value added) is squeezed out of agriculture at a rhythm hardly known since the last great agricultural crisis of the 1930s. Between 1995 and 2005, the net agrarian income of, for example, Dutch agriculture was reduced from 4.6 billion to 3 billion Euros.

Entrepreneurial farming is strategic in this respect. It translates the external squeeze (increasingly tightened by Empire in order to fuel the accumulation of wealth illustrated in Figure 5.4) into an internal squeeze. Individual entrepreneurs believe that this is the way to escape from the effects of the first squeeze; however, the material effect of their actions is that they simply strengthen its negative impact at the level of the sector as a whole.

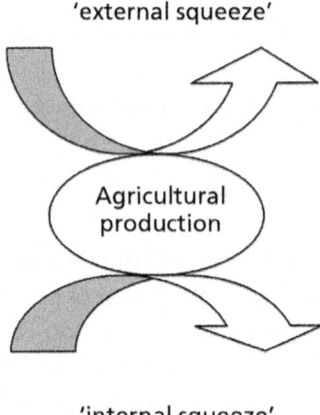

Figure 5.5 *The double squeeze on agriculture*

Source: Ploeg (2006a, p263)

Through the translation of the first into the second squeeze, the entrepreneurial mode of farming increasingly represents conquest as growing numbers of farms are taken over. The advance of entrepreneurial farming also goes hand in hand with a progressive process of multiple degradation. Finally, it impacts upon rural economies as a process of induced impoverishment.

Multiple degradation

Due to expansion being financed substantially through loans, debts are relatively high and affect every available stall in the cowshed. Already during the early 1990s, these debts ranged from some 2300 to 8960 Euros per stall. A modest interest rate of 5 per cent implies that each stall must yield between 115 to 450 Euros per annum. Average debt levels per stall were raised, since then, from 5580 Euros in 1990 to 7240 Euros in 2006 (Alfa, 2007). Wherever high debt levels are present – which is especially the case for entrepreneurial farms – the financial burden translates into the need to produce, at every available place, as high a financial yield or margin as possible in order to pay the interest and capital on loans. Thus, cows are pushed to high milk yield levels – and the more one goes beyond the average gross margin per milking cow (some 2000 Euros), the better. An available place in the cowshed is no longer a use value, no longer a self-evident part of available resources. Above all, place represents here *capital* that must generate extra value. Beyond remuneration for the farmer, it must produce interest payments and redemptions.

This normally translates into the choice of highly productive Holstein cattle, yielding from 8000kg to more than 10,000kg of milk per year. Such cows are the outcome of breeding and selection processes that have significantly altered the social organization of time. Holstein cows produce very high milk yields in their first and second periods of lactation, but thereafter they are likely to diminish. In Figure 5.6, the typical development of milk production over time is illustrated for highly productive cattle and for the more 'traditional' breeds. Highly productive cows are likely to be removed and replaced in the third or fourth year – the more so because the stress they experience is quite likely to develop udder, fertility or other health problems. Currently, the replacement rate in Dutch dairy farming is around 33 per cent. This means that, on average, cows produce for just over three years – while theoretically they can easily go on for far longer periods. However, longevity is not an objective in itself, especially within entrepreneurial farming. Cows will be replaced as soon as their milk yield per year diminishes since they occupy spaces that must yield maximum monetary benefit. Thus, a particular pattern emerges. At one time a cow would occupy a place over a 10- to 12-year period and produce, say, 60,000 to 70,000 litres; nowadays, the same place will be used sequentially by some five cows for only two to three years each.

The irony of this double shift (i.e. higher production per cow per year and shortened productive lifespan) is that within a specific temporal framework (e.g. ten years), up to 40 per cent more animals are needed to realize the same

132 *The New Peasantries*

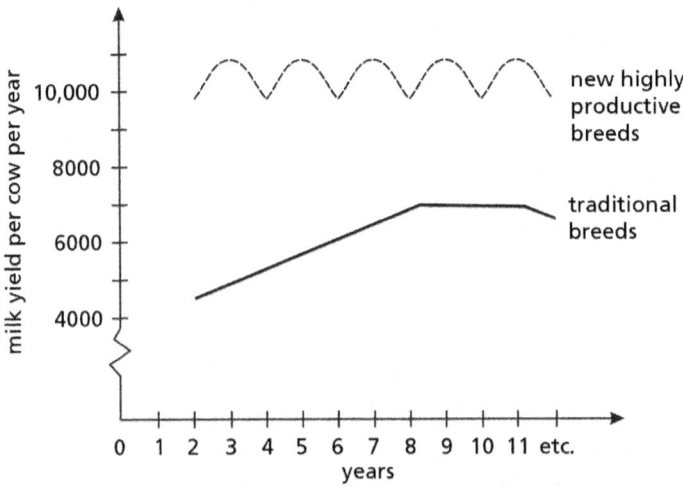

Figure 5.6 *The evolution of production per cow over time*

Source: Original material for this book

total production. Thus, the entrepreneurial mode of farming not only transforms the natural resources (by creating, for example, highly productive cattle and new nitrogen-sensitive meadows), it also reshapes the biophysics of production processes. Together these changes imply that there is a tendency to degrade animals to the status of *throwaway* products. They could produce for many years, some even up to 15 or 17 years (which is what dairy farmers aimed for in the past); but within the new framework of entrepreneurial farming, their lifespan is greatly reduced.

Together with changes in the relevant time dimension and associated increases in milk yields, we find a range of other related adaptations (summarized in Figure 5.7). The use of industrial concentrates is expanding; grassland management is reorganized in order to produce high yields of high-energy and high-protein silage; maize has been introduced; herds per farm have grown; grazing in meadows is being replaced by 'summer feeding' inside cowsheds; the architecture of cowsheds is changing (with cubicle sheds predominating); the techniques for dealing with faeces and urine have been drastically altered ('well-matured manure' has largely disappeared and slurry is taking its place); and the care of animals is completely redefined. All of this significantly reshapes the biophysics of both resources and the process of production built upon it. This reshaping translates into multiple degradations. Cows become far more vulnerable and are materially degraded to things that can simply be disposed of. In the same way, the quality of manure is degraded: it is turned

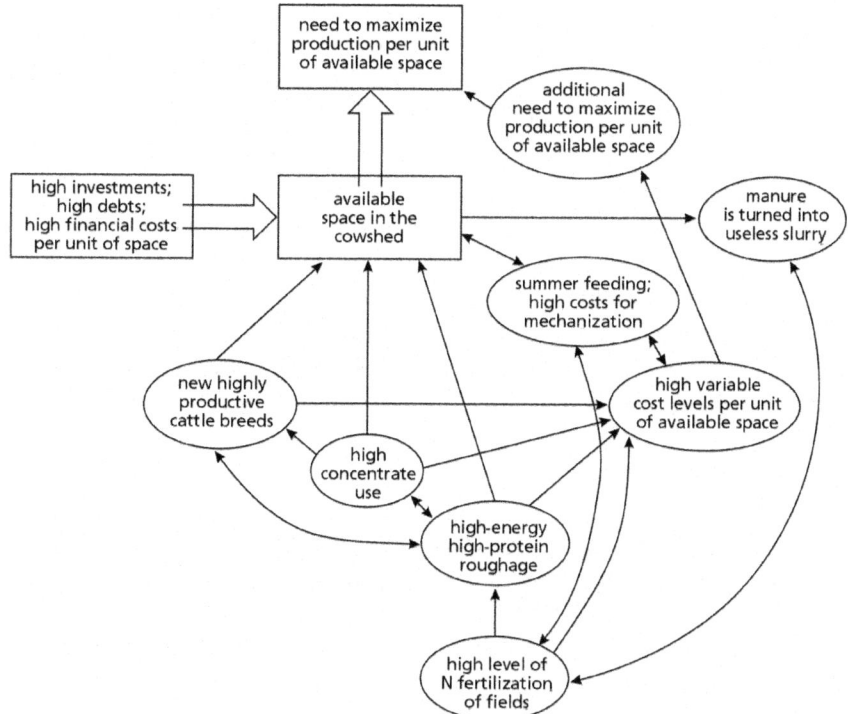

Figure 5.7 *The changing biophysics of production*
Source: Original material for this book

into a waste product that is highly detrimental to soil biology and fertility; in a similar fashion, milk is degraded, among other things, because the CLA (conjugated linoleic acid) levels (i.e. unsaturated fatty acids that have an anti-carcinogenic effect) are gradually diminishing, useful microbiological flora is increasingly filtered out, and fats are homogenized and thus contribute to obesity.

Conquest

Within the rural communities, the entrepreneurial mode of production often translates, at the level of everyday life, as conquest (see Prins, 2006), while in his discussion of similar relations within US agriculture, Marty Strange (1985, p4) refers to 'predatory behaviour'. In order to feed the takeover and concentration of value added, *conquest* and the creation of conditions favourable to it are what is needed. The way *poseiros* operate in the Brazilian Amazon is a striking example of such entrepreneurial conquest. In this case, it is nature that is conquered – and largely destroyed. *Os sem terra* (the peasants of the Amazon) relate in a completely different way to nature (for a full analysis, see Otsuki, 2007). Another example concerns quota trading in The Netherlands. Until recently, the market for milk quotas had two transaction possibilities. Quotas could be bought or leased on a temporary basis. Leasing can help to solve problems of a temporary kind, such as when a father becomes unable to do the job and the son is still studying, or when a new cowshed has to be built. In these circumstances, leasing milk to others is a welcome opportunity. Leasing also occurs when a number of farmers have decided to deactivate their own dairy farm operations: they lease out their quotas and engage, for example, in extensive cattle breeding. With high lease prices, this often turns out to be quite profitable. However, not everybody is allowed to engage in such money-earning behaviour. The Dutch Farmers' Union, Land- en Tuinbouw Organisatie Nederland (LTO), which increasingly operates as a lobby for the large and expanding entrepreneurs, started a campaign to put an end to what they defined as a 'perverse' use of market opportunities. They proposed eliminating leasing. It was thought that by putting an end to it, the 'perverse' farmers would be obliged to sell their quota once and for all and that the augmented supply of quotas would even reduce price levels. The lobby finally proved successful: the legal rules governing the market for quotas were changed. Farmers could now lease out only 30 per cent of their quota. Then an amazing scenario unfolded. Smaller farmers were, as expected, forced to sell their quotas. But, at the same time, large farmers (already possessing, for example, 1.5 million litre quotas) purchased an additional 700,000 litres in order to lease out 660,000 litres (thus staying within the 30 per cent margin), often in small quantities, to others who needed additional quotas (e.g. because during the year their cows had proved to be more productive than was expected). Since the interest rate for loans was somewhat lower than the lease price, interesting extra profits could be generated, while simultaneously extra quotas were accumulated freely. In 1968 Michael Moerman published an interesting book about Thai agriculture. In it he describes how farmers who refuse to obey established rules of reciprocity were socially defined as 'sons of bitches'. Apparently, such a phenomenon is not only limited to Thailand.

Impoverishment

Truly dramatic expressions of induced impoverishment are encountered in developing world countries. The now-widespread process of *ganaderización* is one such expression, especially in Central America. *Ganado* means cattle, and *ganaderización* consequently refers to a slow but persistent change from agricultural activities (sometimes combined with the rearing of animals) to specialized and relatively large-scale cattle breeding.[7] This change is associated with the demise of peasant agriculture and the rise of entrepreneurial cattle ranching.[8]

Currently, the process of *ganaderización* is emerging directly from within the peasant economy itself – that is, a small group of peasants are in the process of transforming themselves into cattle breeding entrepreneurs who control, directly or indirectly, vast areas of grazing land and who organize their ranching practices along new lines and within newly established networks linking local and global in innovative ways. From the point of view of the national and regional economy, this process undoubtedly represents regression: land productivity suffers a sharp decrease; rural employment and disposable income slide and sustainability is often seriously threatened. Nonetheless, for many observers, this type of *ganaderización* represents not only an inevitable, but also a positive, trend towards more feasible entrepreneurial farming and the necessary adaptations of rural economies to global markets.

One of the effects of *ganaderización* is a sudden fall in the total value produced per hectare. In a detailed study in the Sierra de Manantlán in Mexico, Peter Gerritsen (2002) shows that while peasants (involved in arable farming, horticulture and some animal production) produce, on average, a value of 3800 pesos (at that time some US$520) per hectare annually, the more entrepreneurial *ganaderos* only realize an average production of less than 500 pesos per hectare annually. This implies that the spatial expansion of commercial cattle breeding will bring with it a strong downward trend in regional (and national) production levels. While so far we have encountered examples of *relative* decline (i.e. where total production grows less than would have been the case were the peasant mode of production generalized; see Table 5.3), the generalization[9] of the entrepreneurial mode of production results, in the case of *ganaderización,* in an ugly and absolute decline of total agricultural production. The value added per hectare and, hence, the social wealth of the region as a whole characteristically suffers a sharp decline. That is how poverty is *produced.*

In Europe, the relative decline of social wealth as produced and allocated in the countryside can also lead to absolute reductions. Scenario studies show (see, for example, Antuma et al, 1993) that accelerated scale enlargement in the primary sector (also referred to as the 'free trade scenario') will, over a 15-year period, reduce regional sector income to 26 per cent of the initial level. If, however, current trends are simply extrapolated (a 'policy trend scenario' entailing more contained forms of scale increase), then social wealth is likely to decline to 51 per cent of the initial level. Moreover, if current trends were accompanied by an active defence of farming, especially through the integration of additional economic activities (currently known as 'rural development'), the reduction could be limited to 89 per cent of the initial level. This is illustrated in Figure 5.8.

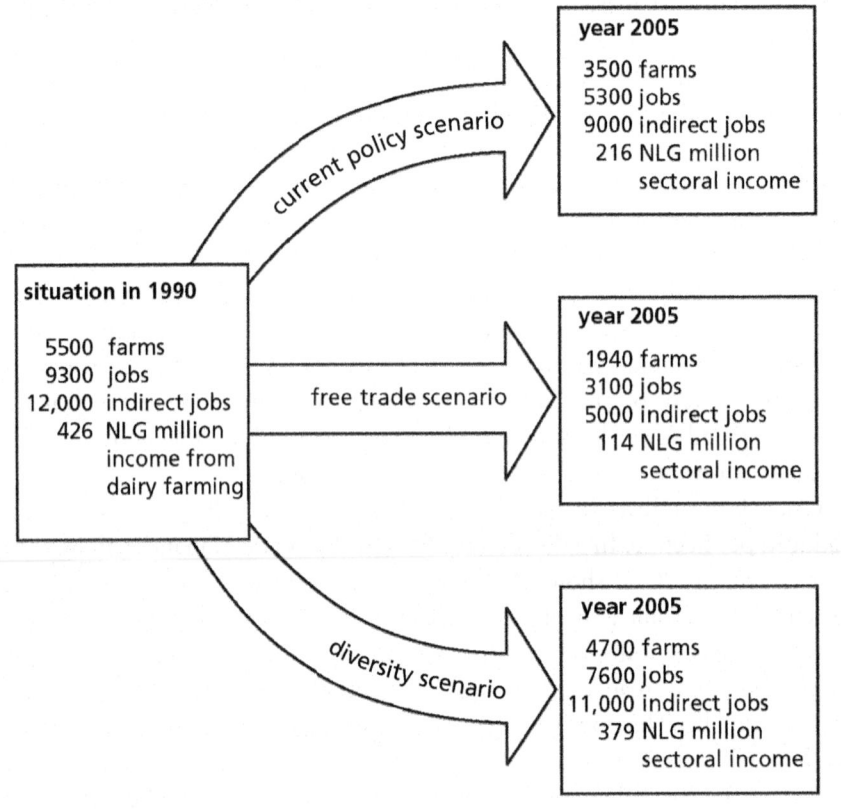

Figure 5.8 *Outcomes of a scenario study that compared different development trajectories (dairy farming in Friesland, The Netherlands)*

Source: Ploeg (2003a, p308)

Heterogeneity reconsidered

During the last 15 years, empirical enquiries have revealed that heterogeneity exists in many different agricultural systems. The patterns of coherence underlying this heterogeneity are what is referred to as 'styles of farming'. These styles represent the material, relational and symbolic outcomes of strategically ordered flows through time. Taken together, they make up a richly chequered range that extends from different forms of peasant agriculture, via highly complex combinations, to different expressions of entrepreneurial agriculture. These different patterns of coherence, or styles, are summarized in Figure 5.9. The vertical axis refers to the domain of production and highlights two contrasting ways of sustaining and augmenting levels of intensity. On the lower

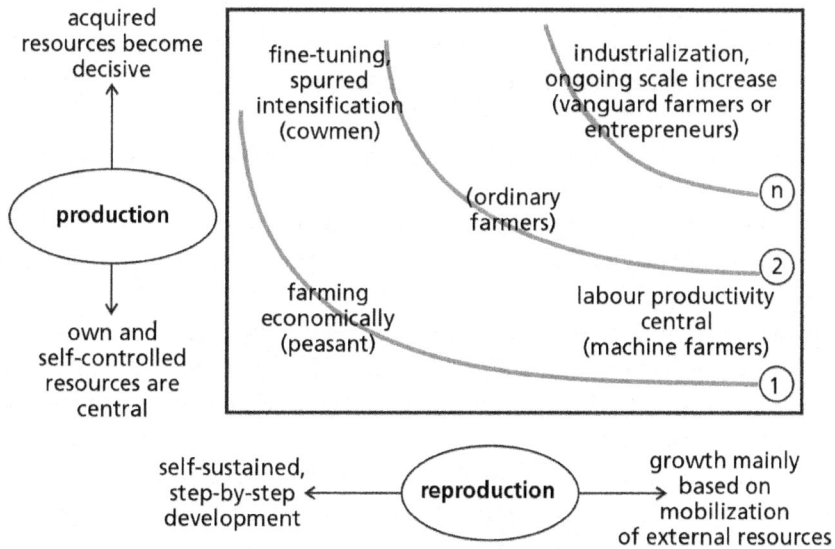

Figure 5.9 *Space for manoeuvre and different degrees of 'peasantness'*
Source: Original material for this book

side, levels of intensity are grounded in available and self-controlled resources. The upper levels of intensity are increasingly based upon external resources. In the first case, we face labour-driven intensification (a process that principally integrates skill-oriented techniques) and, in the second, technology-driven intensification (a process that normally goes together with a reduction of labour input). Evidently, there are many in-between situations in which intensity levels depend upon both the quantity and quality of labour and upon the virtues of specific external resources. The horizontal axis refers to the reproduction of the farm over time. Reproduction may be grounded in self-produced resources or in the mobilization of external resources. Different technological artefacts and especially the way in which they are integrated into the operation of the farm are important here, while again there will be many in-between situations and combinations. Moving along these two dimensions will often imply that the interrelations between specific farm units and markets change and that, in addition, there are changes in the levels of transaction costs (Saccomandi, 1998). The more farming moves away from the bottom-left position, the higher the transaction costs will be – although these may be mediated by institutional arrangements that operate at increased levels of aggregation.

In the space defined by these two dimensions, different positions are possible. In the bottom-left position one encounters the style of farming economically (or low external-input agriculture, as it is defined in many studies realized in developing world agriculture). The immediate expressions of it will, of course, depend upon time and space – that is, they will often differ. The basic ordering principles, though, are the same – regardless of time and space.

Moving from this first position along the vertical axis one arrives at the position of the typical 'cowmen', or '10 tonne grain producers'. This entails a selective use of certain inputs, a high labour input and a meticulous fine-tuning of labour and the production process. These constitute the main ingredients of this particular style.

In the bottom-right corner, we encounter styles that centre on achieving the highest production with the least possible labour input. The 'machine' pops up here as a telling metaphor since it links the two elements. Upper-right represents a position that is normally associated with 'vanguard farms': a strongly increased number of labour objects per unit of labour force, specific technologies that sustain relatively high levels of intensity, and – more than other styles – a process of ongoing expansion.

According to the specificity of time and place, several in-between positions can be discerned, and the associated folk concepts used to classify the different positions are many. The point I wish to make, however, is somewhat different. Different iso-curves can be projected over the space entailed in Figure 5.9. These curves refer to different degrees of 'peasantness' (as put forward by, among others, Victor Toledo, 1995; Jollivet, 1988, 2001). They underline that *'l'agriculteur peut ... décider de devenir "moins exploitant agricole" et "plus paysan"'* ('the farmer can choose to become "less entrepreneur" and "more peasant like"'), or vice versa (Pérez-Vitoria, 2005, p230).

The bottom-left position mainly represents the peasant principle. Here relations, processes, patterns and identities will be ordered according to the peasant mode of farming. At the same time, as Figure 5.9 suggests, this peasant mode flows over into other simultaneously related and different positions: it can move towards further intensification through fine-tuning (*farming gently*), but it might equally evolve into farming roughly. Thus, diversity within the peasantry is conceptually included from the beginning – it is not added *ex post*. Similarly, we encounter, when moving from the bottom-left towards the upper-right corner (i.e. from curve 1 to curve *n*), the field typically occupied (and created) by the entrepreneurial mode of farming. And again one finds differentiation. Depending upon the prevailing conjuncture, entrepreneurial farming will sometimes tend towards one way (technological optimization), and sometimes materialize in other ways (in a 'rougher' type of farming).

One of the strategic elements of research on farming styles has been the thesis that there is no single way (let alone one *superior* way) to produce a reasonable income and promising prospects. There are many ways, each entailing their own specific coherence that can bear good results. This also holds true for The Netherlands during the early years of this first decade of the 21st century. Let me discuss this by referring to some of the main outcomes of a national research project that was inspired by and built upon farm styles research and which tried to explore further their potential. This project, structured as a long-range experiment, was realized at the National Centre for Applied Research in Animal Production at Lelystad, The Netherlands.

Starting from the different strategies encountered in the dairy farming sector, two farms were build there: a so-called 'low-cost farm' and the other a 'high-tech' one (including, among other things, fully automated milking). Both were designed in such a way that one person could accomplish all the work. Both farms were also to achieve a 'comparable income'. To meet these two criteria, the low-cost farm had to produce a quota of 400,000kg of milk, while the high-tech farm needed a quota of nearly 800,000kg. Table 5.4 summarizes a few of the most salient results.

Table 5.4 *Comparison between a peasant and an entrepreneurial approach in Dutch dairy farming*

	Low cost	High tech
Units of labour force	1.0	1.0
Working hours: man hours per year	2500	2490
Hectares of land	32	35
Milking cows	53	81
Milk yield per milking cow	7547	9673
Total milk production (in kilograms)	400,000	783,515
Concentrates per 100kg of milk (in Euros)	3.8	7.5
Calculated labour cost per 100kg of milk (in Euros)	13.0	6.7
Costs associated with technology use per 100kg (in Euros)	5.4	7.1
Production costs per 100kg (in Euros)	34.5	34.7
Realized income per working hour (in Euros)	19.20	16.36

Source: Data provided by Lelystad Centre for Applied Research

The individual differences featured in Table 5.4 are minor and, at first sight, probably irrelevant. However, by combining a range of small differences in a coherent way, decisive contrasts can be produced – which is precisely what Table 5.4 shows. Were the available Dutch dairy quota (10.8 billion kilograms of milk) to be produced by the relatively large-scale entrepreneurial style, there would be space for nearly 13,900 dairy farms. If, however, the peasant style were to dominate, the total number of farms would be twice as high. More importantly, productive employment and the created value added would also be twice as high. For The Netherlands, such a difference is currently somewhat immaterial since there is no widespread rural unemployment. However, there are many other instances within which the indicated contrast would be perceived as strategic, both in Europe (Broekhuizen and Ploeg, 1999) as well as elsewhere in the world.

Two additional questions arise from the comparison summarized in Table 5.4. The first is why farmers (at least some of them) aim for high volumes of production if they can earn the same from only 50 per cent of such a volume. The second question concerns the future: knowing that the gross margin in the entrepreneurial type of high-tech farm is only 50 per cent of the gross margin realized in the peasant low-cost farm, what would happen if, due to globalization and liberalization, off-farm prices dropped significantly?

The moral economy of the agricultural entrepreneurs

Typical of the peasantry is a strong preference for regulating relations within the community through social mechanisms and norms and values – that is, through non-commodity relationships (this also sometimes makes for strong conservatism). Agricultural entrepreneurs largely prefer market mechanisms for the regulation of internal relations. At first sight, the contrast hinges on the difference between a highly visible moral economy in peasant economies, as opposed to the dominance of the market as a regulatory principle in entrepreneurial economies, where a moral economy seems absent. However, I believe this juxtaposition is wrong. The point is that 'the market' represents as much a moral economy as the rules, values and experience of the peasantry. The 'market' is, above all, a *logo* that embraces, unites and hides a widely extended set of rules, perceptions, beliefs, experiences and specifications of internal relations. 'Market discourse' is moral economy *under cover*. What differs between these two highly contrasting moral economies is, among other things, the set of norms that specifies the ways in which farming should relate to empirically existing markets. In the peasant type of moral economy, distantiation emerges as an ordering principle, while the entrepreneurial type of moral economy favours integration. Together with this goes a strongly differentiated evaluation of the effects of market mechanisms as an ordering principle.

When entrepreneurs refer to the market, they evoke a kind of politico-economic programme. When the market is put centre stage, the argument is nearly always that in the future only a few farmers will remain and that the market is a highly selective 'arena' that will exclude many participants. The future is perceived as a scarce commodity[10] and few will survive (although entrepreneurs will hardly use this word; they prefer to speak of those who will *win*). In the moral economy of agrarian entrepreneurs, the 'market' represents an ongoing and harsh contest. Only a few will win, and those who win (and this is an essential part of their moral economy) are to be seen as the 'best'. And being the best, they have the moral right to win. The winning proves their moral superiority. This is reflected in their judgement of those who are not (or are less) engaged in the ruthless competition for the future: they are perceived and devalued as 'stick-in-the-muds'. This attributed inferiority is consequently seen as an expression of moral decadence. Thus, the *others* (those differing from the entrepreneurs) do not deserve, from the moral point of view, any other destination than that of losing and disappearing.

The market is not only a logo for delineating the others and the 'we group' (the group of strong and promising entrepreneurs with a right to participate in the scarce future)[11] – this same logo justifies a particular kind of behaviour. There is a presumed right 'to work according to the market', as the Dutch expression literally goes. Being 'purely economic' is considered a virtue. As discussed earlier, these attitudes have resulted in a wide range of not always very sympathetic forms of conquest. They also result in practices that, in the end, run counter to the interests of the entrepreneurs themselves. As illustrated in Figure 5.10, the Dutch dairy farming sector currently faces very high cost levels. Capital investment per 100kg of milk is more than triple that of its main competitors (the US, Australia, New Zealand, Argentina and Brazil). Total investment for a Dutch dairy farm possessing 95 milking cows is more than

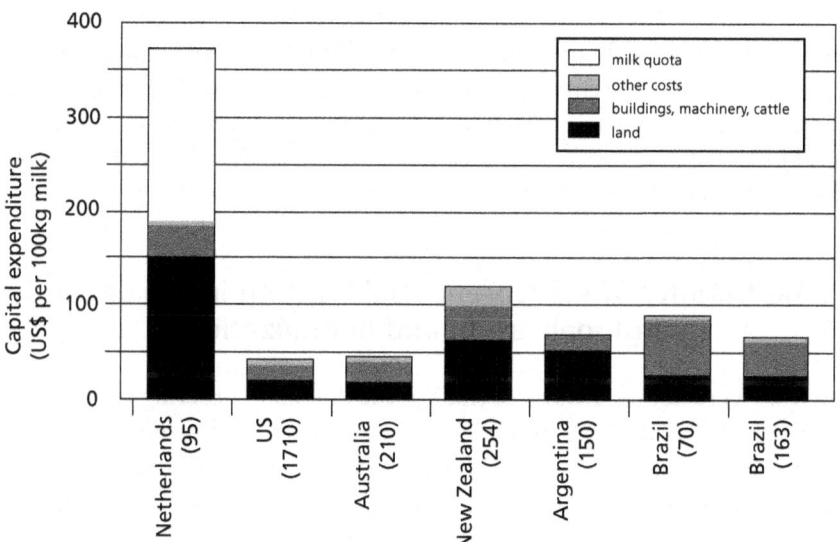

Figure 5.10 *International comparison of investment levels in dairy farming*

Source: Hemme et al (2004)

Note: country names under bars indicate location of dairy farm for which that data applies; number in brackets indicates number of cows on farm.

US$350 per 100kg of milk, whereas in New Zealand for a unit of 254 milking cows, it is only US$125 per 100kg of milk, and in the US (California), less than US$50. These differences are principally due to the extremely high costs of acquiring quota and, to a lesser degree, land. The irony of these differences and the competitive disadvantage that it entails is that they result directly from the strong fixation on the entrepreneurial model underlying agrarian policies and practices in The Netherlands.

Let me enter briefly into some of the most relevant interlinkages in order to show that the entrepreneurial model orders not only practices and trajectories at micro level, but that it significantly shapes a range of contextual elements (located at higher levels of aggregation) as well, the most important of which are the major markets and the functioning of agrarian policies. By (co-)shaping its environment as a set of interrelated 'free markets', the entrepreneurial mode of farming often creates its own pitfalls, such as the competitive disadvantage mentioned above.

During the mid 1980s, the quota system for milk was introduced in all member states of the European Union (EU). However, every member state had to devise its own modality for implementation. In The Netherlands, a fully fledged and 'undisturbed' market model was chosen because it was assumed to fit better than other modalities with entrepreneurial farming. It was thought that an undisturbed market – which contrasts notably with the social, political and/or ecological framing that was applied in many other European countries – would allow an effective transfer of quota from smaller to larger farmers and, thus, for necessary 'structural development'. This, combined with a struggle for the future that many entrepreneurs believe themselves to be engaged in, has led to a continuous rise in quota prices. Thus, the sector created its own stumbling block – that is, it created a context that strongly backfired upon the entrepreneurs themselves.

The fragility of entrepreneurial farming in the epoch of globalization and liberalization

Even if we were not to favour free trade as the dominant ordering principle for society, we would still have to recognize that the coming decennia will probably be characterized by a far-reaching, if not aggressive, globalization and liberalization of agricultural and food markets. This will, without doubt, lead to considerable reshuffling in agricultural production at the global level and to associated price decreases and the reintroduction of frequent price fluctuations. Some people forecast new scarcities (related, for example, to the rise of bio-energy, low levels of grain reserves and rapidly expanding demands in South-East Asia). However, whether such new scarcities will translate into higher off-farm prices remains to be seen – the benefits might well be captured by others.

Nevertheless, the conclusion drawn from the generally expected price decreases is nearly always that in the near future only the highly specialized and large-scale farm enterprises will be able to confront these adverse conditions. Second, it is assumed that the best way of preparing for this harsh scenario is to accelerate scale increase at the farm level as far as possible.

In contrast to this dominant view, I develop here an alternative thesis that centres on the erosion of the entrepreneurial condition discussed earlier in this chapter. Globalization and liberalization, as they are currently advancing, will eliminate the very conditions that are needed for the (enlarged) reproduction of the entrepreneurial mode of production. To realize a spurred scale increase (required to face the expected global competition), high investments are needed which result in high fixed-cost levels. The operation of such large-scale enterprises will require technologies that assume high input levels (among others, energy) and thus relatively high levels for variable costs. Thus, a rather rigid enterprise structure is created while the margins will be low. All of this implies that such enterprises will be highly vulnerable in an epoch characterized by turbulence and unstable prices. The proposed transition will trigger a process that in the end 'bites into its own tail'. While the entrepreneurial condition that was created during the massive modernization project of the 1950s until the 1990s is actively deconstructed, farmers are assumed to engage actively in a transition that assumes precisely the maintenance, if not the strengthening, of such an entrepreneurial condition. Thus, an amazing contradiction is introduced into the debates and practices: farmers invite themselves, and are invited by others, to enter into a contradictory road that probably ends at a Cassandra crossing.

In the spring of 2000, I revisited all the Parmesan farms that provided the main database for the 1979 to 1983 research reported earlier in this chapter. This new visit was meant to reconstruct with farmers from that time (or their sons and daughters) the events that had since taken place. In 2000, the farmers were still in the midst of a deep crisis that had already been affecting for several years the production, transformation and commercialization of *Parmigiano-Reggiano* cheese, a situation that ended only in 2003. During all those years the price for cheese milk was extremely low and prospects were damning. This crisis was, in part, due to the farmers' submission to supermarket circuits and, in part, to competition from the neighbouring *Grana Padana* system. Although the *Parmigiano-Reggiano* system had always experienced cyclical trends (it was never covered by the EU regime of guaranteed prices), the 1997 to 2003 crisis was exceptional. Many observers agree that this period could be perceived as an expression *avant la lettre* of globalization and liberalization. Thus, this revisit allowed for an enquiry into processes that the rest of European agriculture had yet to experience.

My 2000 findings initially proved to be a complete surprise. I had assumed – in retrospect, in a highly naive way – that the entrepreneurial farms would be significantly larger than the farms managed with a peasant logic, the more so since during the 1979 to 1983 period the agricultural entrepreneurs had expressed expectations and plans that implied a considerable further expansion of their farms. And further expansion would, in any case, be the logical outcome of their calculus. What I did not realize at the time was that the reproduction and development of the entrepreneurial mode of production needed a range of specific requirements (summarized as the entrepreneurial condition). In 2000, it turned out that the peasant way of farming had resulted, during previous years, in much more growth and development than that of the entrepreneurial farm. According to available farm accountancy data, the volume of production for peasant units was 510 million lire (at that time more or less 250,000 Euros), while it hardly reached 300 million lire in the second group of entrepreneurial units.[12] In 1980 there had been, on the whole, the same ranking order; but the difference between the two groups was minor (23 per cent in 1980 versus 41 per cent in 1999). Although the entrepreneurs had, in 1980, plans for expansion that far succeeded those of the peasants, in reality, the latter had developed much more. On the farms managed according to an entrepreneurial logic, investments had decreased considerably. This was an expression of rational decision-making. With the ratio of off-farm prices and costs as bad as it was in the second half of the 1990s ('the *margin* is so deplorable'), investing was useless, especially because 'profitability' was considered and judged in a wider framework. The changing land market (and the prospects relating to urbanization) and especially the stock market were considered far more important points of reference. As a matter of fact, several *imprenditori* had invested large amounts in stocks. Others were actively engaged in selling land for house construction.

The reaction of the entrepreneurs to the low prices and eroded prospects was deactivation. The agricultural enterprise was slowly deactivated, while capital was reallocated to other, more promising, economic sectors. Other mechanisms of deactivation considered by entrepreneurs consisted of a change to more extensive forms of agriculture (e.g. meat production and specialized grain cultivation), which would allow for a substantial reduction of labour input. Another mechanism, practised nearly everywhere, was further externalization: the production of feed and fodder within the farm had been replaced by acquisitions, and several farms had eliminated the rearing of calves and heifers – when a cow needed replacing, another one was simply bought on the market. What struck me as well were the new semantic expressions used to capture this changed situation. Several of the entrepreneurs used the expression *agricoltura di salto*, which one may translate as 'hit-and-run agriculture', suggesting a need for the entrepreneur to 'jump' from one opportunity to another. If EU policy or market conditions imply that a particular crop is profitable, then you must jump into it, and as soon as another opportunity arises you must switch to that, and so forth.

The *contadini,* who were also worried and complained bitterly, had reacted quite differently. To start with they had continued to invest (*'fare le spese'*: literally, 'to spend money', but used by the peasant farmers to refer to investments), which, in their farming logic, is an essential commitment to continue to work with *passione, impegno* and *cura*. This applies to themselves but also (and probably even more so) to their successors. During the 1980s this was so self-evident that peasants did not bother to refer to it. The crisis of the 1990s, however, burst on the scene and activated anew this part of their logic. The deeper the crisis bit, the more the need to 'spend' became apparent. *Fare le spese* was needed more than ever, *not* to realize a comparable return on invested capital, but to ensure the continuity of their farm and the promise of the *'bell'azienda'* that it entails. Thus, spending savings and income from pluriactivity and minor loans was not aimed at converting assumed financial capital into an acceptable rate of return. Here we encounter a decisively different process of conversion. The results of former work (and formerly created trust) are converted into contributions to enlarge the reproduction of the farm. The results thus obtained underline the maintenance of the autonomous resource base of the farm and the avoidance of strong dependency relations.

The highly contrasting reactions of the *contadini* and *imprenditori agricoli* are firmly rooted in, and can only be explained by, the different modes of farming from which they emerge. If only average trends at higher levels of aggregation are used, then a kind of general slowdown (a 'recession') is perceived, while the real crisis (the demise of entrepreneurial farming) and the response to it contained in peasant farming are both missed. Moreover, when attention is shifted, in an indiscriminating way, to individual farms, then it is only the presumed (and therefore consequently ascribed) differences in 'entrepreneurship' that come to the fore.[13] Taken together these elements highlight the intellectual poverty of most of today's discourse on farming and agriculture.

But there is more. Farming economically (Ploeg, 2000; Kinsella et al, 2002) never had been absent in the practices of the *contadini*. Yet, now, facing crisis, nearly all peasants embraced and developed this particular strategy. For instance, new cowsheds were built upon the already available infrastructure (often giving rise to ingenious solutions and designs), and construction was mostly done by the farmers themselves or with the help of 'friends from the village'. By thus combining 'spending' with the intention to be as economical as possible, investments could continue under crisis conditions. The same applies to using internal resources as far and as efficiently as possible. In this respect, it is telling that drying hay artificially was now a common practice on all farms managed with a peasant logic. By artificially drying the fodder crops (*alfalfa* being the main one), their quality could be increased while at the same time minimizing quantitative losses, which meant that buying concentrates could be reduced if not completely avoided. It was also revealing that several peasants experimented with making 'good manure' (in Chapter 7 this issue of 'good manure' will re-emerge). From a theoretical point of view, all of these innovations indicate a strengthening of co-production: farming is again based, as much as possible, on nature – that is, on ecological capital.

Many peasants were considering diversification and several had already put specific forms into practice. Young successors, especially, were considering (or already engaged in) agro-tourism, the combination of milk and meat production and marketing the meat through local butchers and restaurants, as well as converting to organic farming.

Comparing technical indicators showed that the way in which peasants organized the 'technical' process of production (the conversion of resources into output) had improved considerably. The contrast with the *imprenditori* was substantial. The replacement of cattle was 19 per cent per year (compared to 30 per cent for the entrepreneurs). Per kilogram of concentrates (produced by the *contadini* themselves, while *imprenditori* mostly acquired this from industrial firms), the peasants produced 3.9kg of milk against only 2.6kg for entrepreneurs. Data like these help to explain why net income[14] as a percentage of total production was only 14 per cent on the farms of the *imprenditori* and 21 per cent on the farms of the *contadini*. The latter produce, for a given volume of production, 50 per cent more income than the former. Thus, a line of defence against globalization and liberalization was *de facto* created.

For several decades entrepreneurial farming represented economic superiority. This was partially virtual – due to the accountancy techniques used for representation and comparison; but undoubtedly it was also, in part, a real superiority. Especially in the so-called *vanguard farms* that combine an enlarged scale with high intensity levels, earnings could be far superior to those realized under other conditions.

Already by the second half of the 1990s, this material difference had started to dwindle, as illustrated indirectly by the comparison between the 'high-tech enterprise' and the 'low-cost farm' discussed earlier in the chapter. However, as perceived by the entrepreneurs involved, even then there remained a decisive advantage in that the large entrepreneurial farm was thought to be a stepping stone for the future. Even when earnings were more or less the same, the entrepreneurs (as well as most of the many experts) firmly continued to believe that in the 'battle for the future' the entrepreneurial farms were far better equipped to survive and win – in particular, because of their dimension and scale.

Today we are witnessing the loss of this last stronghold. It is increasingly clear that large entrepreneurial farms in the most modernized agricultural sector of Europe (i.e. The Netherlands) currently represent the weakest part of the chain. The once strongly convincing expressions of the entrepreneurial mode of farming currently present the worst prospects for continuity – due to the now accelerated breakdown of the entrepreneurial condition. This breakdown is now becoming generalized all over Europe. But especially in The Netherlands, where a considerable part of dairy farming is tuned more than elsewhere to the prevailing entrepreneurial condition, the effects are becoming dramatic. Table 5.5 compares three groups of dairy farms. The first contains relatively extensive farms where milk production is below 15,000kg of milk per hectare per year. Group 2 is an in-between group and group 3 contains the relatively intensive farms (milk production per hectare per year is higher than 20,000kg). This classification coincides globally with the size of the farms. In group 1 the average quota is 560,555kg; in group 2 this is 697,147kg; while in the intensive group the average quota per farm is 787,985kg.

The empirical data summarized in Table 5.5 were collected and analysed by Samenwerkende Register Accountants (SRA), a bureau belonging to the group of private and (formerly) co-operative farm accountancy agencies. These agencies work at the request of farmers and are paid by them. The data, which are always discussed with the farmers and controlled together, excel in their accuracy. But that is not the only difference between them and the national farm accountancy agency Landbouw Economisch Instituut (LEI). A far more important difference is that several of the private and co-operative agencies (the first being Alfa) adopt a Chayanovian approach, while LEI sticks to the neo-classical representation of the farm enterprise. One of the crucial differences is that the former is based on real costs and real expenditure, while the latter is built, in several respects, on *calculated* costs. 'Paid rents' is a typical expression of this difference. It refers to rents *actually* paid to *actually* existing debts. Instead, in a neo-classical approach, an all-encompassing rent is calculated over all capital (whether it is one's own capital or loans) entailed in the farm. Thus, any difference between a highly indebted and a relatively 'free' farm disappears. The same applies to actual repayments versus calculated depreciations, for the temporal horizons assumed for the latter, etc. The consequence is that, in the reports of the private and co-operative agencies, results and tendencies come to the fore that are systematically obscured in the official national farm economy data. Table 5.5 shows that 'final results' per 100kg of milk in the somewhat smaller and extensive farms (group 1) are more than double those realized in the larger intensive farms. The realized cash flow per 100kg is equally superior. When repayments, rents for tenancy, costs associated with the leasing of quota, and private spending and taxes are subtracted, what remains is the amount that can be saved (and eventually re-invested). This again is twice as high in group 1 as in group 3 – not only in a relative sense (i.e. per 100kg of milk), but also in an absolute sense (i.e. for the farm as a whole).

Table 5.5 *Comparative analysis of Dutch dairy farms (2005)*

	Group 1: Extensive farms	Group 2: In-between group	Group 3: Intensive farms	Average
Number of farms	42	15	7	
Total milk production (in kilograms)	560,552	697,147	787,985	631,832
Acreage (in hectares)	48.94	41.03	31.19	44.99
Milk (kg) per hectare	11,454	16,991	25,264	14,044
Total debts per farm (Euros)	668,752	646,349	925,995	718,624
Total debts per hectare (Euros)	13,665	15,753	29,689	15,973
Total debts per kg of milk (Euros)	1.19	0.93	1.18	1.14
Benefits dairy cattle (in Euros per 100kg of milk)	34.87	33.55	32.60	34.32
Other benefits (in Euros per 100kg of milk)	5.83	5.54	2.94	5.11
Bought feed and fodder (in Euros per 100kg of milk)	5.15	5.81	6.54	5.38
Other variable costs (in Euros per 100kg of milk)	4.09	4.33	3.84	4.12
Gross margin (in Euros per 100kg of milk)	31.46	28.95	25.16	29.93
General costs (in Euros per 100kg of milk)	20.54	20.98	19.89	20.39
Remaining results (in Euros per 100kg of milk)	10.92	7.97	5.27	9.54
Paid rents (in Euros per 100kg of milk)	3.49	3.29	1.98	3.43
Final results (in Euros per 100kg of milk)	7.43	4.68	3.29	6.11
Depreciations (in Euros per 100kg of milk)	10.12	11.51	9.24	10.45
Cash flow (in Euros per 100kg of milk	17.55	16.19	12.53	16.56
Repayments (in Euros per 100kg of milk)	2.90	3.78	4.63	3.74
Rents for tenancy (in Euros per 100kg of milk)	1.01	0.64	0.15	0.87
Lease of quota (in Euros per 100kg of milk)	0.42	0.38	0	0.34
Private spending and taxes (in Euros per 100kg of milk)	4.29	3.6	4.36	3.95
Total spend (in Euros per 100kg of milk)	8.62	8.40	9.14	8.90
Savings	8.93	7.79	3.39	7.66

Source: SRA (2006, p3)

Notwithstanding the fact that farms of group 3 are, on average, 40 per cent larger (in terms of quota), their savings remain at a level of 26,713 Euros per year for the farm enterprise as a whole, while this is 50,052 Euros for group 1.

Currently, the economic performance of farm enterprises structured and managed according to the logic of the entrepreneurial mode of farming is caught in a regressive process.[15] This is due to the interplay of a deteriorating entrepreneurial condition and the ongoing expansion at farm enterprise level to which entrepreneurs feel 'morally' obliged (and which is also a material need built into their farms). However, there will be, probably sooner than later, a moment when the *fuite en avant* can no longer be continued – precisely because economic performance is unsatisfactory. Then, a widespread deactivation will follow. It is important to underline that such regression will not be a generalized one. It will primarily affect the entrepreneurial mode of farming. In a recent publication of the Alfa accountancy agency (that previously belonged to the Dutch Christian Federation of Farmers and Horticulturists, or NCBTB) entitled *Telling Figures*,[16] the question is asked whether 'scale increase and intensification are accompanied by improvements of the margin'. The answer is 'no' (Alfa, 2005, p18). 'Notwithstanding their smaller dimension, the lower labour productivity and the low milk yield per cow, the extensive farms realize a result that is 11,500 Euros higher' (Alfa, 2005, p18).

In the past, further expansion was the way for entrepreneurial farms to escape from the squeeze on agriculture. In another edition of *Telling Figures* (Alfa, 2006), an interesting comparison is made: between farms that are not growing (growth is less than 5 per cent per year), farms that grow slowly (between 5 and 25 per cent) and a third group of farms that are *quickly expanding* (more than 25 per cent in the 2000 to 2004 period). These latter farms have the largest quota per farm. The final results in 2005 were: +8.10 Euros (per 100kg of milk) for the non-growers; +5.50 Euros for the slow growers; +0.80 Euros for the quick growers. Forecasts show that this will be, in the year 2010, +5.80 Euros for the non-growers; +1.80 Euros for the slow growers; and a *negative result* of –4.10 Euros per 100kg of milk for the quick growers. That is how the demise of entrepreneurial farming proceeds.[17]

In short, globalization, the associated price decreases and, especially, the frequent price fluctuations will indirectly provoke a repeasantization: first, because globalization eats its 'own children' (the entrepreneurs); and, second, because globalization and its consequences can only be responded to in a resilient and sustainable way through peasant modes of farming. This applies especially to the newly emerging ways of peasant farming that I will discuss in the following chapters.

6
Rural Development: European Expressions of Repeasantization[1]

Introduction

Over the last decade and a half, Europe has witnessed a widespread process of repeasantization. This process mainly expresses itself qualitatively. It involves enlarging autonomy and widening a resource base much narrowed by previous processes of specialization that followed the script of entrepreneurship. Repeasantization is also about fine-tuning that allows for new, often highly pleasing, gains in productivity. In short, it is about making agriculture again more peasant like. The degree of peasantness (Toledo, 1995) is increasing, resulting in the creation of new relations that concern both society and nature and which allow for a new 'embedding' of farming. Although repeasantization stems from many different sources, it is triggered by, and develops as a response to, the squeeze imposed on farming, and the marginalization, deprivation, degradation and growing dependency that go with it.

Taking into account the international situation, one might argue that repeasantization is a typical European response to the global squeeze. While South-East Asia fights this by putting extremely low levels of remuneration centre stage (thus simultaneously contributing to its reproduction), and the US, Brazil, Australia and New Zealand do so by increasing the scale of farming (thus also contributing to deepening the squeeze), Europe leans towards an alternative route that centres on the strengthening and further unfolding of multifunctionality (OECD, 2000; Huylenbroeck and Durand, 2003; Groot et al, 2007a) – that is, one and the same set of resources is used to generate an expanding range of products and services, thus reducing the costs of production of each single product (Saccomandi, 1998) and simultaneously augmenting the value added realized on the farm.

Repeasantization is far from being the only existing developmental tendency to be noted in Europe; neither are contrasting trends exclusive to other continents. Within Europe, repeasantization occurs alongside further industrialization and deactivation. The simultaneous presence of contrasting and, in a way, mutually competing development trajectories creates a complex 'battlefield' in which different interests, prospects and projects compete.

However, as a growing range of impact studies show, repeasantization tends to be, within Europe, the most important trajectory in terms of the number of farms and farmers. Also, in view of the changing conditions regarding energy, quality of life, water scarcity, etc., it is probably the most convincing. But it is, at the same time, a highly contested development pattern. European farmers are enlarging the peasantness of their farms, and reconstituting themselves as *new* peasants – not as 'yesterday's peasants', but as peasants located at the beginning of the third millennium.[2] What remains the same, however, is that current forms of repeasantization are barely understood by most scientists and politicians, which has perhaps been the case throughout the ages.

Mechanisms of repeasantization

There is a growing body of literature[3] describing the new modes of farming blossoming throughout Europe (and elsewhere, albeit to a lesser degree). This newly emerging morphology is often referred to as an outcome of 'rural development'. In itself this is not incorrect and it points to a complex interface between rural development policies and practices. As observed by Deirdre O'Connor et al (2006, p2), rural development is 'an increasingly important, but often misinterpreted phenomenon' (see also Wiskerke, 2001). That which – at the level of immediate empirical phenomena – is denoted as rural development is, from an analytical point of view, principally the outcome of an underlying process of repeasantization (Ploeg et al, 2002d). Rural development practices are, in other words, the outcome of a massive, grassroots-driven endogenous process of change, which was already gathering momentum before the first outlines of rural development policies were formulated.[4] Currently, supra-national, national and regional rural development policies interact, in complex and often highly contradictory ways, with rural development as a broad range of peasant-like responses to the squeeze on agriculture. I will come back to these complex interactions, and the interests underlying them, in the final section of this chapter.

Current forms of repeasantization might be explained analytically by starting from the notion that farming is always a process of conversion (of inputs into outputs) based on a twofold mobilization of resources. Resources can be mobilized from the respective markets (and, thus, enter the process of production as commodities) or they might be produced and reproduced within the farm itself (or within the wider rural community). This implies that 'outputs' can also be oriented in two ways: towards output markets or towards reuse (perhaps after socially regulated exchange) within the farm.

Facing large commodity markets, which are increasingly controlled and restructured by powerful food empires, many farmers have started to diversify their output in a range of ways. New products and services are produced, while simultaneously new markets and new market circuits are created (see number 1 in Figure 6.1). Thus, *multi-product farms* emerge that contain new levels of competitiveness and simultaneously entail more autonomy. Often, diversified output is combined with on-farm processing and the construction of new short links to consumers. Parallel with this first tendency (and often neatly intertwined with it) is, number 2, a shift away from main input markets, a shift that has also been called *farming (more) economically* (Reijntjes et al, 1992; Ploeg, 2000). In other words, the process of production is increasingly based upon resources other than those controlled by agro-industry such that autonomy is further enlarged. In the corresponding transition, number 3, the *regrounding of agriculture upon nature* plays a central role. Under the same rationale,

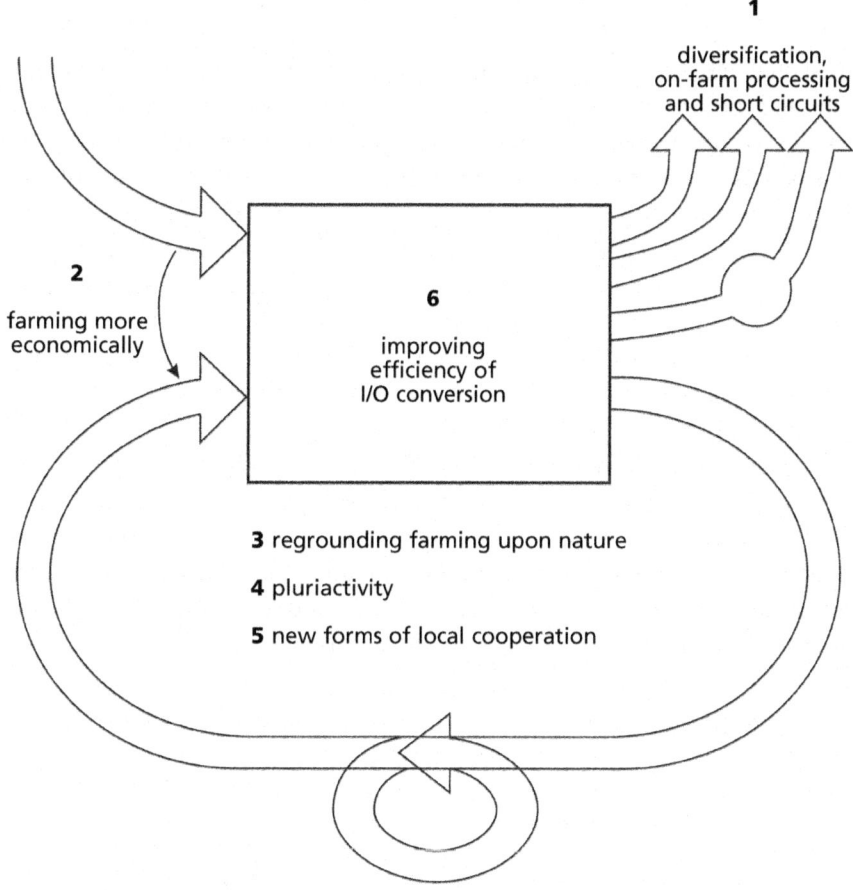

Figure 6.1 *The choreography of repeasantization*

Source: Original material for this book

number 4, pluriactivity and, 5, new forms of local cooperation are rediscovered and further developed. They also allow for de-linking agriculture from direct dependency upon financial and industrial capital. Within the core of the production process there is, 6, an increasing reintroduction of *craftsmanship* (an organic unity of mental and manual labour that allows for direct control over and fine-tuning of the process of production). This reintroduction is associated with the development and implementation of a new generation of skill-oriented technologies (Bray, 1986) and often results in the continuous production of novelties (Swagemakers, 2002; Wiskerke and Ploeg, 2004; Wolleswinkel et al, 2004).

It is important to stress that the changes indicated above should not be seen as mere additions to an unchanged way of farming. Neither do they concern only the details of farming. Taken together, they tend to represent both theoretically and, in practice, a 'structural' adieu' to the script for entrepreneurial farming that, until recently, dominated agriculture. They also represent a rupture (or at least a partial one) in the specific division of labour that linked farming to agro-industry, banks and the expert systems. Diversification, for instance, runs counter to the centrality of specialization in both entrepreneurial scripts and modernization theories. It is also at odds with the prescriptions articulated by agro-industry for farmers. The same applies to on-farm processing and the elaboration of short circuits that directly link the production and consumption of food. Diversification in farm processing and direct marketing represented, especially in the early stages of the repeasantization process, numerous small 'rebellions'. It was a deviation from the rule – it ran counter to vested routines, interests and identities, and those who did implement it were thought of as incapable farmers who were simply trying to find refuge in inappropriate ways out.

The same can be said about the other moves indicated in Figure 6.1. Farming economically (move 2) was perceived by many as a step backward, especially when combined with a regrounding of farming upon natural resources. And within the modernization paradigm, which centres on the embodiment of scientific progress in industrial inputs and new technologies for farming, such a combination would be seen as an outrage. Pluriactivity was likewise assumed to be something for the periphery; and new forms of local cooperation were thought to be unnecessary as long as the state and the key farmers' unions ordered the sector properly. The former were, in any case, thought incapable of 'competing' with the latter. Improving the efficiency of farming (creating new 'frontier functions': move 6 in Figure 6.1) was understood to be the exclusive role of science and associated expert systems. Again, a central role for farmers in such developments was understood as regression and if there was a role for farmers, it could only be miniscule.

Considered in isolation, then, the many empirical changes associated with the current process of repeasantization may, indeed, appear miniscule and almost irrelevant. However, as soon as analysis moves beyond the level of single units of production, a widespread and radical *re-patterning* of both the social and the natural world is revealed. The relevance of on-farm processing of milk into cheese, yoghurt and other products and the direct marketing of it, for instance, resides not only in the newly made cheeses and other products, but in the fact that it entails a redefinition of the interrelations between farming and agro-industry. It redefines the farm as being limited to the delivery of raw material only, into being a new multifunctional unit that relates in several new ways to society and nature. It also implies a redefinition of identities (of farming men and women and farmers' wives alike), as well as the creation of new networks that link with consumers (thus also redefining consumers; Miele, 2001). On-farm processing and direct marketing might also entail a reversal in the tendency for value added to decrease continuously. This may also reshuffle (at least potentially) the market for cheese and yoghurt and influence the high degree of monopolization that currently characterizes the dairy market.

Thus, a paradigm shift starts to unfold. This is also reflected, albeit indirectly, in the widespread discomfort and critique expressed by the entrepreneurial pole in the farming sector. Expressions of repeasantization are experienced as 'betrayals', as forms of inappropriate behaviour, and as blocking the free flow of resources badly needed for further expansion of entrepreneurial farming.

This paradigm shift entailed in the process of European repeasantization has never been clearly articulated at institutional levels. This is because it runs counter to too many institutional interests associated with previous modernization processes. Admitting that such a far-reaching shift is occurring would imply that vested positions, scripts and routines need reconsidering. It might also damage the aura of 'always being on the right track' (indispensable for expert systems and agrarian policy). Hence, the shifts shown in Figure 6.1 and the resulting multifunctionality are represented as something additional to farming, while the agricultural sector as a whole is conceptualized in terms of co-existence, meaning by this that alongside 'productive farming' there are other 'rural development' types of farming.

Such interpretations miss some essential points. First, 'classical' activities such as milking, vegetable production, etc. are not *separated* from new activities (such as on-farm processing, direct marketing, landscape management, energy production, or whatever) but *combined* with them, and the better this combination is organized, the more it delivers (Saccomandi, 1998). Second, earnings from 'old' and 'new' activities cannot be separated in order to compare them; it is their *unity* that matters. Third, within the co-existence view it becomes difficult to explore the many problems that changing farmers face and to understand the many new contradictions that arise.

Be that as it may, European farming is experiencing a far-reaching, complex and, as yet, unfinished process of transition that is unfolding along several different dimensions, and is located at several mutually interacting levels. At the grassroots level, this process of transition is shaped from an analytical point of view, as illustrated in Figure 6.1. It also indicates that repeasantization is, in a way, a 'process of borderline shifts' (as Ventura and Milone, 2004, phrased it): it is about moves that flow over the traditional boundaries of the specialized farm enterprise, just as it is about flows that translate, through newly created networks, to other levels. Together with others, I have referred to this transition as repeasantization (Ploeg and Rooij, 1999; Ploeg et al, 2000; Prodi, 2004; Johnson, 2004; Pérez-Vitoria, 2005; Hervieu, 2005; Sevilla Guzman, 2006, 2007; Valentini, 2006; Ventura and Milone, 2007).

The first reason for doing so relates to the fact that several, if not all, of the indicated moves translate into increased autonomy and, sometimes, to the creation of new moves.[6] Often, the creation of more autonomy and increased room for manoeuvre, decision-making and learning is the explicit intent of the processes summarized in Figure 6.1. It is a struggle for autonomy in a world strongly and increasingly characterized by dependency patterns and processes of marginalization and deprivation.

Second, increased autonomy materializes in a reconstitution of the resource base of the farm: it is broadened and diversified, and combinations allowing for new productive activities are wrought (OECD, 2000; Brunori et al, 2005; Caron and Cotty, 2006). It also means that more or less forgotten resources are rediscovered. Manure and soil life are excellent illustrations. I return to these rather mundane issues in the next chapter. Relevant in this context is that labour again becomes a central resource within the resource base as a whole. This applies both quantitatively and qualitatively. The heavily tailored labour processes that emerged during the epoch of modernization (and which are often imposed by food empires) are actively countered and replaced by others that allow for more overview, flexibility and quality, and greatly reduce stress. This is especially related to the growing influence of farmers' wives and their changing background (Rooij, 1992; Rooij et al, 1995; Bock, 1998). Due to these new relations, the 'art of farming' (Columella, 1977) is being rediscovered and materially reconstructed, albeit in highly differing degrees.

Third, the shifts indicated in Figure 6.1 identify another strategic feature (see the debate between Goodman, 2004, and Ploeg and Renting, 2004): they tend to enlarge the value added produced on single farms, as well as in the agricultural sector as a whole. Contrasting and competing processes, such as accelerated scale increase and deactivation, diminish the value added in the sector. Value added is actively decreased by entrepreneurial farming and by its expansion, while peasant-driven rural development augments it both at the individual farm level and at the level of the sector as a whole. This is due to the way in which the latter relate to other farms and to non-agrarian sectors of the rural economy. In this respect, it is also important that the so-called multiplier of rural development activities is considerably higher (and more localized) than that for entrepreneurial farming (Heijman et al, 2002). Entrepreneurial farming progresses through the takeover of other farms, while remaining strictly within the boundaries defined and imposed by food empires. Peasant-like farms progress, not through takeover, but through the creation of new additional wealth, and in doing so they actively cross the borderlines imposed by different empires – even when such crossings are tabooed as infractions.

Fourth, the same shifts that move today's agriculture beyond the model of the highly specialized agricultural enterprise are also reconnecting farming again to society, nature and the interests and prospects of the direct producers. While the model of entrepreneurial farming only contributes to further deepening the current agrarian crisis (see Figure 1.4 in Chapter 1), repeasantization may potentially bridge the many chasms that have, in the meantime, been created.

The fifth reason for describing the ongoing transition of European agriculture as a process of repeasantization relates to the fact that, in practice, rural development unfolds as a *struggle against state apparatuses, their regulatory schemes and agribusiness* (Marsden, 2003). It involves a struggle for autonomy, the creation of new value added and survival, rather than, as some assume, a more or less straightforward implementation of European Union (EU) schemes and the associated rhetoric. I come back to this crucial feature in the last section of this chapter.

The transition occurring has some specificities that likewise point to its peasant-like nature. It is not governed from any central locus of control; instead, it is endogenous and somewhat anarchic. It does not offer a global solution for a range of different local problems and situations, but is evolving as a growing range of diversified local solutions for a general problem (i.e. the squeeze on agriculture). And, finally, it does not proceed as a mega-project (as one big sweeping and all-embracing change or rupture that could evoke havoc; see Scott, 1998); but as a wide range of interconnected steps (that increasingly extend through time and space), which together compose, in a way that is constantly fluctuating, the overall and, indeed, massive change that is transforming agriculture and the countryside.

Magnitude and impact

Currently, some 80 per cent of European farmers are actively applying one or more of the above-mentioned responses that together compose the European process of repeasantization. A general overview (based on a 1999 survey with n = 3264 in six European countries)[7] is provided in Figure 6.2. This depicts the 'new forms of heterogeneity' (Oostindie et al, 2002, p218) that result from rural development (RD) as a grassroots process of repeasantization.

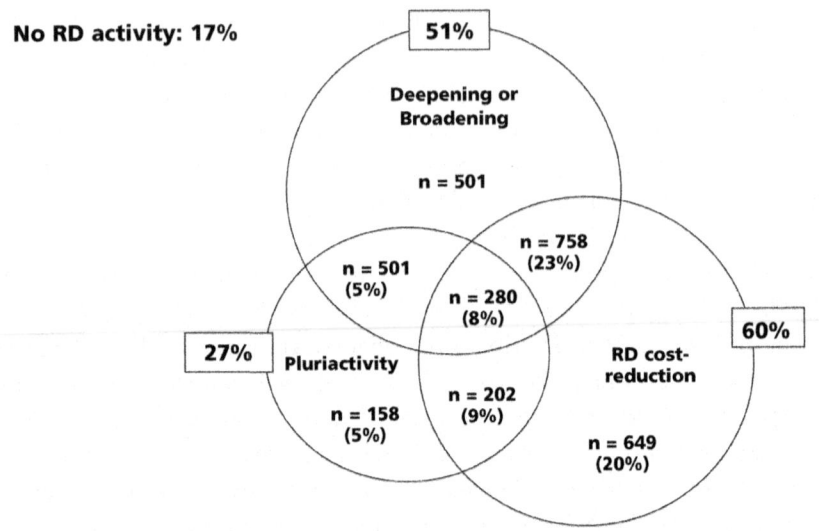

Figure 6.2 *Newly emerging expressions of repeasantization*[8]

Source: Oostindie et al (2002, p218)

Figure 6.2 shows that among professional farmers[9] more than half (51 per cent) are actively pursuing activities that imply 'deepening' and 'broadening'. Deepening refers to those activities that augment the value added per unit of produced product. Typical expressions are organic farming, high-quality production, production of regional specialities, on-farm processing and direct marketing. Broadening refers to adding non-agricultural activities to the farm (again, raising the value added at farm level). Well-known expressions are the (paid) management of nature, biodiversity and landscape; energy production; agro-tourism; provision of care and other services; and a wide range of more traditional rural services. Thus, more than half of the professional farmers are engaged (some from ancient times onwards, and others more recently) in what is shown as 'flow number 1' type of repeasantization (see Figure 6.1). It is important to stress that this type of repeasantization is neither a return to the past, nor the desperate construction of a last lifeline (as argued by Rabbinge, 2001). It is about peasants of the third millennium: On average, they work

93ha compared to 74ha for those not involved in this kind of diversification, and operate with an average of 3.8 labour units (mainly family labour), against 2.5 for those who have not diversified. And they are relatively younger.

Being engaged in new forms of cost reduction (moves 2 and 3 in Figure 6.1) that greatly contrast with cost reduction constructed through scale increase constitutes a second important, though far less visible, domain in which repeasantization unfolds. A total of 60 per cent of all farmers are actively involved in this domain.

Finally, we need to consider pluriactivity. Once considered an expression of a disappearing peasantry, it is again present as a mechanism through which the peasantry reconstitutes itself anew: 27 per cent of them are involved in it.[10] There is considerable overlap between the domains distinguished in Figure 6.2, which contributes notably to the creation of this new heterogeneity. In this emerging panorama, farmers strictly following the entrepreneurial script (i.e. those who, according to the survey, are not involved in any of the three domains) are becoming a minority of 17 per cent.

The importance of the peasant moves in Figure 6.1 goes much further than simply affecting the morphology of the countryside. One important aspect is that it is difficult to understand or theoretically represent a reality increasingly re-patterned in a peasant-like way with tools and concepts that belong to an entrepreneurial mode, nor will agrarian and rural policies function well if they are based on a fundamental misunderstanding. I will come back to this point later.

As discussed in Chapter 5, the search for, and construction of, additional value added is an important characteristic of the peasant economy. Through deepening and broadening activities, 3414 million Euros and 2458 million Euros, respectively (i.e. a total *extra net value added* of 5.9 billion Euros) (Ploeg et al, 2002c), are added to the agricultural sectors of the six key countries (1997 data).[11] This is twice as much as the *total* agrarian income of Dutch agriculture. If the latter is referred to as an 'agricultural giant', one cannot but conclude that, in the meantime, another 'giant' has been born. Alongside this impact of deepening and broadening, it is calculated that 'farming economically' (i.e. in the peasant mode) contributes another 5.7 billion Euros (for the six countries together) to farm family income.

Let me briefly comment here on a few aspects. First, we are dealing here with (relative) increases of value added and, hence, agrarian incomes that are achieved without enlargement of total agricultural production (which would have detrimental consequences both for developing world agriculture and for the environment in Europe). It is basically about raising the value added/gross value of production (VA/GVP) ratio at the level of primary production itself.[12] Second, we must take note that the new domains in which repeasantization is unfolding are not separated from classical production systems such as those of milk and potatoes. They combine as multi-product farms. This implies that thanks to the additional value added created, considerable parts of European agriculture survive that would otherwise probably disappear. At the micro-level, this is seen in the dairy farmer who is able to continue and even develop his farm due to earnings obtained from, for example, the agro-tourist facilities on the farm and from the income earned by his wife in the nearby village.

What applies to Europe as a whole might also be encountered at regional and local level. De Wolden is a small rural region in the province of Drenthe, located in the north-east of The Netherlands. It is a very rural area. Of a total of 25,000ha, 16,000ha are farmed; the rest is basically woodland and heath. Agriculture provides 30 per cent of total employment in the area. Official statistics indicate more than 600 farms, though in practice there are less than 400 (such misrepresentation of the number of farms is a generalized problem in The Netherlands). Total agrarian income in the Wolden area is 6 million Euros (based on 2003 data). However, on these same farms another 2.5 million Euros is earned through all kinds of old and new broadening activities (and a few deepening activities), while salaried work undertaken elsewhere contributes a further 5 million Euros to total farm family income. Thus, strictly speaking, agricultural activities contribute less than half (i.e. 44 per cent) of the family income as a whole (Oostindie and Broekhuizen, 2004; similar relations have been reported in LEI, 2005; LNV, 2005; Immink and Kroon, 2006; and Knickel, 2006).

The quality of life in rural areas

As far as impact is concerned, peasant-driven rural development may also contribute to the quality of life in rural areas as perceived by rural dwellers. In a recent Italian research programme (Ventura et al, 2007) a multiple-level approach was used to examine the contribution of multifunctional agriculture to quality of life. The first level took in the different municipalities. Italy has more than 8000 municipalities, 6356 of which are classified (according to Organisation for Economic Co-operation and Development (OECD) criteria that reflect population density) as rural and semi-rural areas. Within this latter category additional distinctions were made (see Figure 6.3). These were introduced to obtain a better understanding of the spatially differentiated process of counter-urbanization. As nearly everywhere in Europe, Italy, too, has experienced considerable '*repeuplement de la campagne*', as Bernard Kayser (1995) phrases it.

A first group of municipalities were defined as marginal areas, where the average net income per inhabitant was below the average for rural areas in the region as a whole. A total of 3075 municipalities were classified as marginal.

Rural Development: European Expressions of Repeasantization   161

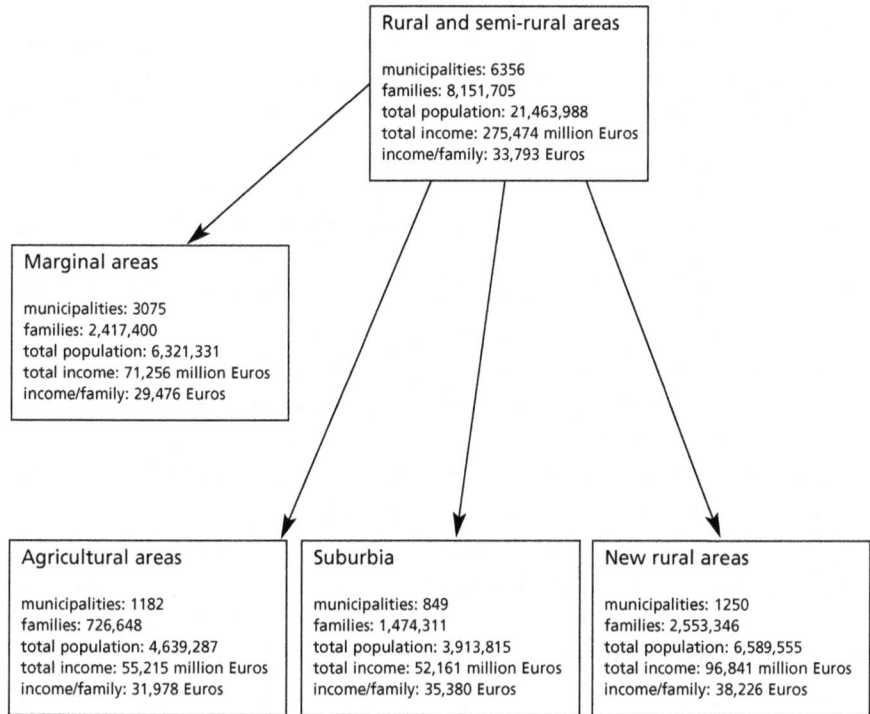

Figure 6.3 *Differentiation of rural and semi-rural areas in Italy*
Source: Ventura et al (2007, p48)

They were especially, though not exclusively, located in the southern parts of the country and on the isles. The remaining non-marginal areas were further divided into three mutually contrasting categories: first, areas in which agricultural activities were relatively more important than in other areas – that is, statistically, the number of people directly engaged in farming was higher than the regional average. In practical terms, when more than 10.8 per cent of the economically active population of a municipality is directly engaged in primary agricultural production, the municipality is defined as a 'specialized agricultural area'.

In the other two categories agriculture was of secondary importance. In one, its importance was not only low, but rapidly declining (the Agricultural Censuses of 2001 and 1991 record the decline as exceeding 36 per cent). This category of municipality is referred to as *suburbia* and is characterized by a relatively high inflow of commuters. The third category of municipalities, while having a relatively low presence of agriculture, lacked the sharp decline in agriculture. A remarkable number of municipalities even showed an overall absolute growth in agrarian employment and a revival of their rural economies due to the development, for example, of high-quality wine production. This category was classified as 'new rural areas'. The rationale underlying this classification is clear. With a more or less generalized decline of agriculture (resulting from globalization and liberalization), the countryside is *not* converted, in a uni-linear way, into a 'generalized' marginal area. Instead, different development trajectories are likely to unfold, thus giving rise to different kinds of spaces (see also Murdoch, 2006). Hence, alongside a diminishing number of specialized agricultural areas, there are not only marginal areas and suburbia, but also new rural areas where agriculture plays a new role (for well-documented cases, see Ventura and Milone, 2005b).

Together with this differentiation of space, rural dwellers themselves can be distinguished into different categories. A representative survey (n = 1445) revealed that 58 per cent of respondents said that they were rural dwellers by tradition and were tied to the area through family relations. Another 10 per cent indicated that they lived in the rural area out of 'necessity'; and the remaining 32 per cent said that they made a 'conscious choice' to live in a rural area.

If we now tie together people and places, a pattern emerges that is summarized in Figure 6.4. It shows that people who have the possibility of choosing where they live tend to locate themselves in the new rural areas. The latter are apparently attractive when seen in terms of the 'repopulation of the countryside'. Whether this is actually associated with the distinctive features of these areas and, if so, which features, was checked in a second level of analysis that explored the interrelations *within* municipalities.

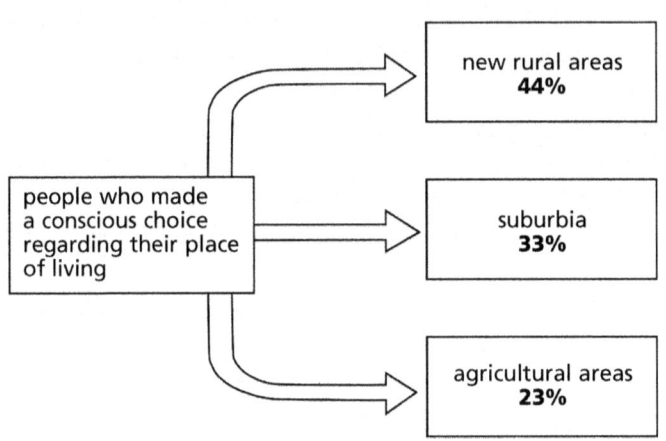

Figure 6.4 *Where do people move to?*

Source: Ventura et al (2007, p53)

A theoretical model, which centres on the concept of social capital,[13] underpins the level 2 enquiry into the quality of life in rural areas. Following Putnam's (1993) well-known account of Italian civic culture, social capital is mostly understood as a complex set of interlinked and well-functioning networks that tie people together through sets of shared norms and beliefs. This definition is close to that of the World Bank:

> *Social capital refers to the norms and networks that enable collective action. Increasing evidence shows that social cohesion – social capital – is critical for poverty alleviation and sustainable human and economic development.*[14]

However, following Long (2001), I believe that these definitions overstress the solidity of 'shared' norms and values:

> *Social networks are infused with a multiplicity of partial connections, exchange contents, normative repertoires and multiple markers of morality. They are never fully integrated or organized around an unambiguous set of values, rights and obligations. They are entangled and ambivalent.* (Long, 2001, pp132–133)

Consequently, social capital was measured at level 2 of the enquiry by extended evaluations given by rural dwellers of the multiple (and partly connected, partly disconnected) networks in which they were actually involved.[15] Such networks might centre on the education of children (a network involving teachers, buildings, local and regional authorities, school kitchens and cooks, other parents, etc.), political life in the area (local meeting points, debates, organizers, etc.) or organizations for voluntary work of a social and religious nature, etc. It was assumed that a positive evaluation would indicate good integration of individuals within such networks[16] and therefore well-developed social capital. Social capital, of course, is never simply 'out there'; it only emerges when networks are actively used. It was assumed that a fair development of these networks and active participation in them would lead to positive judgements of the particularities of a place – namely, that it was better than other places. If people participated in many networks in their area they would obviously get to know many local people, including those who care for children, for the elderly, etc. In other words they would know how things were dealt with locally, how to contact people and how to obtain information about local affairs; when and where culturally relevant encounters took place; and so forth. In synthesis, this kind of social capital would then produce a *sense of belonging* and contribute to a *positive evaluation* of place. Moreover, if the difference between the urban and the rural was relevant in some or all of these respects, then such differences would also be reflected in the empirical research.

Social capital assumes 'collective assets' (Lin, 1999, p41) – that is, resources that are embedded in relevant networks. In the research under discussion, these were operationalized in two ways: first, as the availability and quality of a wide range of services (health, schools, public transport, post offices, sport facilities, etc.); and, second, as the attractiveness of surroundings (mainly landscape, nature, accessibility, absence of pollution, etc.). Without services and the associated meeting points, social capital cannot easily develop. More and better-quality services will translate into more social capital and thus contribute directly and indirectly to quality of life.[17] It also applies that *rurality* (Ploeg, 1997b) or the quality of the rural environment will be a crucial element that contributes to the quality of life in general, and might even be the main motive for moving to the countryside (Kayser, 1995; Kinsella et al, 2000).

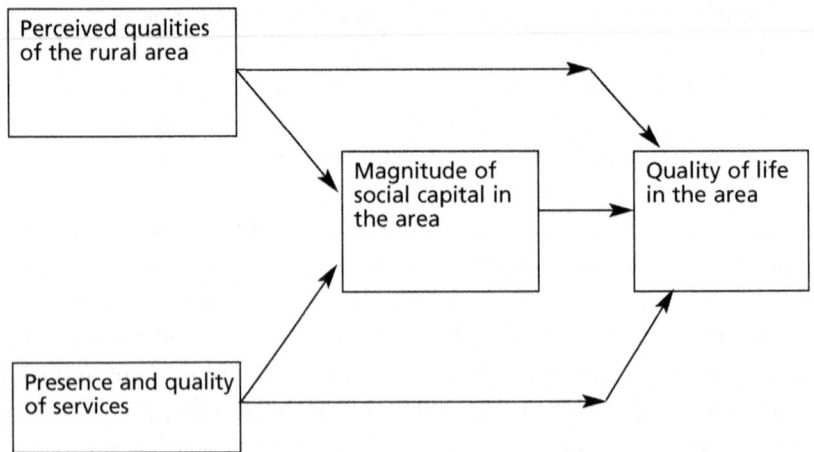

Figure 6.5 *The theoretical model underlying the enquiry into quality of life in rural areas*

Source: Ventura et al (2007, p56)

Figure 6.5 summarizes the theoretical model sketched so far. An important feature is that it might also be read the other way around – that is, it helps to clarify and specify the very notion of quality of life. Quality of life embraces three dimensions: the social, the economic and the physical. The social dimensions involve social cohesion and social networks that link people and allow them to obtain a grip on their own situation. The economic dimension refers to the availability and quality of services and productive activities.[18] And the physical dimension concerns, among other things, the landscape, its qualities, its accessibility and the capacity to maintain it.

Figure 6.6 contains a path diagram that summarizes the main empirical findings of the research based on this theoretical model. Without wishing to go into all implied technicalities,[19] the figure shows, in synthesis, that social capital indeed translates into quality of life as perceived by rural dwellers. It does so through two paths: one that begins with a set of networks that apply to the community as a whole, and the other that departs from the networks directly linked to family and children. The path diagram also shows that social capital is, in turn, related to ('is explained by') a range of relations that connect to farming and, especially, to multifunctional agriculture: the more positive the judgement of agriculture's role in the making and maintenance of the qualities of the rural area, the more social capital there is (β = +0.25 and +0.24, respectively). And the more multifunctionality developed in local agriculture, the more social capital is strengthened (β = +0.11 and 0.07, respectively).[20]

The importance of newly emerging multifunctional farm units resides not only in the products, services and the associated value added that they render; it also resides, and perhaps more especially so, in their contribution to social capital (and, thus, to quality of life in the rural areas). Apart from the relevance of a tasty new cheese, there is the relevance of the encounters and exchanges implied by this new cheese. Multifunctionality generates new networks that become important ingredients of social capital. On the other hand, new rural dwellers might constitute, in their turn, an attractive (additional) market for renewing farmers (see the positive, albeit small, β linking the amount of rural dwellers consciously choosing the area with the magnitude of new multifunctional agriculture). This is how symbolic supply and demand meet and subsequently translate into material exchanges. The new goods and opportunities provided by these multifunctional farms turn out to be the vehicle for new exchanges, new growth and new networks that start to sustain both these exchanges and growth.

In this respect, it is also telling that multifunctional agriculture positively feeds into the general provision of services and their quality (β = +0.18). Indeed, it provides shops, sporting facilities, recreational and tourist facilities of all kinds, and employment opportunities – that is, it creates, in synthesis, part of the economic dimension of the quality of life, just as it, in the same way, strengthens the physical and social side.

Returning to the classification of rural areas, it is striking that the specialized agricultural and suburbanizing areas negatively feed into social capital (β= –0.24 and –0.08, respectively) and into the magnitude of multifunctional farming (β = –0.12). It is in the new rural areas where multifunctionality is especially blossoming and where the new rural dwellers (motivated by the attractiveness of the area) become the carriers of quality of life in the countryside. Thus, repeasantization translates into strengthening the beautiful 'Greek' side of farming – even in Italy.

166　*The New Peasantries*

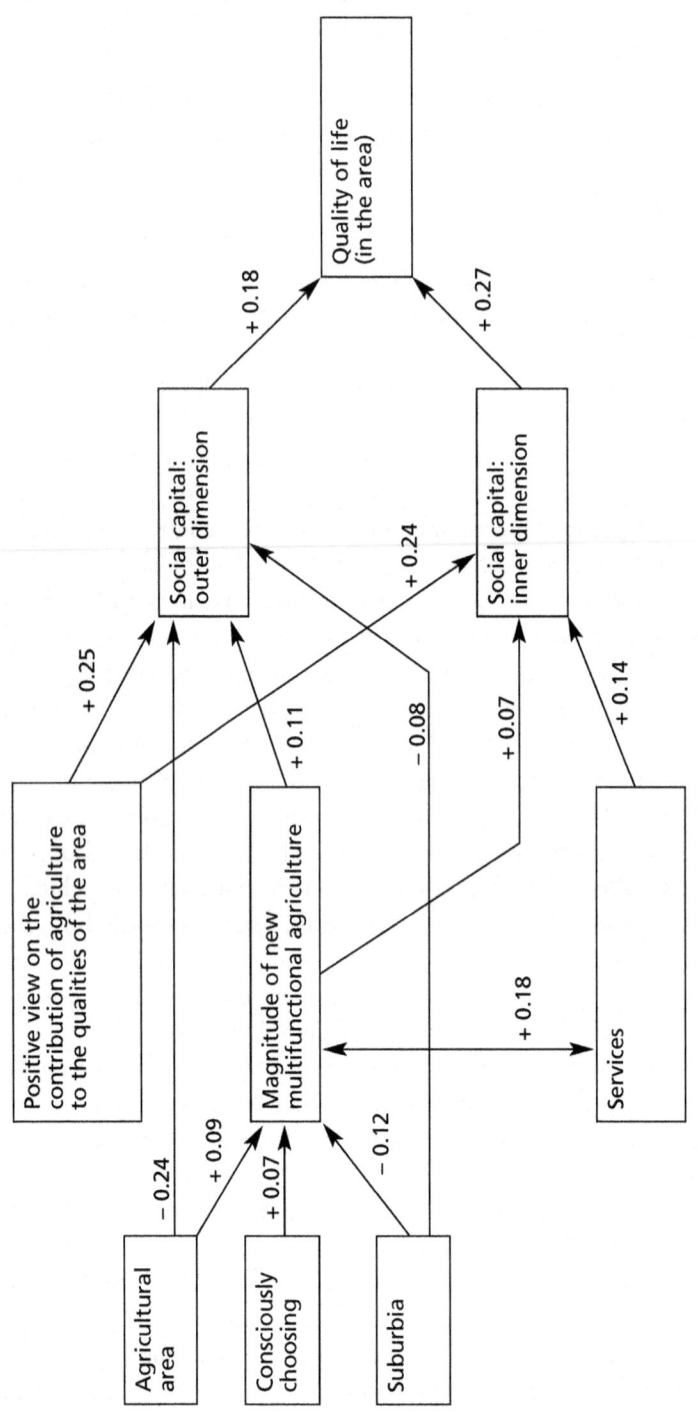

Figure 6.6 *Explaining the quality of life (overall path diagram)*

Source: Ventura et al (2007, p83)

Newly emerging peasant types of technology

Zwiggelte is a small village not far away from the De Wolden area mentioned earlier. It is a rather sad village, one of those villages from which 'God disappeared' (Mak, 1996). There is a lot of arable farming, focused for decades on the production of 'factory potatoes', a sub-sector that represents a chronic crisis. Dairy farming occurs, as well as some intensive pig and chicken breeding and, thus, a large manure surplus; there is much forest, a huge recreational park with bungalows, some small- and medium-sized enterprises and, finally, an interesting piece of 'archaeology': a pumping station once used to channel the formerly abundant reserves of natural gas into the national (and international) delivery system. Currently, the pumping station is no longer used as the gas reserves have been depleted.

A group of seven farmers (already with some experience of new and innovative ways out of the seemingly perennial standstill experienced in Zwiggelte)[21] took the initiative to look for alternative opportunities. A first, somewhat implicit, design principle centred on the combination of the following elements to:

- build as much as possible on local assets;
- strengthen them through the selective introduction of specific external elements; in order to
- create a new and productive combination of these assets.

In this way, once more or less useless assets could be actively converted into productive resources, without degrading or putting into disuse other resources.

The proposal developed by these seven farmers is explained in Figure 6.7. In the first place, it shows elements already mentioned (forest lands, manure surplus, the archaeological remnant, etc.). It also highlights a second design principle aimed at creating *new and as yet non-existing connections*. The first connection (not widely known at the time) was between manure surplus and energy production. However, the efficiency of a straightforward conversion of manure into energy was deceptively low. Here, a second connection turned out to be decisive. The group came to know of a new technology – developed in Germany – that considerably increases efficiency by fuelling the process with carbon. After a study tour to Germany (the second connection) they concluded that this method could well be applied to their own situation, especially when a third and fourth connection was made – through maintaining the forest they could 'harvest' much of the required carbon, and agricultural waste could also be used. Conversion of manure enriched with carbon produces gas. This provoked the fifth connection: the ancient pumping station could be brought back into use for introducing the gas directly into the delivery system. In order to convince the company (Gasunie) that controls gas distribution, a sixth connection was created and used: the Petten Research Institution (ECN) was asked to make a chemical and physical analysis of the gas produced. It turned out to have the same characteristics as natural gas; hence, it could be introduced without any inconvenience into the delivery system. Being keenly astute, the peasant farmers of Zwiggelte immediately realized that one of the implied risks would be an almost complete dependency upon the *Gasunie* network. Thus, a seventh connection was studied: the possibility of converting the gas, with a turbine, into electricity and channelling it into the regional distribution network (*Nuon*). By doing so, they could create flexibility: energy could be channelled according to the terms of trade to Nuon or to Gasunie. However, the conception of a new pattern that promised to turn more or less useless assets into productive resources did not stop there. Producing electricity out of gas also generates a lot of heat, which is normally lost. Connection number eight was thus invented: channelling the heat towards the bungalow park and its swimming pool so that the open-air swimming pool could be used for a more extended period in the year, making the park more attractive. A ninth connection explored was the direct delivery of electricity (through a new cable) to the local small- and medium-sized enterprises; connection number ten concerns the use of the realized value added within the local community.[22]

Although there were several other connections in the Zwiggelte proposal, the main point is clear. Innovation proceeds here as re-patterning, as the making of new connections. I have focused mainly on the material aspects; but evidently each and every step involves negotiation, renegotiation and, possibly, the creation of new *institutional* relations. What we are looking at here, then, is a techno-institutional design (Rip and Kemp, 1998) for re-patterning a specific set of relations that form part of the social and natural worlds.

Re-patterning is inherent to whatever change takes place. It is also a striking feature of Empire as an ordering principle (see, for example, Figures 3.5 and 4.1). There are, nonetheless, certain decisive differences that one can elaborate upon in terms of (additional) design principles – which, in the end, might materialize into alternative ordering principles.

Alongside the two design principles already specified – building upon available assets and creating new connections – are three others that are crucial. One is that the realization of the final objective is not external to the local situation (as in Figure 3.5) but resides within it: local needs and local resources are moulded, remoulded and combined (through complex processes of conceptualization and materialization) in order to create local solutions for global problems (such as the squeeze, embodied here in the misery of 'factory potatoes', manure surpluses, etc). A key feature of such local solutions is that they not only enlarge total value added, but place it in the locality that produces it. Peasant innovation thus potentially contains an ordering of the world that runs diametrically contrary to that entailed in Empire.

The second principle maintains that the exchanges assumed to occur with every newly established connection are conceptualized and materialized, in the first instance, as *conversions* – and not primarily as profitable transactions. The *entire* set of newly established connections (as sketched in Figure 6.7) may render new wealth; but it is not expected that every *single* conversion will, or ought to be, a transaction that renders profit. Indeed, if the latter was a necessary condition, the transition as a whole would definitely be impossible. Peasant inventiveness implies an ongoing reflection on, and material reshifting of, boundaries between commodity and non-commodity circuits (see also Ventura and Milone, 2005a). Put somewhat differently: the 'market', or a set of interlinked markets, should not be understood as an ordering principle that will shape and reproduce the required connections. On the contrary, organizing the new constellation as a market (i.e. through a set of market relations) would be tantamount to its annihilation. The conceptualized constellation is linked only to output markets *at the end of the 'chain'*. Relative autonomy, an

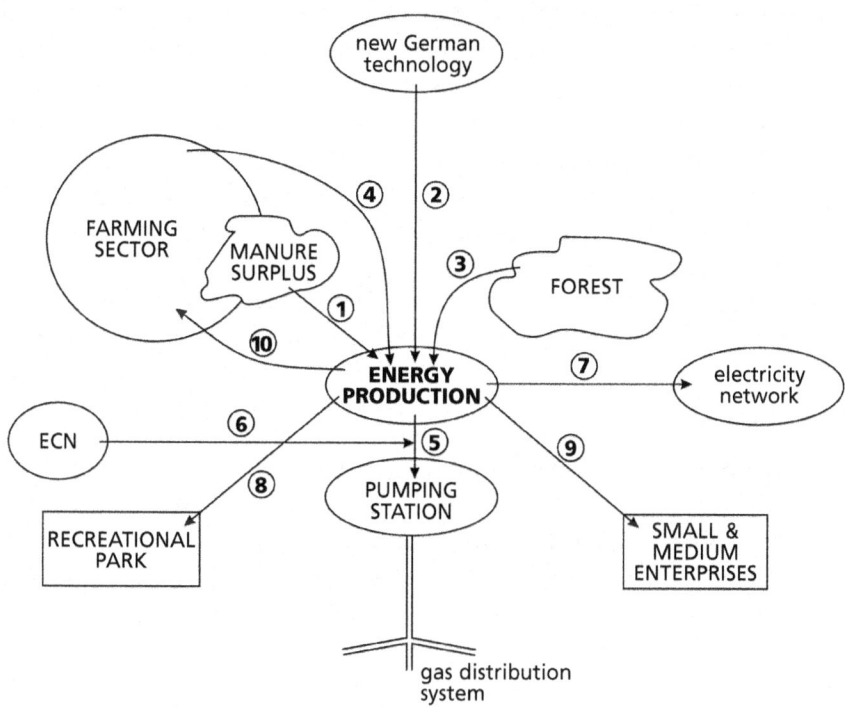

Figure 6.7 *Re-patterning resource use in Zwiggelte: An illustration of peasant inventiveness*

Source: Original material for this book

important feature of the peasant mode of ordering (and graphically illustrated in Figure 2.5 in Chapter 2), is, thus, made to 'travel': it is actively moved from the farm to a higher level of aggregation (i.e. the new constellation that aims to produce energy). Relative autonomy is, in synthesis, incorporated within the new energy-producing technology and, simultaneously, reproduced by it.

Another decisive feature of the peasant mode of ordering, *craftsmanship*, also 'travels' through peasant innovation. It travels to the design of new technologies, thus making for skill-oriented technologies that critically depend upon the centrality of labour and associated skills: the more capable the labour force, the better the productive results.[23] As opposed to mechanical technologies in which labour is mainly an extension of the machine, in skill-oriented technologies labour *governs* the process of production, which also implies that ongoing and cumulative improvements can be created.

Francesca Bray (1986), in her study of 'rice economies', introduces the distinction between skill-oriented and mechanical technologies. She conceptualized the first as being composed of a combination of relatively simple technical devices with a highly skilled, well-experienced and knowledgeable workforce. On the other hand, mechanical technology represents highly sophisticated technical artefacts that could even be set in motion by unskilled labour. This distinction is well reflected in the empirical domains that she focuses upon (the Far East and the US). The point I want to make here builds on Bray's work but goes a step further – that is to say, within the domain of *today's* peasantries, especially (though not exclusively) in Europe, a new combination is emerging: complex and sophisticated instruments are combined with highly skilled labour. I refer to such combinations as skill-oriented and sophisticated technologies.

Let me discuss a few additional examples of these new technologies in order to explore some of their specificities. The examples are drawn from Italy.

Olive oil

Olive oil production basically has two stages. The first centres on the pressing of olives in order to release the oil, while the second basically comes down to cleaning the obtained product through filtering: impurities such as small particles of the crushed nuts need to be removed. Filtering is mostly done with water, which then has to be removed from the oil. Currently, a new technical device is available for this second stage: it is a centrifugal system that separates the oil–water mixture into two circular layers (see Figure 6.8) – one at the centre containing the oil, the other, the outer layer, containing the water and dirt. This separation results from differences in specific gravity and the high-speed revolutions of the centrifuge.

The complication here is that the oil has to be taken out while the machine

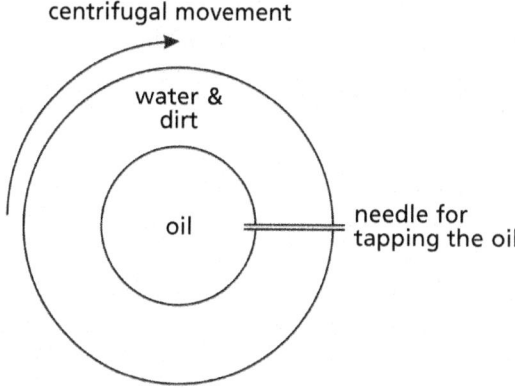

Figure 6.8 *Centrifugal filtering of olive oil*

Source: Original material for this book

is rotating at high speed. This occurs through a 'hollow needle' that taps the oil. The art of operating this technology then consists of locating the needle exactly at the required 'depth': too deep means that oil is lost; too shallow and the oil produced will still be contaminated with water. Adjusting the needle (by one tenth of a millimetre) has to be coordinated with a range of other factors – the revolutions per minute, the volume and composition of the oil, the flow and velocity with which it enters the cylinder, etc. Fine-tuning – that is, the simultaneous and mutually interdependent adjustment of a broad range of variables (i.e. managing many switches at the same time) – thus becomes central. This is based on the simultaneous observation and interpretation of a range of indicators, such as the quality and colour of the subtracted oil and the colour and composition of the outflow of water. This requires thorough knowledge, considerable experience, strong nerves and the capacity to translate the constantly changing cycles of observation and interpretation into the required adjustments. This is what I refer to as skill.[24] A skill consists partly of a language without words; it partly is knowledge that cannot be expressed in precise, unambiguous and quantifiable concepts (see Darré, 1985). It can only be obtained through long periods of apprenticeship, training and experience. In so far as it is knowledge, it is clearly experiential or practical knowledge.

However, the description of the centrifuge and its running hardly describes the technology exhaustively. In order to obtain the best possible oil, several additional conditions must be met. Among them are the following:

- The processing of the oil has to be a continuous flow (frequent stops and interruptions must be avoided).

- The time lapse between the harvest of the olives and the subsequent processing must be 12 hours at maximum.
- The olives must be ripe at the moment of harvest.
- The olives must be picked from the tree (not collected from the ground).

Without entering into all of the associated complexities and technicalities, it is clear that these conditions imply a complicated, extended and, at the same time, flexible social organization of time, space and work. Which olive yards are to be harvested at what moment and in which sequence? And how is the labour for harvesting and transport to the processing plant to be organized so that the most optimal and 'seamless' flow of olives through time and space (as well as the best quality of oil) is obtained?

This is what skill is: the capacity to (re)organize and coordinate time, space, labour, technical artefacts, flows and quality standards, while taking into account the capriciousness of soil, weather and other unpredictable factors. I have often watched the working of such small units for olive oil processing and have always been impressed by the initial nervousness, especially during the fine-tuning at the start of the campaign. Young farmers can become very tense when thinking about the conversion efficiency of the process and the quality of the oil obtained. Within their household economy these are often very important concerns. Then it is up to the *frantoiolo* (the man running and adjusting the machinery) to reassure the nervous ones and thus avoid damaging interruptions.

Technology is not only about interlinking artefacts and governing material flows – it is as much about interlinking people in specific ways in order to obtain the right kind of conditions and flows. Thus, skill is all about being able to overview, observe, handle, adjust and coordinate extended domains of the social and the natural world. This is done by building upon the specificities of different elements of the social and natural world. It is probably in this latter aspect (building on encountered and/or created specificities) that the main difference between skill-oriented and mechanical technologies resides. The latter cannot easily handle (or build upon) specificities. Continuous adjustments are neither feasible nor desirable. If you produce coca cola, then only coca cola comes out of your plant. A better or worse coca cola or even somewhat different coca cola is unthinkable and would be immediately perceived as a disaster. As objectified patterns of 'through flow', mechanical technologies assume a standardized inflow, as much as they produce a standardized outflow. They cannot deal with specificity or variation. Specificity is a deviation, a threat and even, potentially, a destructive factor.

If skill-oriented and sophisticated technologies are to be interpreted as an ongoing process of engineering, creation and discovery, then it follows that mechanical technologies represent formalization, routine procedures, endless repetition and, in a way, standstill: things are done the way they always have been. Endogenous improvements are hard to perceive.

Milk

In the case of milk and, more generally, dairy products, standardization (i.e. the elimination of deviations and specificity) occurs among, other things, through the pasteurization of raw milk. Raw milk is a living product that contains a range of micro-organisms, many of which are essential to its taste, smell, flavour, quality and health, while others may be potentially harmful. Through pasteurization this living milk (as it is called in Italy) or raw milk (as it is called in The Netherlands) is literally turned into a 'dead substance' to be rebuilt into different outputs such as butter, yoghurt, consumption milk, and so forth. Pasteurization is an intrinsic feature of mechanical technologies currently used in the dairy industry. Instead of building on nature (and the specificities that it implies), living nature is eliminated (in the discussion of Parmalat, I gave an example of twofold and even threefold pasteurization in order to turn old and improper milk from far away into 'fresh milk' again). Pasteurization is not *per se* required. As a matter of fact, many exceptional cheeses, such as *Parmigiano-Reggiano* (Parmesan cheese, as it is commonly known) are made out of raw living milk – but then the making of such cheeses critically involves skill-oriented technologies. In the latter, the highly skilled *casaro,* or cheese-maker, is the nucleus of an extended *actor network* that links well-developed fields with alfalfa crops, extended irrigation systems, highly skilled farmers, special cattle breeds (the milk of which contains high levels of KKB caseins), summer feeding techniques, on-farm drying of hay, a well-outlined *disciplinare di produzione* (the manual that specifies the best way in which to make good cheese milk), roads, small milk containers and a radius of activities of not much more than 20km (in order to avoid the decomposition of living milk). The same network also explicitly excludes many other elements such as maize, grass silage, and uni-feed technology for cattle feeding. Likewise, such a network, which is composed of interlocking and interacting human and non-human agents, contains considerable flexibility, which allows for the making of further specificity (in the case of *Parmigiano-Reggiano*, this materializes in special cheese made out of milk from traditional red cows, special cheese from the mountain areas, organic cheese, etc.). Within this network, craftsmanship and skills are crucial in order to deal with the highly variable material entities and for dealing with social aspects of the network – just as in the case of high-quality olive oil. Pasteurization is excluded (and not required) in the specific network that characterizes the production of *Parmigiano-Reggiano* and similar cheeses. If the milk is bad or contaminated the *casaro* would probably recognize it immediately and reject it (or use it as feed for pigs). If it were to go unnoticed, the result would be foul-smelling, fermenting and gas-producing cheeses that would explode and be impossible to commercialize. Hence, the constellation as a whole has its safety valves.

Consumption milk has no such safety valve (at least none built into the final product). Nonetheless, especially in the north and centre of Italy, there is a very interesting and widespread revival of *latte vivo* (living milk). New miniaturized and automated technical devices for controlling and bottling fresh milk and its consequent distribution through new extended networks are important and, indeed, sophisticated elements in this respect. The milk is distributed from the farm to a range of schools, hospitals, shops, public restaurants, etc. It is transported every morning (after milking, cooling and bottling) in refrigerated vans to points of distribution where it is put into cooled display cases where the public can buy it (see Figure 6.9). What remains of yesterday's milk is then taken back (normally it is used for making yoghurt, butter, etc.). However, what is central is the fine-tuning of the farm as a whole. Cows are milked in conditions of extreme hygiene, the milking equipment is chosen with care, and cleaning procedures and controls are painstakingly precise. Cows must be free of stress (and the associated vulnerability of disease), while feed and fodder must meet the highest criteria for cleanliness and health. Individuals who milk the cows must be able to observe and correctly interpret any change in cattle behaviour, and so on and so forth. In short, the farm, as a whole, as well as the network in which it is embedded, is turned into a well-functioning and well cared for *organism* that produces fresh milk that meets the highest standards of quality and safety.

I also refer here to the socio-technical network in which the farm is embedded because, apart from technology, knowledgeable consumers are needed. If consumers should take fresh milk from one of the cooled display cases, leave it for hours in a car parked in the sun and then put in the refrigerator at home – or if they use the fresh milk after, say, two days – then the efforts made at farm level are in vain. The network needs consumers who are not only capable of judging and appreciating the specific virtues of fresh milk, but who are willing to treat it in the required way.

Central, then, to the technology entailed in this 'organism' and the associated network are the following characteristics (see also Roep, 2000; Rip and Schot, 2001; and Ventura and Milone, 2005a):

- It produces a *quality* level that cannot be reached in industry (i.e. by applying mechanical technology). Craftsmanship and technology that allow for its full employment and development (i.e. a skill-oriented technology) are decisive. This skill-oriented technology is developed and strengthened through the use of sophisticated technical devices (as with high-quality olive oil) and/or by turning the farm into a highly sophisticated 'organism'.

Figure 6.9 *New technological devices*

Source: Servizi Commerciali Allevatori (2005, p1)

- It is by definition a *localized* technology: it builds on local factors (i.e. on specificity) and thus also results in specificity.[25] More generally, skill-oriented and sophisticated technologies (what one might call 'peasant technologies') are actor networks that build upon and adapt to the specificity, variability and also to the unexpected (Remmers, 1998) in the social and the natural world, while, at the same time, translating them into higher quality levels. Here resides the central difference vis-à-vis mechanical technologies: the latter neither build upon nor adapt to, but essentially *subordinate*, nature and the social world. This is done by requiring them to fit into previously assessed standards, schemes and procedures.
- It entails complex back and forward exchanges of information and judgements[26]: between, for example, grassland quality, weather conditions, etc., and the treatment and quality of the milk. Together these exchanges allow

for flexibility within the process of production in *sensu strictu*. Somewhat deviating inflows are not defined and treated as *waste* – but are, instead, adapted. Likewise, the inflow is shaped and reshaped[27] in order to meet as well as possible the requirements of the process of transformation (or, for that matter, the expectations of the consumers). Thus, the flows of communication are essentially two sided: they flow back and forwards (which is in neat contrast to mechanical – i.e. industrial – technologies).
- Apart from this 'technical' flexibility, there is also flexibility in another sense: milk, for instance, that is not commercialized within 24 hours is simply taken back and converted into other products.
- Skill-oriented technologies tend to augment the value added per produced unit. They thus tend to raise the level of value added for the unit of production as a whole.
- Skill-oriented technologies generate and are simultaneously dependent upon knowledge. The generated knowledge has specific features. It is, to use the expression coined by Henry Mendras (1967), *art de la localité* – that is, knowledge of the specific. Likewise, skill-oriented technologies are apt as a context, if not a highly appropriate tool, for *learning*.
- And, finally, they are *open* constellations. They are accessible to everyone who is knowledgeable (or wants to become knowledgeable). Anyone may try to improve them, after which the created novelties can travel to any other location.
- In short, peasant technologies are relevant not only for the directly involved producers, but potentially for society as a whole. Their relevance far exceeds the 'quality niches' so far discussed (although even in these niches a potentially powerful approach to sustainability is germinating). Let me now discuss this by returning to the issue of bio-energy since this directly relates to the global need to find and elaborate upon alternatives to oil-based energy.

Bio-energy

Many different technologies are deployed to produce bio-energy, each simultaneously a way of patterning both the social and the natural. But what does the peasant way of producing energy amount to?

In its most simple form, it starts from slurry available on the farm and, for example, a part of the maize harvest. Together these elements might be used to fuel a process of anaerobic fermentation and subsequent burning of the methanol gas resulting from it. These processes can be located in relatively simple, small and interlinked technical installations (such as a fermentation tank and a generator) – and they result in electricity and heated water. The electricity can be sold to the network (and be bought back for the same price for use on the farm itself), while the heated cooling water can be used to heat neighbouring houses, horticultural glasshouses, etc. Thus, on the output side, there is some flexibility (just as in the previously discussed Zwiggelte proposal). Currently, flexibility is further enlarged because the production of bio-energy renders a 'bio-energy certificate' for which major players on the energy market will pay big money since they are legally obliged to produce up to 5 per cent alternative energy, which they prefer to externalize through such certificates. And, second, the process renders, in the end, a powder containing a high level of organic nitrogen that makes good fertilizer.

In this process of bio-energy production, inflow can also be variable. A typical smart feature encountered in practice is that maize production on the farm is partly oriented toward cattle feed, and partly toward fuelling the fermentation process. This *dual-purpose* construction allows for the following. The best parts of the ensilaged maize go to the cattle, the worst to fermentation. Since one never knows for sure what the quality of silage will be (partly due to weather conditions, but also inherent to natural processes within the silage), one can use both the best and the worst of it.

Thus, three elements emerge. First, energy production is neatly interwoven with, if not constructed upon, farming. Second, energy production is fuelled with waste products (slurry, low-quality maize silage, but also dried grass, straw, shredded wood, or waste products from the local bakery, etc.). In other words, waste products are converted into new resources, which, in turn, are converted into new values – in this case, energy. And, third, the conversion process (of waste into energy) is characterized by flexibility both at the input and output side of the farm.

This, together with its other characteristics (see Box 6.1), allows for its denomination as 'peasant technology', the more so since this decentralized small-scale peasant way of producing bio-energy (also encountered in the biodiesel production in Germany; see Knickel, 2002) contrasts sharply with *centralized* production (as occurs in the case of bio-ethanol). In the latter, huge industrial plants for conversion of maize, sugar cane and soya are the central elements, while the delivering farms are restructured or reduced to being large appendices that provide the centre with cheap raw material for conversion into energy. Since countries such as Brazil and the US already produce huge amounts of bio-ethanol (at relatively low prices), farms face a highly competitive market, which will probably induce large-scale mono-crop systems at the level of primary production. In the 'peasant way', where energy production is mostly grounded on waste products (and embedded in patterns that allow for flexibility), such competition is felt less severely and less directly.

Thus, the social and natural worlds become patterned in highly contrasting ways. This is especially so since an average industrial plant for centralized energy production requires at least 100,000ha of agricultural land dedicated to the production of the required raw material. In most European countries, this can have far-reaching consequences on, for example, the landscape and biodiversity. Apart from this, there will be high pressure on the transport systems to bring the raw material to the central plant with a consequent energy leakage due to the structure thus created. In comparison, the peasant way of producing energy, which entails direct delivery to the already existing electricity network, cannot be characterized as other than smart. Transporting electricity is far more intelligent than transporting maize.

Box 6.1 *Features of the peasant mode of energy production*

Features of the peasant mode of energy production include the following:

- on-farm conversion, basically grounded on the use of waste products;
- small but highly efficient units, directly connected to widely extended networks already existing for energy transport;
- high degrees of flexibility both in the use of inputs and in the outflow;
- relatively high levels of value added at the farm level;
- a high degree of decentralization and low vulnerability of the system as a whole;
- a high degree of multifunctionality: farms are not reduced to simple providers of raw material for energy production;
- social benefits in terms of the maintenance, instead of the destruction, of scenic landscapes, biodiversity, employment, etc.

Repeasantization as social struggle

Repeasantization is a process of transition that unfolds at many levels, along several dimensions, involving many people. Like every process of transition, it runs counter to existing techno-institutional regimes and interests, and thus generates a wide range of contradictions. Repeasantization is one of a broader set of three competing transitional processes (see Figure 1.3 in Chapter 1) and this generates a range of additional contradictions. At the same time, repeasantization is a massive and widespread process that is triggered and spurred by the interests and prospects of the farmers involved; in this sense, it constitutes a social struggle. Repeasantization is about facing problems, opposition, adverse interests, hostile opponents and fierce competition. And it is also about the endeavour to go beyond them, about struggling against the tide in order to go forward. This applies to many moments, many places and many levels, some of which I will now briefly discuss.

I have participated several times in panels mandated to choose prize winners in contests for the 'best' or the 'most exiting' innovations in the countryside.[28] This is how I came to know the Zwiggelte case. What struck me about the different processes that I witnessed was that practically all of the innovations proposed had emerged from a developed and often radical critique of the state of the art, or the dominant routines in particular domains of society. The feeling that things 'could and should be done better' was time and again central.

Nowadays, many aspects of rural life are subjected to highly detailed regulatory schemes. These sets of generalized rules are often at odds with the diversified and dynamic nature of agriculture – and several other economic activities in the countryside – and they therefore result in various kinds of friction. They are especially detrimental to the materialization of important ideas that often form the starting point of new rural development trajectories. In fact, in the European survey drawn upon earlier of the farmers involved in 'deepening' and 'broadening' strategies, some 69 and 61 per cent of farmers, respectively, indicated that 'restrictive regulations' constituted the major constraint to realizing their new activities (Ploeg et al, 2002c, p227).

The European survey also made an inventory of the main driving forces implied in the creation of new peasant-like types of farming. Respondents referred to the *area* (and the specific qualities that it entailed), to *personal* skills and interests, to the availability of labour within the *family* – in short, to *local* factors – as being the main drivers for change. However, it is precisely the local that becomes uncomfortable, if not awkward, within the framework of global and highly formalized rules – the more so when local deviations become the starting point for new development trajectories. As a matter of fact, many innovators live, formally speaking, in a situation of 'illegality' (Morgan and Sonnino, 2006, p19).[29] In short, peasant-driven rural development occurs, in practice, as resistance.

One of the most intriguing aspects of ongoing social struggles is certainly the range of contradictions that emerge at the interface between widespread rural development practices and the new rural development policies that now function at supranational, national, regional and local level. This is elaborated upon in Figure 6.10.

Rural development *policies* were born at the supranational level (especially during and after the European Conference on Rural Development, held in Cork, Ireland, from 7 to 9 November 1996) as an explicit endeavour to go beyond the limitations of the Common Agricultural Policy (CAP).[30] However, the European principle of subsidiarity ('define policies and programmes at the lowest possible level') stopped as soon as responsibilities were shifted from Europe to the member states. Member states developed rural development policies that embodied the principle of strict control exercised by various state apparatuses, implying a high degree of formalization.[31] This ran counter to the often informal (Bock, 1998), flexible and necessarily open-ended experiments

180 *The New Peasantries*

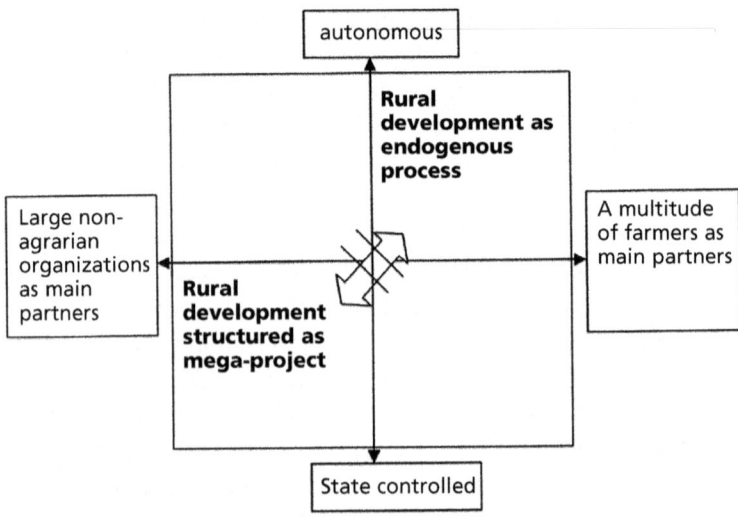

Figure 6.10 *Rural development as a contested and fragmented process*
Source: Original material for this book

of many of the autonomous initiatives stemming from the countryside. It was equally at odds with the heterogeneity of the social and natural world. One cannot take hold of nature and landscape, or prescribe for their management through central bureaucratic programmes – not even in a country as small as The Netherlands (I return to this problem in Chapter 8).

Alongside this first dimension, there is a second that concerns the issue of 'red tape'. State apparatuses managing rural development find it far more attractive to deal with a few large organizations (in many circumstances, preferably non-agrarian ones) than with an endless multitude of small farmers.

Thus, an interface is born that is characterized by strong frictions and a set of far-reaching contradictions. At this interface, rural development policies and practices sometimes interlink; but, ironically, they mostly enter into contradiction with each other. Consequently, the design of new intermediating institutions becomes a major challenge. In Chapter 7, I discuss the construction of one such new institution.

7
Striving for Autonomy at Higher Levels of Aggregation: Territorial Co-operatives[1]

Introduction

This chapter focuses on the construction of new mechanisms for creating autonomy at levels of aggregation that go beyond single peasant units of production. Central to the argument is the case of territorial co-operatives: an institutional innovation born more or less simultaneously in several places throughout the north-west of Europe at the beginning of the 1990s (although pleas for them were already voiced during the late 1970s and 1980s). Territorial co-operatives can be highly effective mechanisms of supporting repeasantization. They are also strategic in the attempt to overcome the current agricultural crisis (see Figure 1.4 in Chapter 1) because they involve new forms of self-regulation. They re-link the farming and rural population as active and knowledgeable participants to processes of rural development and agrarian transition. This is particularly important politically since the more conventional ways of expressing and negotiating interests through agrarian syndicates and corporatist frameworks have failed to produce cohesion and practical results (Frouws, 1993; Frouws et al, 1996; Hees, 2000).

To underpin my argument I draw heavily on the case of the Noardlike Fryske Wâlden – the North Frisian Woodlands (NFW). The NFW is one of the most prominent Dutch examples of a territorial co-operative. Today, it has some 900 members, most of them farmers and rural dwellers with some land, but also non-agrarian members. It covers an area of around 50,000ha that includes large spaces dedicated to nature. Within the area, almost 80 per cent of all farmers are affiliated to the NFW. As impressive as this may appear, one should not forget that it began some 15 years ago as a very small and highly vulnerable initiative. Since then, however, the NFW has developed in a sturdy and persistent way. It became a major field laboratory and has on several occasions affected important aspects of Dutch agrarian policy. It has also stimulated major new scientific insights into the nature of co-production (Sonneveld, 2004; Reijs, 2007).

The NFW co-operative is located in the north-eastern part of the province of Friesland in the north of The Netherlands. The area is characterized by an attractive historically created hedgerow landscape (see Figures 7.1 and 7.2) with a beautiful variety of open and closed, high and low-lying, wet and dry areas (Schaminee et al, 2004). It was recently declared a 'national landscape' deserving special attention and support from national and regional government. Landscape and farming practices are also the main reasons for highly developed biodiversity in the area that embraces both flora (Weeda et al, 2004) and fauna (Swagemakers et al, 2007).

What are territorial co-operatives?

The agrarian crisis of the 1880s was, to a degree, provoked by deteriorating relations between farming and markets. Butter fraud, usury, a lack of transparency and failing market power were but a few expressions. This triggered a first wave of agricultural co-operatives. These emerging co-operatives were not aiming for major changes in the markets as such (nor could they ever have done so). Their target was basically improved articulation between farming and markets (Dijk, 2005).

Currently, interrelations between the state and the farming sector have become strongly disarticulated. The state imposes regulatory schemes that are increasingly felt as inadequate, if not asphyxiating. Mutual distrust is a 'structural' feature (Ploeg et al, 1994; see also Ploeg, 2003a; Breeman, 2006). This

Figure 7.1 *An overview of the hedgerow landscape*

Source: Schaminee et al (2004, p17)

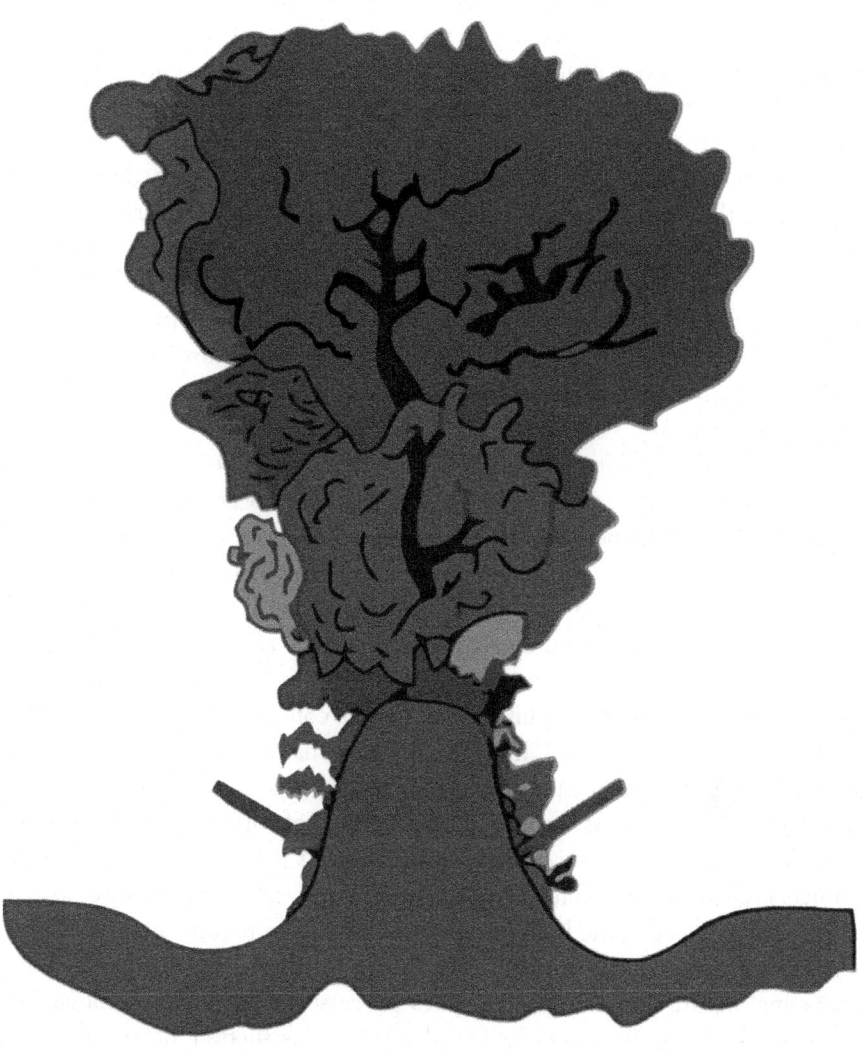

Figure 7.2 *The anatomy of a hedgerow*

Source: Boer (2003, p20)

disarticulation has triggered a new form of rural cooperation that has materialized in what are increasingly referred to as territorial co-operatives,[2] which aim to radically improve relations between farmers and the state by introducing new forms of local self-regulation and new strategies for negotiated development (Ploeg et al, 2002a). This second wave of cooperation that is currently gaining momentum is in line with the generally accepted European Union (EU) principle of subsidiarity. It also reflects the strong democratic traditions of north-west Europe. It reduces the transaction costs associated with current rural and agricultural policy programmes (Milone, 2004), while augmenting their reach, impact and efficiency. Territorial co-operatives could therefore, in synthesis, be a perfect complement to agricultural and rural policies. Yet, despite this, the 'marriage' between state apparatuses and the newly emerging territorial co-operatives is turning out to be an unhappy one. This is due to the fact that the ways in which ministries of agriculture (and associated bodies) relate to farmers are increasingly structured in a hierarchical (i.e. an Empire-like) way.

Theoretically speaking, territorial co-operatives may be understood as the interlocking of three emancipatory moves that aim to go beyond particular impasses. The first is the search for, and construction of, *regional cooperation*, which aims to integrate within farming practices activities that are oriented towards protecting the environment,[3] nature and landscape (Wiskerke et al, 2003b, p3). The background is twofold. Regulatory schemes imposed by the state are highly segmented: for example, one set of rules pertains to nature values and their protection, while another set is concerned with the reduction of ammonia emissions, etc.[4] In turn, this internally segmented (and often contradictory) set of prescriptions is disconnected (due to its particular design) from farming practice. As a consequence, the sets of rules materialize as innumerable limitations imposed upon farming (WRR, 2003). In addition, biodiversity, landscape and high-quality levels of resources such as water and air cannot be produced at the level of the single farm; they require a regional scale, both from a material and social point of view. Protecting the environment and 'managing' nature and landscape implies processes of learning, exchange and cooperation. Thus, it is widely felt that the construction of sustainability requires regional cooperation, and that this is the only way of successfully redressing the many frictions and limitations inherent in the sets of general rules defined by expert systems and the state (Stuiver and Wiskerke, 2004).

A second move concerns the search for, and construction of, new forms of *rural governance*. This move emerged from the early 1990s onwards (Dijk, 1990; Marsden and Murdoch, 1998; Ploeg et al, 2000) and produced, within different arenas, a broad spectrum of expressions (Hees et al, 1994; Horlings, 1996; Wiskerke et al, 2003a) in which the principles of responsibility, accountability, transparency, representation and accessibility (Schmitter, 2001) became important beacons for gaining legitimacy. In this respect, it was concluded in an Organisation for Economic Co-operation and Development (OECD) report that 'the farmer-led co-operatives [out of which the NFW grew] are consistent with both Dutch institutional and democratic traditions'. It also noted that 'from the government's perspective, the emergence of these groups has proved a useful vehicle for mobilizing farmers' commitment to environmental protection, and for finding ways to shift more responsibility over the implementation of environmental policy to local communities' (OECD, 1996; see also Fischler, 1998, and Franks and McGloin, 2006). Central to the newly emerging forms of rural governance entailed in territorial co-operatives is the principle of institutionally defined exchange. Territorial co-operatives accept the general *objectives* regarding landscape, nature and environment (and often promise to go beyond these objectives) on the condition that they receive room for manoeuvre ('freedom', to echo Slicher van Bath, 1978) to define for themselves the most adequate *means* of reaching the objectives.[5]

Third, territorial co-operatives represent a move away from expert systems towards the innovative abilities of peasants. Territorial co-operatives are thus also *field laboratories* (Stuiver et al, 2003, 2004): they are places where the most adequate *means* of locally resolving new global problems (such as the environmental crisis) are developed, tested, implemented, evaluated and further improved upon.

In the new territorial co-operatives, the moves mentioned are tied together into one new institution, which crucially builds upon and at the same time strengthens the social capital available in the territory. Equally important is the network of interrelations with other regional, national (and sometimes supranational) institutions. Through such a network new services, products and additional room are created and delivered that otherwise would be difficult to achieve.

A brief history of the North Frisian Woodlands

The birth of what later became the NFW is a perfect illustration of institutionally defined exchange. During the early 1990s, a national law (first known as the ecological guideline and later specified as the law on ammonia and animal production) was implemented to protect valuable nature from acid rain. Natural elements that were declared acid sensitive were hemmed in by restrictions that meant agricultural activities could no longer be expanded. The new regulations implied that in a dense hedgerow landscape such as that in the NFW, all agricultural activity would have to be 'frozen'. A general standstill, if not a massive regression, would have been the result.

The proposition provoked considerable anger in the area. A main argument raised was that farmers had created this landscape (from 1850 onwards) and had always been the ones to take care of it (see Bruin and Ploeg, 1992). It was seen as wrong and unjust to turn this landscape into a noose that would literally strangle farming in the area. Some were prepared to eliminate hedgerows and other acid-sensitive elements in the area to avoid such a danger. Luckily, another way out was found through an institutionally defined exchange. The municipality (and the province) promised *not* to declare the many hedges as acid-sensitive elements in exchange for a promise from farmers to maintain and protect the hedges, ponds, alder rows and sandy roads of the area. For this reason, farmers created the Eastermars Lânsdouwe (VEL). Thus, state objectives were secured, but through other, more appropriate, means. This first association was formed in the spring of 1992. The second, Vereniging Agrarisch Natuur en Landschapsonderhoud Achtkarspelen (VANLA), followed in the autumn of the same year. Later, another four were created in the surrounding municipalities, and together, during the course of 2002, these six associations and co-operatives created the overarching NFW co-operative.

The creation of the first two nuclei involved difficult bargaining (for an extensive description, see Ploeg 2003a): the expectations of participating farmers and the surrounding institutions had to be brought in line. A solid contractual base for reciprocity had to be constructed, without one of the parties concerned feeling the victim of any opportunistic behaviour by the other. The effective grounding of the co-operatives took further shape when a contract was worked out and signed with van Aartsen, the then minister of agriculture.[6] Through this contract the co-operatives obtained legal room to develop and test several novelties: a large programme for the maintenance of alder rows[7] and another to construct a new peasant trajectory towards sustainability. To create room for these programmes, specific measures and exemptions from legal obligations (such as the injection of slurry into the subsoil) were granted.

Following this agreement, VEL and VANLA developed their own locally adapted modules for landscape and nature management, enrolled the majority of farmers in it and thus started a large programme that greatly improved the area. At the same time, they designed an 'environmentally friendly' machine for manure distribution (a machine appropriate for small fields surrounded by hedges and alder rows), and succeeded in engaging nearly all farmers in the management of nutrient accountancy systems (MINAS), which were not yet legally required. With the widespread application of this accountancy system, environmental progress (included in the contract with the minister) could be monitored adequately.

Strategic in producing effective environmental progress was a twofold modification for participating dairy farms: the use of chemical fertilizer was strongly reduced while slurry was rebuilt into 'good manure'. The effects of this typical peasant approach (reducing external inputs and improving internal resources) were impressive: within a few years the frequency curves representing nitrogen losses per hectare changed completely (see Figure 7.3). During 1995 to 1996, the largest category still realized losses of 360kg to 400kg of nitrogen per hectare; this was reduced to 200kg to 240kg of nitrogen per hectare by 1998 to 1999. The average loss per hectare decreased from 346kg per hectare in 1995 to 1996 to 236kg per hectare in 1999 to 2000 (see Figure 7.4). During 2001 to 2002, nitrogen losses were further reduced to an average level of 150kg per hectare. The fact that several farms managed to go far below that level was taken as an indication of future potential.

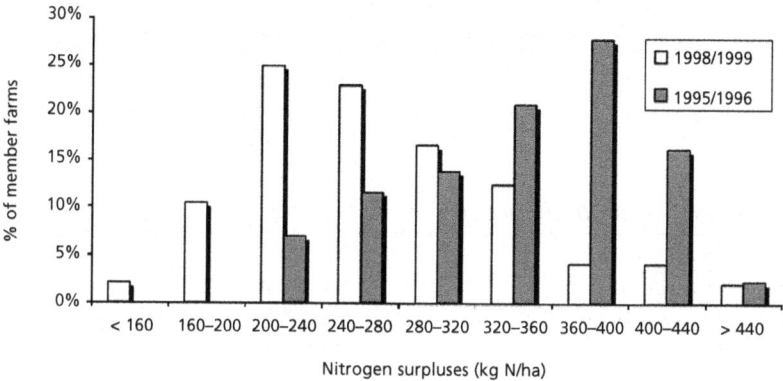

Figure 7.3 *Distribution of nitrogen surpluses among VEL/VANLA member farms*

Source: Atsma et al (2000, p23)

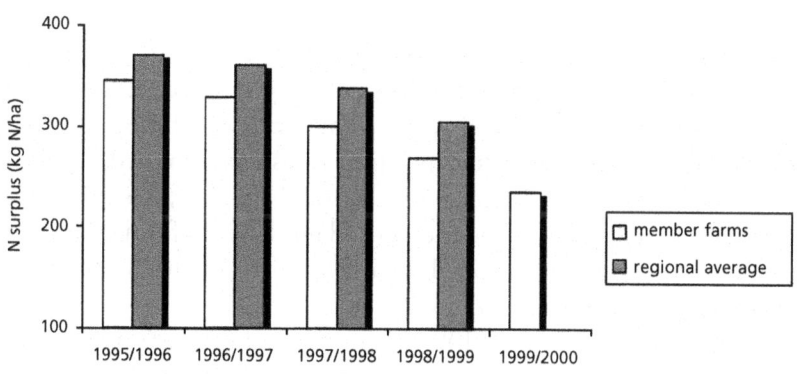

Figure 7.4 *Nitrogen surpluses on VEL/VANLA member farms compared to the regional average*

Source: Atsma et al (2000, p13)

The NFW also became involved in a wide range of activities for maintaining and improving landscape and nature. Table 7.1 offers some quantitative data. Approximately 80 per cent of the area as a whole is covered by some form of nature and landscape management. Nowhere else in the country is there such a high coverage. Regional cooperation was crucial in this respect (Eshuis, 2006) and, as a result, it was possible to achieve qualitative improvements of landscape and biodiversity far beyond those to be gained from single units of production. With a cooperative approach, the management of landscape and biodiversity could be lifted to the level of the territory as a whole. The cooperative management of nature and landscape creates an additional flow of income into the regional economy of 4 million Euros per year. During 2004, the average farm participating in the programmes for nature and landscape management gained an extra value added of 11,000 Euros (Heijman, 2005).

Table 7.1 *Some quantitative data on the management of nature and landscape*

Management of field boundaries	900ha
Protection of meadow birds	12,000ha
Farming with natural handicaps	3700ha
Protection of geese	3000ha
Hedgerows	344km
Alder rows	860km
Ponds	430 units
Wooded copses	9ha
Pollard willows	457 units
Associated fencing	1085km

Source: Original material for this book

Nature, landscape and environmental data were related to what, in the end, became a large and promising new research programme carried out by scientists and farmers together.[8] This programme not only shifted several of the boundaries between science and practice; it also transformed, albeit slowly, several boundaries within science itself. At the same time, this slowly growing research programme created new levers and mechanisms for the further development of local self-regulation. I return to some of the outcomes of this research in the last section of this chapter.

From 2003 onwards, the NFW greatly enlarged the fields in which it operated or planned to operate. Figure 7.5 provides a synthetic overview of these fields, with the central overlap highlighting potential synergy. The NFW also prepared the text of a 'territorial contract' in which a range of institutional partners declared that they would actively engage in bringing about the goals of the working plan of the NFW.

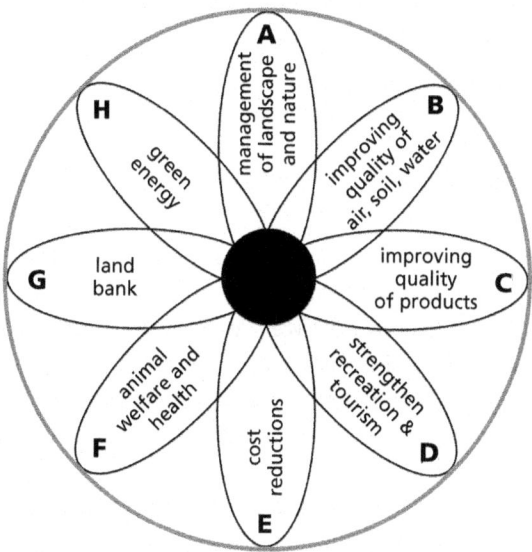

Figure 7.5 *The outline of the new North Frisian Woodlands plan*
Source: NFW (2004, p17)

The working plan contained 30 specific projects, which together covered many aspects of the regional economy and its sustainability (see NFW, 2004). Among those who signed the territorial contract were the provincial government, the ministries of agriculture and spatial planning, the district water board, the five municipalities, the environmental federation, nature organizations and Wageningen University. This agreement has resulted in the creation of a new territorial board in which the NFW and other partners meet twice a year to discuss problems associated with implementing the working plan. In the longer term, it is assumed that the regional economy will be strengthened, sustainability improved and self-regulation expanded upon. If this succeeds, it will represent numerous reversals of existing trends. However, to date, it is turning out to be a heavy struggle that is sometimes perceived as an illusory endeavour that runs against the tide (but that was also the case for the initial establishment of the first nuclei of the NFW some 15 years ago).

A highly interesting feature of the process that finally led to the territorial contract is the mission statement that the NFW co-operative formulated after several rounds of consultation. This lists ten commonly shared values that reflect the history of both the area and the co-operative (see Box 7.1). They also reflect the interests, prospects and the emancipatory ambitions of its people (as reflected, for example, in the claim for 'our own entitlement'). Taken together, they represent the strong social capital that has been forged in the 15 years of successfully expanding the NFW from its first vulnerable nuclei to a now solid and well-rooted territorial co-operative. The commonly shared values are each condensed in Frisian catchwords.

> **Box 7.1** *Commonly shared values as specified by the North Frisian Woodlands in its mission statement*
>
> 1 *Mienskip (community).* As a community we are proud and strongly aware that over the past 100 years we have managed to shake off the yoke of poverty. In the province of Friesland we are known as *wâldpyken*, as headstrong people, and we are proud of that. We resolve the conflicts that emerge within our community ourselves, and usually manage to find our own area-specific solution to problems that come from outside. Our sense of community and togetherness is strong, and we expect others to respect this.
> 2 *Lânsdouwe (the unity of man and land).* Our area is characterized by an attractive, varied and living landscape: a unique unity of man and nature (*lânsdouwe*). The landscape has been made by our forefathers. At present, the guardianship of North Frisian Woodlands' (NFW's) farmers is complemented by the active involvement of bird protectors and other volunteers, thus securing the further development of nature and landscape. We expect to be given an active role in the management of nature, landscape and environment. For this, a lively and land-based agriculture is indispensable.
> 3 *Kreas buorkje (farming gently).* We, the members of the NFW co-operative, are fully aware that the unique unity of man and nature brings about a special responsibility. Farming should be carried out in a responsible and sustainable way: *kreas buorkje*. Given our historical and current experience and skills, we, more than anyone else, are the obvious entity to ensure that this continues.
> 4 *Eigen gerjochtigheid (our own rights and entitlement).* The Northern Frisian Woodlands is our area, shaped by us and our forefathers. It is our *eigen gerjochtigheid*. Therefore, we are entitled to participate in all planning and decision-making that concerns our area.
> 5 *Wy kinne en dogge it better (we perform better).* The NFW association is made up of six grassroots organizations that have up to 15 years of experience in managing programmes for nature, landscape and environment protection. Furthermore, we have a strong cooperative tradition of communally owned dairy factories, commons, village associations, study groups, voluntary land consolidation schemes, and mutual help. We have proven that we can develop and successfully implement high-quality area development plans. We have demonstrated that our way of farming is environmentally sound, and that we sustain nature far better than is achieved through state-imposed regulation. Using our knowledge, our skills and our cooperative tradition, we perform better than would be possible with any generic approach: *wy kinne en dogge it better!*
> 6 *Wissichheid (reliability).* In agreements with other parties, the NFW co-operative is a reliable partner. We offer perseverance and reliability (*wissichheid*) and we expect our partners to be trustworthy.

> 7 *Stadich oan foarût (progressing slowly but steadily)*. History has taught us that the battles we are fighting can last a long time. This is why our aim is steady progress, step by step (*stadich oan foarût*). Sometimes, however, we make big leaps too. Step by step or with big leaps, the common interest of the area is always paramount.
> 8 *Net allinnich (not alone)*. We are convinced that challenges should be faced together, not alone (*net allinnich*). We have recently built, and maintain, fruitful coalitions (at local, provincial and national level) with politicians, environmentalists, conservationists, scientists, and with water management boards and farmers' lobby groups. The NFW association will continue along this path.
> 9 *Tinke oan'e takomst (caring for the future)*. In these times of globalization we stand up for the future and put future first – the future of the area and its future inhabitants (*tinke oan'e takomst*). Thus, future generations can also continue to nurture the area, be proud to live and work in it, and enjoy it together with others.
> 10 *Mei wille en nocht (with satisfaction and joy)*. We, the members and governors of the NFW co-operative, have worked over the years with joy and satisfaction (*mei wille en nocht*). This has increased resoluteness in the area. We wish to continue in this way and our organization will play an important role in helping us to manage our own affairs in our own area.

From an analytical point of view, these commonly shared values might be interpreted as constituting a moral economy. In this respect there are – notwithstanding the many differences in context and time – remarkable similarities with the shared values of the peasant community of Catacaos, formulated during the early 1970s (see Chapter 3). One similarity is the centrality of *community* (i.e. the historically rooted sense of community that is neatly interwoven with the interests and prospects of those who are part of it – whether it is a *mienskip* or a *comunidad*). And other important features, echoed in both declarations, are the notions of constant *struggle* (within and against an often hostile environment) with the community as its main vehicle, and the focus on potential *superiority*: 'we perform better'.[9]

These and other continuities indirectly point, I think, to a general bias in current social science research, which seems no longer capable of adequately recognizing and theoretically representing (rural) communities. Since Tönnies's (1887) stress on the shift from *Gemeinschaft* to *Gesellschaft*, and especially, since the predominance of modernization theories from the 1950s onwards, rural communities have, theoretically, ceased to exist and in this respect the social sciences have been blinkered. Ironically, this also applies to social research in Peru, where it is often assumed that peasant communities only exist in the 'traditional' Andes and not in the 'modernized' coastal regions. This accounts for the dearth of peasant community studies in the latter.

Novelty production

As indicated earlier, territorial co-operatives may materialize, in part, as field laboratories. This is especially the case with the NFW. Within the co-operative, and in close cooperation with a group of scientists, a range of novelties has been explored and linked to other novelties that together make up an important 'web' – that is, a well-integrated constellation of interconnected changes that have far-reaching and multidimensional impacts on farming practice and transition. Novelties are, in a way, *deviations from the rule*, which might have been deliberately created or are simply the unexpected outcome of the messiness of life (Richards, 1985; Remmers, 1998; Wiskerke and Ploeg, 2004; Flora, 2005). Thus, novelties can be new practices, new artefacts or simply changed definitions of a particular situation or task. A key element is that they entail a promise, which often implies that things can perhaps be done better. Of course they may also prove to be a failure, or will take time to become fully comprehended. Novelties 'infringe' existing codes of conduct or rules for understanding things. Materially, novelties also generally turn out to produce some kind of rupture. They are not incremental and thus differ from innovations. They may build upon already available elements, connections and their specific ordering (as outlined in Chapter 6), but at the same time they imply a reordering of elements, connections and overall patterns (see Ploeg et al, 2004b). Thus, novelties occur as *change agents in disguise*: as *undercover agents*. In addition, they emphasize the importance of the local, which contains but also produces them. Novelties are, in other words, *hidden* in the local and may have to be identified and unpacked in order to allow them to travel to other localities. Furthermore, as novelties represent a deviation from the rule, this implies that the rules may have to be changed or at least 'softened' in order to allow 'things that should not occur' to occur. This is the theoretical side of the equation. The material side involves the idea that a rule represents some kind of institutionalized code of conduct (it is an integral part of the wider sociotechnical regime, as depicted in transition studies; see Rip and Kemp, 1998). Space has to be deliberately created for a novelty to develop, and the creation of such room might run counter to existing infrastructure, interests and/or laws. Hence, a central role must be accorded to various forms of strategic niche management since it is through such forms that the local is strengthened vis-à-vis Empire-like threats to annihilate it.

The NFW co-operative represents a locality in which many novelties have been encountered and actively developed. 'Good manure' is probably one of the most telling, yet also most contested. The background to this particular novelty is located in the former modernization process that deeply restructured farming practices and resources. 'Well-matured manure' was once a highly valued resource and the making and use of it were closely embedded in local cultural repertoires. However, in accordance with the modernization trajectory, well-matured manure was converted (unintentionally) into a waste product – a 'nuisance to be got rid off' (Eshuis et al, 2001). But, as many farmers put it: 'Once you have a nuisance on your farm, it has the habit of repeatedly re-emerging.' The loss of organic matter in the subsoil, the increased need for high levels of fertilizer and deteriorating grassland conditions are but a few examples of this 'reproduction of misery'. For some farmers this worrying state of affairs triggered a multifaceted search to create, once again, good or at least improved manure. Thus, the emphasis on good manure began as a critique of inefficiency and loss. It also started from the careful observation and interpretation of heterogeneity: the fields of a particular farmer yielding far more production than the fields of others in the same neighbourhood thus became an important point of reference. Could he perhaps be using somewhat different, and maybe improved, manure on his fields? And if so, how was this different manure constructed? One thing was evident among the farmers involved: good manure is not an isolated artefact; it is the outcome of a particular repatterning of the process of co-production, as is illustrated in Figure 7.6.

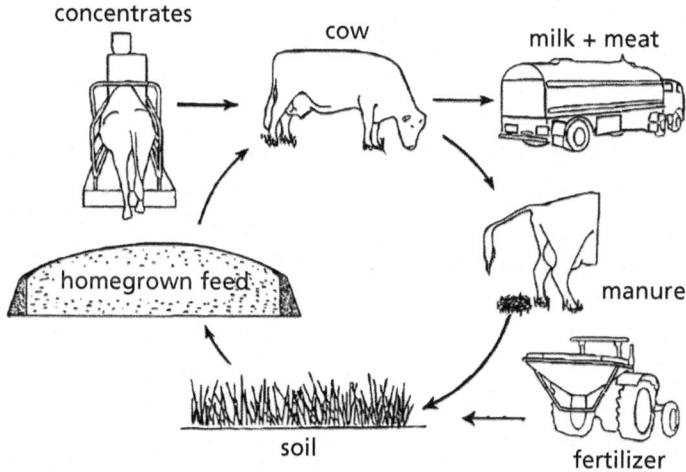

Figure 7.6 *The cattle–manure–soil–fodder balance*

Source: Verhoeven et al (2003, p150)

Good manure is slurry with a high carbon:nitrogen ratio and a relatively low ratio of ammoniacal nitrogen (and, consequently, an elevated proportion of organic nitrogen). These and many other features are now (after almost 15 years of research) well known, documented and scientifically explained (see, for example, Verhoeven et al, 2003; Goede et al, 2003, 2004; Reijs et al, 2004,

2005; Sonneveld, 2004; Reijs, 2007). However, prior to this, such insights were lacking and farmers' opinions differed widely (Eshuis et al, 2001). The only expectation was that manure could be improved upon. It was hoped that rebalancing the soil–plant–animal–manure cycle (see Figure 7.6) would have positive outcomes (Verhoeven et al, 2003), especially since former modernization strategies had focused almost exclusively on one component of the whole constellation (namely, the cow), thus creating many frictions and setbacks.

In the beginning, good or improved manure represented a novelty. It was different in terms of composition, outlook, smell and effects. It also differed in its history – that is, in its making. However, exponents of the Dutch agricultural expert system considered good manure to be a *monstrosity*, something that, according to available insights, should and would not work, the more so since farmers of the NFW area proposed applying it in 'a good way': by surface distribution instead of the legally prescribed injection into the subsoil.

Nevertheless, it is now abundantly clear, at least with respect to the NFW, that when combined with good application techniques, good manure translates into improved soil biology (Goede et al, 2003), which in turn (i.e. through the augmented autonomous nitrogen delivery of the soil) allows for the production of more and better fodder with less chemical fertilizer (Ploeg et al, 2006; Groot et al, 2007b). Delaying the date for mowing (which for many reasons, including psychological reasons, is quite difficult for farmers) also contributes to obtaining better fodder (characterized, technically speaking, by a high degree of fibre and a low level of protein, which is diametrically opposed to fodder produced according to the modernization model). Feeding cattle with this improved fodder leads to less stress in the dairy herd, fewer veterinary interventions, longer life, and to milk containing more protein and fat (and probably more conjugated linoleic acid, or CLA) and, finally, – again – good manure (Reijs, 2007). Thus, the cycle is closed and a new self-sustaining balance is created that tends, from an environmental point of view, to be superior to the model imposed by the state: nitrogen losses and ammonia emissions are far lower (Groot et al, 2003, 2007b; Sonneveld and Bouma, 2003, 2004; Huijsman et al, 2004; Sonneveld, 2006).[10]

The construction of this now widely accepted and scientifically supported re-patterning of the social and natural world, contained in the micro-cosmos of the dairy farm, took many years to develop. It also took the concerted action of 60 dairy farmers belonging to the first nuclei of the NFW and a multidisciplinary group of scientists who could operate outside the vested routines and interests of the expert system as a whole. The approach has spread, like ink dots, all over the country, especially since it impacts positively upon the economy of the farm unit. Monetary expenditure is reduced and additional benefits are realized through extended longevity and improved milk content. Figure 7.7 summarizes some of the main findings through the identification of four categories. The first is comprised of farmers who during 1997–1998 had already reorganized their process of production in an integral and rebalanced way. The graph shows that during that particular period they obtained the highest margin per 100kg of milk. Through further fine-tuning this balance, they succeeded in augmenting the margin even further (line 1). This represents a clear deviation from the general trend in those years, when dairy farming in Friesland as a whole (line 4) and a control group within the NFW area (line 3) showed decreasing margins. While the squeeze exerted on agriculture is reflected in downward pressure on margins, the production and development of novelties becomes a response to this negative trend. Finally, we have line 2, the category of farmers who initially (i.e. in 1997 to 1998) only partly re-patterned their production, but who later succeeded in rebalancing it further. They also show an increase in margins.

The comparative study summarized in Figure 7.7 was based on 37 dairy farms, of which 17 belonged to categories 1 and 2. Compared to the average trend, these 17 farms realized together an *additional* gross value added of 250,000 Euros. If their example were to be followed throughout the North Frisian Woodlands, the gross value added realized in the area would rise by some 14 million Euros – that is, the room for further endogenous growth is considerable.

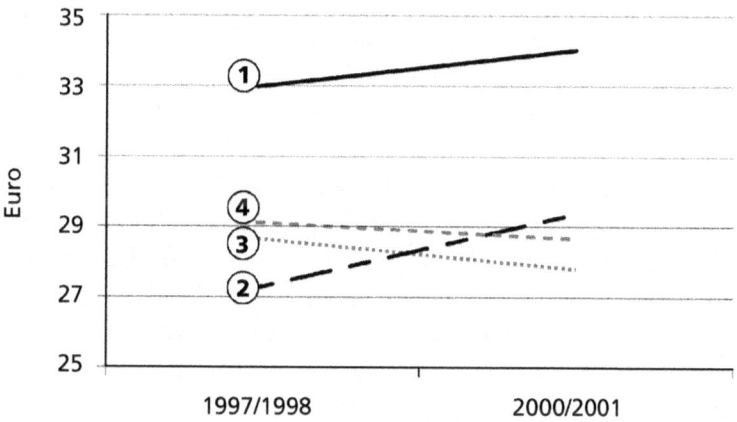

Figure 7.7 *Development of margins per 100kg of milk (in Euros) for several groups*

Source: adapted from Ploeg et al (2003); main findings were later confirmed by Groot et al (2006a) and Reijs (2007)

As a whole, this process of change centres on the reactivation of ecological capital, the regrounding of farming upon it, and the simultaneous strengthening of the local as a self-organizing space (Friedmann, 2006). A remarkable feature of the same process is that it has throughout been the object of sharp criticism (and denigration) by the prevailing agricultural expert systems – and by directors of the large-scale dairy industry (Friesland Foods), as well as by leaders of the Dutch Farmers' Union (LTO). Their opposition can be understood in so far as the transition achieved in the field laboratory was a kind of practical critique of the scientific models that the experts had developed to inform and help shape agro-environmental state policies. The slowly consolidating novelties showed *de facto* that several of the generically imposed rules and prescriptions were not needed (i.e. things could be done differently). Through a careful rebalancing of farming, identical if not superior results could be achieved. This made it possible to go beyond the contradiction between 'environment' and 'profitable farming' that expert systems had introduced into the core of national agro-environmental policies (Ploeg, 2003a).

Agricultural experts also saw the results obtained by the NFW approach as a threat since it implied that expert systems did not automatically have a monopoly on knowledge, truth and the best solutions (a monopoly that is crucial for their institutional reproduction, even in economic terms). There is, in the cosmos of these expert systems, simply one scientific truth: the one produced by themselves. Consequently, the epitaph commonly used to characterize their opponents (farmers and associated scientists) is that they are 'unscientific'.

Underlying all such fuzz there is probably a more decisive feature – namely, that through its relatively successful novelty production, the NFW and the other groups that followed their lead have tended to *escape* from the controls imposed by the state, expert systems and the farmers' unions. This escape, of course, is not meant to be an opportunity for augmenting levels of pollution. Because environmental progress translates into economic gains (as implied by the new balance), farmers themselves will secure effective control and avoid 'leakages'. Furthermore, the territorial co-operative as a whole has an interest in and mechanisms for securing a satisfactory use of the new potentials. The point is, however, that centralized control (i.e. an imperial type of control) becomes superfluous if not impossible if it has to deal with many different localities. This also explains the sometimes fierce opposition from the farmers' unions: a breakdown of imperial control would necessarily mean many self-regulating localities, thus reducing the relevance of centralized bargaining.

Although the technicalities of the different bodies of critique have been amply discussed elsewhere (e.g. Ploeg et al, 2006, 2007; Groot et al, 2007b), it remains important to summarize here some of the reasons that explain why most expert systems find it difficult to deal with novelties. From a methodological point of view, this is due to the fact that complex constellations (such as those shown in Figure 7.6) are mostly divided into isolated segments (the cow, manure, land, etc.) and consequently studied in isolation (preferably by means of controlled experiments). The scientific division of labour strengthens this tendency. Thus, only partial relations will be encountered, while more complex interactions – for instance, of the *multiple conjunctural causation* type (Ragin, 1989) – will remain unexplored (and therefore declared non-existent). What is equally relevant in this respect is that much research proceeds through the construction of averages. Deviations are filtered out by definition: statistically, they simply represent noise. The fact that some of these deviations might entail the promise of new ways forward is neglected. Currently, this situation has become even more worrisome since empirical research is increasingly replaced by modelling. Modelling implies the introduction of *already known and accepted general rules* into (the understanding of) every locality and then the prospect of re-patterning the local according to the general rules and algorithms entailed in the model. Thus, deviations are ruled out even in practice itself.[11]

From a theoretical point of view, the problem is that within expert systems, resources (whether they are land, manure, cattle, labour force or machinery for automated feeding) are essentially conceptualized as *things on their own*. Thus, it is assumed that every resource has *immanent* properties and modes of behaviour. These have been explored and unravelled through scientific research and reformulated into laws that are understood as governing the behaviour of each and every resource. Consequently, context no longer matters. It might impact negatively (by blocking full development of the laws), but it cannot contribute in any positive way. The best thing to do is to reorganize the context (i.e. the range of all possible localities) in such a way that it optimally corresponds with the potential entailed in the resources that are central to the system (see also Ploeg, 1993). Thus, the local – as the place where resources are *combined* in specific ways that simultaneously *remould* these resources (i.e. by changing their properties and modes of behaviour) – is annihilated, at least theoretically. Chapter 8 returns to this issue.

An important aspect of novelty production in the NFW co-operative is that it is an unfolding programme in which the first novelties provoke additional ones. Thus, novelties are strung together in a growing wickerwork that often develops in unexpected directions through which single novelties are consolidated. Figure 7.8 shows how the initial novelty – good manure – translates into a range of interconnected novelties. One might consider this web as an unfolding programme that is multiple layered: it involves and reshapes the practice of farming; it constitutes the core of the activities of the co-operative; and, third, it translates into scientific research of the unknown. Is it possible, for instance, to go beyond an autonomous nitrogen delivery capacity of, say, 200kg per hectare per year on sandy soils? Is it possible to monitor the environmental qualities of an area with new smart systems that go beyond the current segmentation and atomization? Under what conditions might self-regulation become a vehicle of transition? These are some of the new questions triggered with the growth of the web that might result in new solutions and generate new novelties.

As indicated in Figure 7.8, good manure became translated, among other things, into an 'outstretched hand'. This is political jargon used in The Netherlands for a major correction to the Manure Law that allows for *local* exceptions to a *global* set of rules imposed on farming (see Chapter 8). Here it is important to note that the web of strung-together novelties extends beyond the (geographical) boundaries of the area of the NFW. It branches off into agrarian policy-making, into science (provoking new designs, such as the 3MG mentioned in Figure 7.8; Sonneveld, 2006), and into changed soil biology 'beneath' the area, modified value flows in the regional economy and an enlarged 'goodwill' for farming.

Figure 7.9 depicts another web relating to the management of nature and landscape, and which focuses on organizational novelties.

Central to this web are the session days, the survey commission and the field guide. These are very familiar, if not old-fashioned, terms used in the countryside. They entail (and at the same time conceal), however, a completely new way of patterning relations between farmers and state apparatuses, while introducing simultaneously new connections between farming and its natural environment. A session day once meant attending a meeting, mostly held in one of the local pubs, where every farmer had to provide required information about his farm to an extension officer who subsequently introduced the data into national statistics. This was known as 'May counting'. Today, it is the other way around. The NFW co-operative has trained a group of farmers' wives in the mechanics of the state programme for nature and landscape management. The women know exactly what the administrative requirements are (as well as the pitfalls and dangers). During a session day, the women sit behind a table and interested farmers come to discuss their farms with them. Together they make the best match between the possibilities, needs and limitations of the farm and the opportunities, conditions and timetables of the programme. In this way, an interface is bridged that would otherwise imply high transaction costs and carry the danger of failure and associated government-imposed penalties. Thus, behind the apparent routine of a session day lie drastically altered relations between state, farmers, landscape and nature.

The same applies to the field guide and the survey commission. The field guide (produced in close cooperation by the NFW, local people knowledgeable in history, landscape and biology, and some NGOs) describes the best way of managing hedgerows, ponds, alder rows and other landscape elements (Boer, 2003). It includes a flexible system for rating well-maintained, beautiful and rich hedgerows, distinguishing them, at the same time, from those less biologically diverse or less cared for. A crucial detail is that the rating is flexi-

Striving for Autonomy at Higher Levels of Aggregation 199

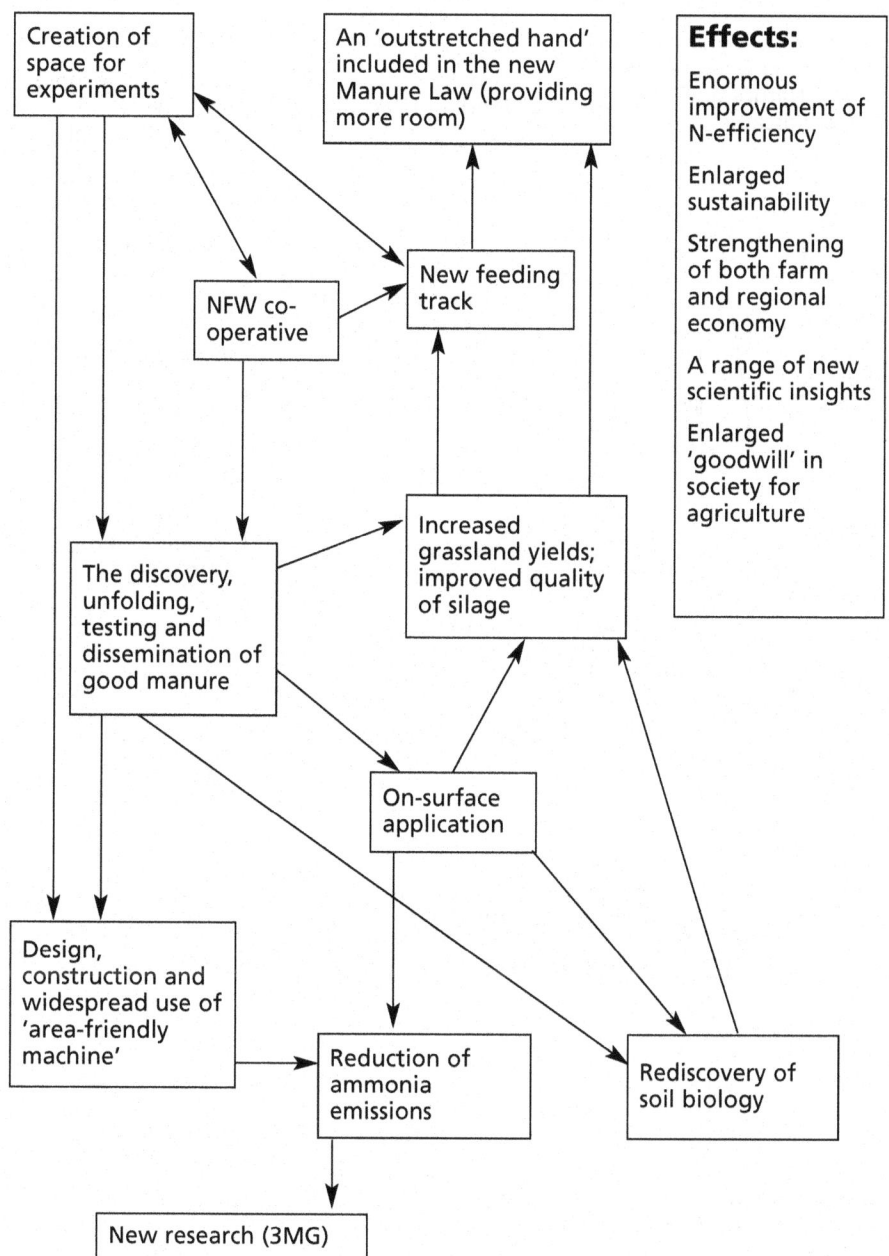

Figure 7.8 *A web of interconnected novelties*
Source: Original material for this book

200 *The New Peasantries*

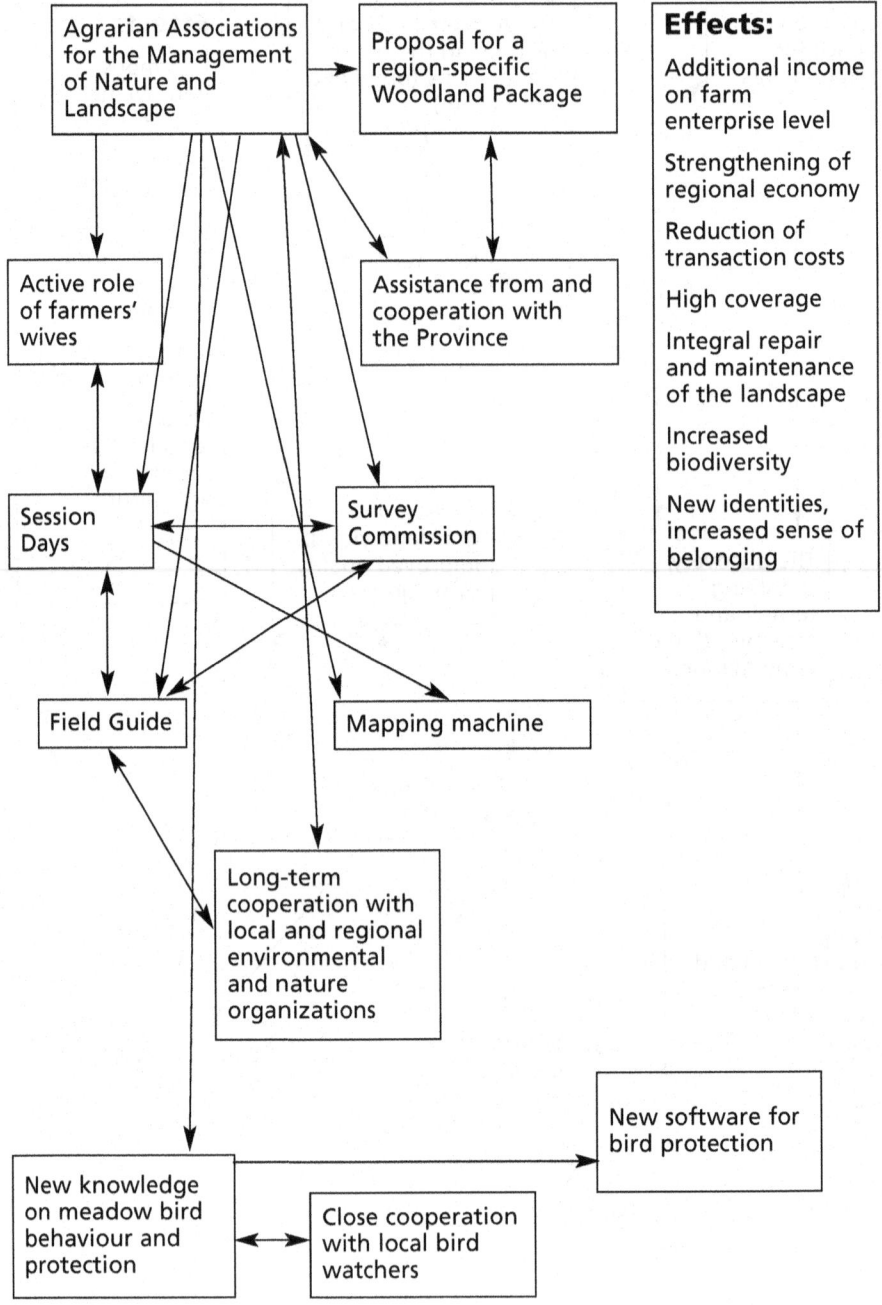

Figure 7.9 *A second web relating to the management of nature and landscape*
Source: Original material for this book

ble. A hedgerow might be beautiful for different reasons since there is a wide range of ways of making a hedgerow beautiful and rich in biodiversity. Thus, the field guide introduces the notion of quality, while also allowing for flexibility. The importance of this seemingly small detail will be highlighted in Chapter 8.

The survey commission (composed of local experts and farmers) regularly visits and inspects farms participating in the nature and landscape programme and, by applying the criteria given in the field guide, comments on the quality of the different landscape elements and perhaps offers some advice. The effect is twofold. The committee introduces the notion of quality in a positive way into the practice of nature and landscape management;[12] at the same time, a kind of countervailing power is created. If the state inspection services observe 'infringements' (as increasingly happens), the farmers can turn to the judgement made by the survey commission to counter the negative assessment. This is especially useful when in court. The core of the web shown in Figure 7.9 (the triangle of session days, field guide and survey commission) has now become a far wider network that also includes new artefacts such as a 'mapping machine'[13] and a try-out of a global positioning system (GPS) registration of bird nests.[14] This has resulted in new insights into and new knowledge of managing the habitat of meadow birds (Swagemakers et al, 2007), and has consolidated working relations with nature conservation organizations and the province.

Dimensions of strategic niche management

Taken together, the many novelties, their careful coordination and their active development in widening webs have had major effects. These are, in the first place, a reversal in interrelations between economy and the environment. Within the NFW, the classically dominant zero-sum game has been converted into a realignment of the two, which results in considerable synergy. A second reversal concerns the change from mutual distrust to negotiated cooperation between farmers and state organizations. The third relates to the change from the single farm to the territory as the unit of operation, thus allowing issues of landscape, biodiversity and environmental quality to be discussed and dealt with *at the required level*. A fourth and last reversal is the cultural reversal among participating people. Where once certain despair reigned, it is now steadfastness (as expressed in commonly shared values), hope and sometimes anger that tend to dominate.

These major reversals could not have been realized if there had not been a keen management of the NFW as a strategic niche (Kemp et al, 1998, 2001; Rip and Kemp, 1998; Hoogma et al, 2002; Moors, 2004). The same applies to the novelties discussed, many of which would not have been developed and dovetailed with others without the protective room provided by the NFW co-operative (Wiskerke, 2002; Roep and Wiskerke, 2004). Figure 7.10 summarizes several dimensions of niche management as it occurred in the case of the NFW. These dimensions underlie the different episodes described in previous sections. Together they describe the multidimensional nature of the linkages between the NFW and the surrounding socio-technical regime (Wiskerke et al, 2003a; see Ventura and Milone, 2005b, for an application of the same dimensions to another case: the rise of the Montefalco district in Italy).

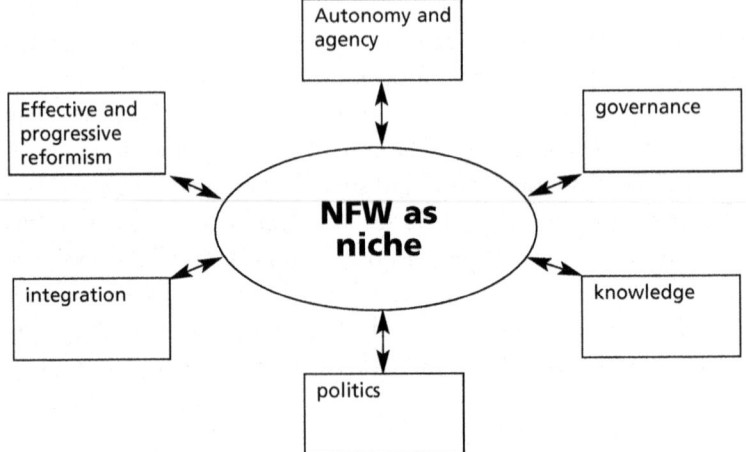

Figure 7.10 *Dimensions of strategic niche management*

Source: Roep et al (2003)

Governance refers here to the capacity to play simultaneously on different chess boards and to coordinate the variously located 'moves' into an adequate and progressively evolving flow over time. Governance is about negotiated development in arenas where different schemes of state regulation interact, as well as about the creation of exemptions for (or ways of dealing with) measures that impact negatively on the local situation. It is also about creating and providing a smooth-running internal organization and effective functioning of technical services for members. Above all, it is the coordination of these different domains that is central to this type of governance. If governance is successful, it results in room to develop and tie together promising novelties, thus producing a double capacity to deliver or, as local discourse has it, 'to do it better' than can be done through the unmediated imposition of regulatory schemes (Eshuis, 2006). Figure 7.10 characterizes this as *effective and progressive reformism*,[15] which refers not only to intentions, possibilities and projections, but above all to newly induced *practices* (hence, reforms) and to the associated *results and outcomes* evaluated as superior to those normally realized (hence, effective). Effective reformism refers to the capacity to get things done: it results in a positive record that strengthens the coalitions which are strategic for governance and the politics associated with it.

Integration refers to the need to glue different activities together in a seamless pattern. It implies going beyond the many dissimilarities and discontinuities entailed in generic and segmented regulatory schemes of the central state. Integration can also occur within a wider network by coordinating local activities, for example, in such a way that they fit into provincial programmes. Together, integration, effective reformism and good governance are appealing. This appeal resulted in an early visit to the NFW by the crown prince and the award of a prestigious prize for innovation presented by the Ministry of Spatial Planning and Environment. In their turn, such symbols greatly help to strengthen governance, and this is how synergy functions.

Knowledge is another crucial dimension. In a society that considers itself to be knowledge based, it is increasingly the case that only those things 'proven' to function well are allowed. Returning for the last time to 'good manure', it should be noted that the main argument against it was that it 'had not been proven'. Thus, a timely construction of new knowledge (or at least the timely design of appropriate research) becomes crucial, not only in the arena where the NFW meets state apparatuses, but also for participating farmers. As indicated earlier, novelties must be 'unpacked' to be understood and further developed. And the more these novelties cross the borderline that separates the known from the unknown, the more attractive they become for scientists.[16] Niches such as the NFW are the places where 'the music is being played'. On the other hand, it is also the case that crossing these borderlines often turns this dimension into a 'battlefield of knowledge' (Long and Long, 1992).[17]

In this context, *politics* refers to the capacity to involve, engage, mobilize and use the support of 'others' in order to create, defend and expand the room needed in which to manoeuvre. Its creation and maintenance (i.e. the creation of a strategic niche) has been far from easy, and has in no way been a smooth and uni-linear process towards expanded self-regulation. In retrospect, what was decisive – from the formation of the first nuclei onwards – was that the NFW could repeatedly involve the Standing Committee on Agriculture of the Dutch parliament in order to correct Ministry of Agriculture decisions. This committee intervened several times on behalf of the NFW. Good relations have been maintained with members of parliament and a broad range of political parties. This goes back, in part, to the fact that several of them had been invited to the area, had played a role in internal discussions and institutional steps that had been taken. The very existence of the NFW represents an important point of reference for the parliamentarians to critically examine the overall development of policy proposals emanating from the ministry – especially as the NFW represents how 'things can be done better'.

The link with parliament is also strengthened through the strong interconnections between the NFW and various other local and regional entities and political bodies, including the province of Friesland, the Friesian Nature Movement, the *Fryske Gea* and Landscape Management Friesland. Due to this wide support, the co-operative emerges as a mediating point *par excellence* – and that is what politicians need. Political support is time and again of strategic importance. The development of the NFW has frequently been threatened, if not blocked. It was only through the mobilization of a large support network (in which members of parliament occupied a pivotal position) that these blockades could be lifted in time.

A final dimension to be briefly touched upon here is that of *autonomy and agency* (see also Wijffels, 2004) or, in other words, how the *peasant condition* (see Chapter 2) *is translated to a higher level of aggregation*. The more effective the actions taken with regard to the above dimensions, the more autonomy and effective agency are likely to result; and, in their turn, autonomy and agency can be expected to feed back to generate more effective reformism, more knowledge and better governance.

Design principles

The activities that occur along the different dimensions are deviations from established routines and, hence, require new design principles (Ostrom, 1990, 1992). The latter are mostly derived from local cultural repertoires and the experiences acquired over time by the co-operative itself. Let me now try to make explicit some of these design principles, which are the informal rules through which *agency* expresses itself.

A first design principle is the *exploration of relevant heterogeneity* in the territory. Within the NFW, people are convinced that many improvements are already available, albeit hidden – hence, there is no need to 'reinvent' them. What matters is to find, unwrap, test and combine them. This principle was very important in the case of good manure, but equally so with regard to the construction of 'area-friendly' machinery for slurry application. Here the relevant heterogeneity included Germany, from where a specially engineered pump was obtained. The exploration might also embrace history, as occurred with the reactivation of the nearly forgotten 'session days'.[18] A second principle centres on the creation, use and development of *autonomy*. It is considered important to move forward on the basis of already available resources and assets (both material and social) and, wherever possible, to avoid new patterns of dependency. Third, changes should be structured in such a way as to render *practical benefits*; and a fourth principle stresses the need to *combine*, in an intelligent way, the many available sources for change, whether they are novelties, local resources or actors (see Geels, 2002, who proposes what seems like an identical principle, although within a completely different context).

An intriguing aspect of the four design principles mentioned so far is that they are, from a theoretical point of view, very much *peasant* principles: modest but, in the end, powerful, especially when activated in a place where there is some space for manoeuvre. They are principles that are associated with the peasant development of single units of production. However, here they are made to operate on another level: that of the co-operative and its specific trajectory aimed at strengthening and transforming regional agriculture.

In the patterning of the interrelations between the NFW and its institutional environment, one might discern some additional design principles. These are based, in the fifth place, on the awareness that it is useless to fight an impossible battle – that is, it is far better to engage in *collateral moves*. If, for example, civil servants can't be convinced of the need for certain changes, then it is more effective to nurture relations with parliamentarians instead.

A sixth principle stresses the importance of *using shifting interfaces* within the state apparatus. Thus, if it is not possible to change a particular regulatory scheme at odds with local conditions, then it is better not to waste energy trying, as the Dutch saying goes, to 'kick a dead horse into the future'. However, in the rare moments when such regulatory schemes are to be redefined or decentralized – perhaps from the national towards the provincial level – there may be opportunities to introduce changes. Anticipation, or 'knowing where the hares are running', becomes strategic.

A seventh principle concerns the *nature of the created interrelations*. They are likely to be flexible and movable, and therefore allow for heterogeneity and further unfolding. The field guide, for example, does not specify one particular hedgerow model to be applied to all existing hedgerows. Instead, it defines a wider range of objectives and possible interventions that can be applied and interpreted in specific localities: a mechanical application is neither possible nor desirable. Thus, a flexible whole emerges that allows for different hedgerows that together compose a beautiful landscape. Here it is also important that remuneration is coupled with quality, and that further development is possible and immediately stimulated.

In itself, this might all appear self-evident. However, as Chapter 8 will show, flexibility, heterogeneity, further unfolding and an active role for the farmers involved are *far from evident* in a world increasingly structured according to Empire. Important relations and connections are becoming more and more 'frozen' as they are squeezed into a single model for regulations that straitjacket the many practices encountered in real life. Expert systems and state bodies play a decisive role in the production and implementation of straitjackets that not only exclude heterogeneity but likewise block development. The creation and subsequent implementation of *fixed interrelations* play, within this context, a strategic role, which I try to unravel in the chapters that follow. However, before entering into this discussion, it may be helpful to underline that such interrelations need not necessarily be formulated as fixed. By means of a careful application of tacit or scientific knowledge (or a combination of the two as we find in the case of the NFW), many relations that are more or less fixed (for whatever reason) can be made *mobile* and, therefore, *movable*. Let me illustrate this important point with two examples from the NFW co-operative.

The construction of movability[19]

Production and landscape ecologists at Wageningen University in The Netherlands have recently developed, in close cooperation with the NFW co-operative and Landscape Management Friesland, a local and highly flexible model that might help to further improve the already created synergy between farming and landscape. I will briefly discuss this model because it shows that modelling *in as far as it departs from and centres on the local* can render results that strongly contrast with the Empire-type of modelling which currently dominates the agrarian scene. An important feature of the model is that it concerns the 'connectivity' entailed in specific landscapes. Here, *connectivity* is taken to be the whole set of connections present in the landscape that make it possible for different species (birds, insects, bats, deer, butterflies, weasels, etc.) to travel through it. Thus, connectivity is one of the important carriers of biodiversity (see Figure 7.11).

The model also relates to farming and landscape management. On the one hand, landscape elements (that together result in a specific mode of connectivity) are the outcome of *historical* land use; on the other, they interact with *current* and highly dynamic forms of land use. Thus, farming can easily enter into contradictions with the given landscape structures (fields may be too small, small corridors between one field and the next are lacking, etc.). However, legal frameworks (at national, provincial and municipal levels) are highly static. They simply forbid the cutting or removing of a hedgerow, or the linking of a pond to a small canal, or whatever. Thus, protection tends to introduce a general standstill. Changes are seen as so many threats to biodiversity. However, with the introduction of connectivity as a central concept, it becomes possible to go beyond this paralysing situation.

A new computerized model called Landscape IMAGES (where the second word is an acronym for 'interactive multi-goal agricultural landscape generation

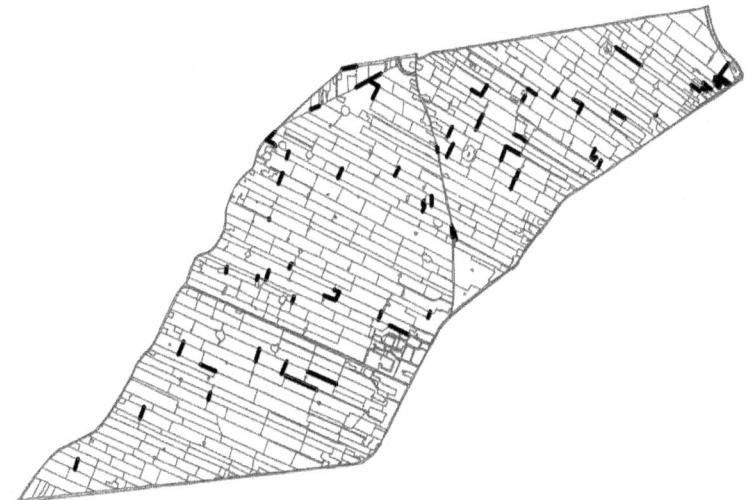

Figure 7.11 *Improved connectivity suggested by Landscape IMAGES*
Source: Groot et al (2007a, p62)

and evaluation system'; see Groot et al, 2007a) is loaded with the specific structure of the local landscape and with the 'travel requirements' of important local species. On the basis of this, one can calculate the loss of connectivity due to a particular intervention; at the same time, it indicates the potential gains associated with compensating measures elsewhere. The latter might include, for example, bridging certain open spaces that are hard to cross for specific species (see bold lines in Figure 7.11). Thus, it is possible to go beyond the classical zero-sum game. Flexibility for farmers is increased while connectivity is improved. In synthesis, the structure of the landscape is made 'movable' in order to improve farming efficiency *as well as* landscape quality and biodiversity.

The same model also includes productive activities on the farm (starting from agronomic data) and the associated economic performance (Groot et al, 2006b, 2007a). Simultaneously, the model allows for the introduction of a broad range of nature-oriented activities on the farm (e.g. less-productive grassland containing a rich flora). Normally the interrelation between the two (farming and nature) is conceptualized in terms of a contradiction – as the *fixed* iso-curve presented in Figure 7.12.

However, with new non-deterministic modelling that uses a mode of ongoing Pareto optimization – in which obtained output is fed as input into the next round of calculations in order to reach new and better solutions – the 'tragedy' contained in the normally assumed iso-curve can be superseded (see Figure 7.13). After repeated calculations (up to 12,000 times) new farm models emerge that improve both nature and economy. Thus, new *moves* become

208 The New Peasantries

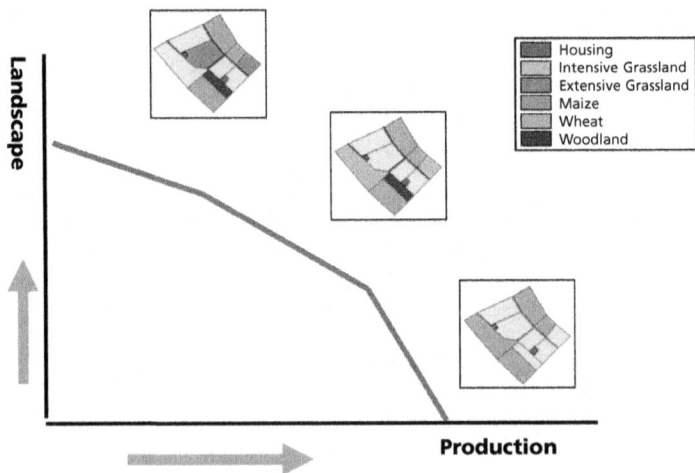

Figure 7.12 *Nature and economy as mutually exclusive categories*
Source: Groot et al (2007a)

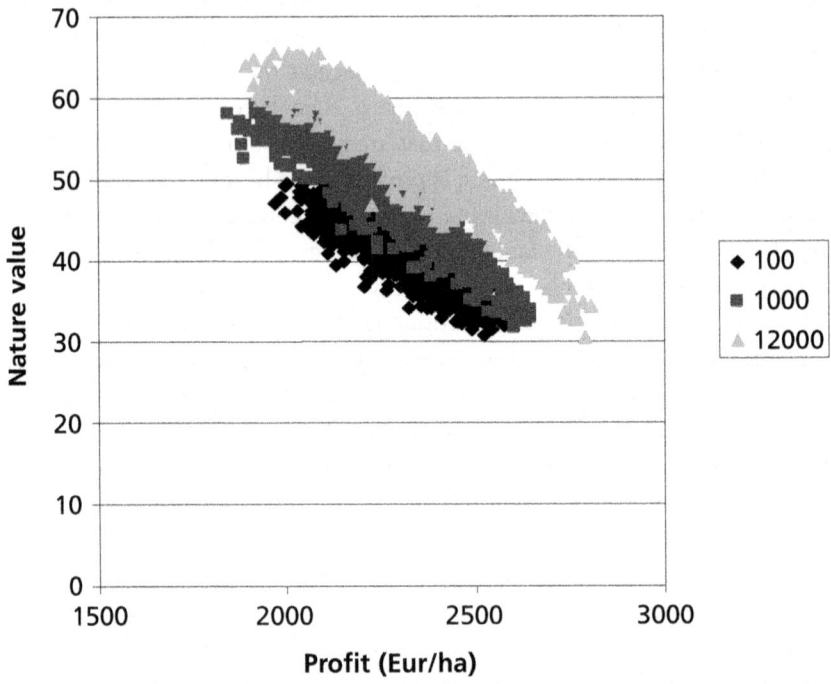

Figure 7.13 *Pareto optimization*
Source: Groot et al (2006b)

possible: moves that will render improved incomes as well as enriched nature. Rephrasing this in more theoretical terms, one may talk about the reintroduction of the co-production principle (see Chapter 5) in farming. Thus, Figure 7.13 shows how the peasantry of the third millennium reshapes farming and nature in a well-integrated way.

8

Tamed Hedgerows, a Global Cow and a 'Bug': The Creation and Demolition of Controllability[1]

Introduction

There are many regulatory schemes currently being used to shape and control agriculture. Here I discuss two that relate to nature conservation and the creation of sustainability (or, more specifically, nitrate levels in groundwater). Both examples are about supranational European Union (EU) policies and their application in a specific locality: the Northern Frisian Woodlands (NFW). The reason for this discussion is twofold: it allows me in Chapter 9 to look at some specific features of Empire and in Chapter 10 to deal in general terms with the central contradiction between Empire and peasantry as contrasting and conflicting modes of ordering.

Control is a central feature of Empire (Colás, 2007). In order to impose control, the subjected world (whether the social or natural world) must be made controllable. Thus, *controllability* is a key word of this chapter. The making of controllability is intimately allied to the Janus-faced nature of science. Science may be a vehicle *par excellence* for the Empire-like conquest of the world. However, science can also be subversive, allowing the order implied by Empire to sometimes crack.

Taming hedgerows

Figure 7.2 of the previous chapter shows the anatomy of a hedgerow as found in the Frisian Woodlands. Through the national programme for the management of nature and landscape (the *Programma Beheer*), these hedgerows are subjected to an extended range of generic requirements. Only when farmers meet these requirements do they get paid for their care (which involves considerable labour) or avoid the penalties imposed for not meeting them.

I focus here on just one of the many requirements: the obligation to protect hedgerows by constructing a fence on both sides of them. These fences must be a series of posts located some distance from the hedgerow, connected by barbed or electric wire. This more or less coincides with current practice since farmers do not want cows damaging them; but they do want the cows also to be able to browse[2] and therefore locate the fence quite close to the hedgerow. So far, so good. The problem, however, is that the hedgerows normally provide an abundance of wild blackberries, which local people like to pick, and which are an important element of biodiversity. Thus, the wires that accompany the hedgerows are often covered, more in some spots than others, by the quickly growing bramble branches. In the late autumn or early winter the farmer cuts them back in readiness for the next spring cycle to begin anew.

However, the presence of these brambles gave rise to a most unbelievable problem, which under normal circumstances would not exist, but in this case made NFW farmers almost decide to give up nature management: the rules forbid bramble branches from hanging over wires. *If the wires are covered, the controlling agent cannot observe whether or not the legally required wire is present or not.* In fact, if brambles cover the wire, an infringement occurs that warrants formal sanctions. And many farmers have actually paid, year after year, substantial fines for this particular infringement. This provoked the question as to whether the ministry wanted farmers to use herbicides on hedgerows and effectively kill all the bramble.

I have joined several delegations of the NFW co-operative in its attempt to change this blatantly absurd prescription, which is at odds with the dynamics of both nature and farming. These delegations have argued in vain for years, in different arenas, at different levels and in different places. And when there was, at last, a change, it was not due to arguments but to shifting interfaces. Today's regulatory systems are characterized by a one-sided permeability. Directives, rules, requirements, procedures and protocols easily pass through, from the top downwards; flows of information, comments and critique that go the other way (i.e. from the local to the national and global level) are, however, almost completely blocked.

Technically speaking, this one-sided flow and its persistence boils down to three things. The first is arrogance and distrust. Civil servants believe themselves to be in charge and do not welcome proposals to change established routines, especially when they are articulated by outsiders. The 'not-invented-here syndrome' only strengthens this attitude. Second, one has to acknowledge that the introduction of change is difficult, precisely because of the way in which we have organized the world. Managing nature (as simple as it might look in the first instance) involves a range of institutional entities. To begin with, there is the policy unit of the ministry that designs the overall lines of nature management: *directie natuur*. Then follows the *directie regelingen*: the unit that is mandated to implement the designed policies. This is followed by yet another level, the Dienst Landelijk Gebied (DLG) and sometimes the

Algemene Inspectie Dienst (AID), which control whether farmers have correctly followed the prescriptions. And finally there are the institutions that evaluate the programmes themselves.

So a first set of connections regulates the translation of policy objectives into required practices, control and evaluation. A second set is related to the financing of the required practices. This involves almost everywhere the principle of co-financing. Part of the funding is from the EU, part is nationally funded, and from 2007 onwards, yet another part is funded regionally. Thus, an institutional density is created that tends to exclude flexibility. Policy formulation, implementation, control, evaluation and the associated financial flows have materialized in different subsystems at supranational, national and regional levels – subsystems that are connected through formal and highly detailed information and communication technology (ICT)-based definitions on what is to be done and how one should proceed. This implies that reconsideration of a specific rule such as the management of hedgerows in the Northern Frisian Woodlands, for example, would necessarily imply *an overhaul* of all other implied systems. Adapted rules for control have to be matched with adapted rules that specify objectives and modes of regulation. This entails adapting the rules for evaluation and at the same time getting the adaptation accepted by 'Brussels'. It must also be accepted, operationalized and translated at regional level. The transaction costs could be enormous.[3] Consequently, the system as a whole tends towards rigidity.

Through this multiple and systematized interconnectedness, everyday life, such as caring for hedgerows, becomes a true Gordian knot: it can no longer be unravelled. Simultaneously, this knot turns into a nightmare for the farmers participating in the programme for nature conservation. The civil servants defining the new programme do a good job. Those who implement it equally provide an adequate service. And those who have to monitor and control it must make sure they avoid wasting public money. Everybody is doing his or her best – the final result, however, is havoc.

After the many disappointments suffered during the attempts to improve this situation of the badly judged blackberry shrubs, the NFW co-operative decided to take the matter to court.[4] This move, and the strong critique provided by the regional deputy for rural areas, Anita Andriesen, ended by finding a typical Dutch solution: the deviant behaviour of the hedgerows was to be *tolerated*. Charges for infringements would no longer be enforced. To arrive at this point had taken seven years! This is another relevant feature of Empire-like patterning: even the simple things of life are transformed into highly complicated operations.

In the years to come, nature management will be decentralized from the state to the provinces. Thus, the NFW co-operative has been invited by provincial authorities to prepare a 'Woodland Package' for nature management in the area. This new package is to reflect the peculiarities of both nature and farming in the Frisian woodlands. It will also allow for the introduction of quality, differentiation, flexibility and adequate remuneration.[5] The field guide and survey commission will probably play an important role in this new package. None of this implies, at least in this realm, that self-regulation has been definitely established. The basic grammar (of the new rules now to be developed by the NFW co-operative itself) is still centrally defined. The pivotal centre of such a grammar was once the additional costs and foregone production; but now it is the implied labour that is central. This provokes new contradictions since the most valuable places in terms of scenic beauty and biodiversity are obviously those that should be touched as little as possible. According to the new grammar, these most valuable places are those that will receive very little or hardly any payment. Thus, ongoing struggles will certainly continue to accompany nature management activities.

The global cow[6]

There is seemingly nothing more local than cows that have been bred to fit local pasture lands. The same applies to fields that have been developed, over time, to meet the nutritional needs of these cattle. The 'only problem' is that such balances (and the practices through which they are created) cannot readily be controlled by outside agencies. This is not only because of the autonomy of the farmers involved but also due to the large *heterogeneity* of the created balances. Ironically, this 'only problem' turns out to be the *major* problem within the context of current agro-environmental schemes, especially since the latter are increasingly moving in an Empire-like direction. Empire requires controllability. Thus, the well-balanced unity of fields, farmer and cattle emerges as an iniquity – regardless of the level of sustainability it achieves.

The resulting impasse is resolved in an Empire-like way by completely moving away from local specificity and the balances entailed. Through complex modelling, a 'global cow' has been constructed that specifies the so-called nitrogen-excretion per cow. However, this cow is an abstraction. It is a virtual or a *global cow* – and similarly the calculated level of nitrogen excretion is merely an *average*, a *global standard*, that will more often than not *deviate* from concrete situations. Nonetheless, it is this *global cow* that emerges as the main instrument for control. Having assessed a maximum level per hectare for nitrogen from cattle, control comes down to simply counting the number of animals in order to know whether a farmer is within or beyond the centrally defined levels of sustainability. Through computerized data systems, the required control can thus be realized from one remote locus of control.

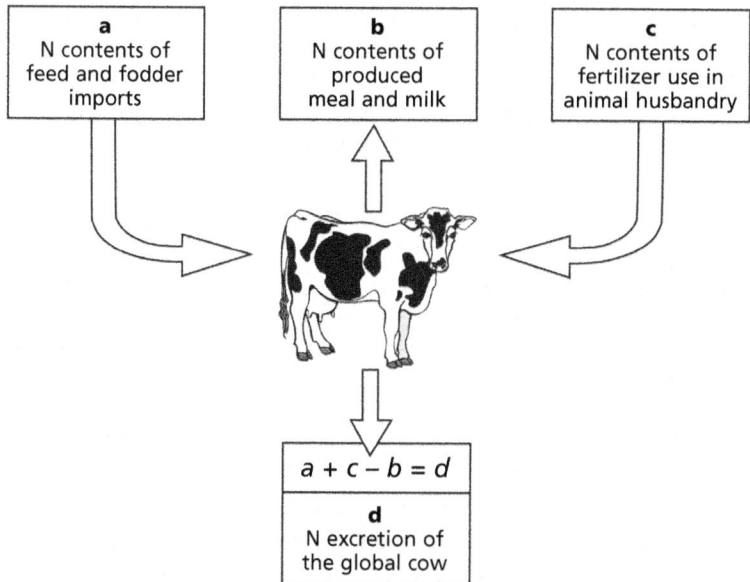

Figure 8.1 *Calculating the nitrogen excretion of the 'global cow'*
Source: Original material for this book

Figure 8.1 summarizes the global cow. For the country as a whole, all flows of feed and fodder – these are often, indeed, global flows – are taken together and converted into the corresponding amount of nitrogen. The same goes for grassland and maize production in so far as it is located within The Netherlands: it equally represents a specific nitrogen flow from the fields towards the cattle. This flow is heavily associated with fertilizer use. On the other hand, there are two outflows. The first relates to milk and meat production. Both products have well-known nitrogen contents, which allows for a specification of this flow in terms of nitrogen. Thus, a specific amount of nitrogen is imported into the national herd and a specific amount is exported. The remainder[7] then is contained in the last flow: the manure produced. When divided by the total number of cattle, this results in an average nitrogen excretion per cow of some 114kg of nitrogen per year.[8] Subsequently, a 'global algorithm' was developed (see Box 8.1) that further specifies the global cow: it differentiates the average nitrogen excretion according to milk yield and urea content of the milk. This latter parameter is somewhat problematic. It is not related in an unambiguous way to the nitrogen content of slurry. However, it was *easily available* (just as milk yield) since it is registered in the dairy factories. Even the scientists developing this model were not very happy with it. Controllability, however, took priority over precision. Furthermore, using the urea content of the milk as a parameter is also problematic because it is very difficult for farmers to get and keep it at a desired level, especially when the herds are pasturing outdoors. Thus, a wet summer can heavily penalize farmers.

Box 8.1 *The 'global algorithm'*

$N = 0.95 \times 0.8825(136.7 + 0.0094(m - 7482) + 1.8(u - 26))$

where N = nitrogen excretion (kg per cow per year)
   m = milk yield (kg per cow per year)
   u = urea (mg per 100g milk)

The global cow and the global algorithm built into it have become cornerstones of the Dutch Manure Law. Knowing that a maximum of 170kg of nitrogen per hectare of animal manure can be distributed over the fields (or 250kg per hectare in the case of derogation), the total 'legal' number of cattle (given the surface) can be calculated. Since the number of cattle, acreage, milk yield and urea content are precisely known at the level of the main databases, control and sanctioning are made easy. If there are too many head of cattle and it cannot be proven that manure is delivered to arable farmers in a legally specified amount and way, the number of cattle must be reduced.

The Manure Law has an interesting *lookalike* aspect. It appears as though the Manure Law definitely resolves the environmental problem in Dutch agriculture – at least the problem of huge nitrogen surpluses. Behind this appearance, though, things might be completely different. In fact, the Manure Law aimed to solve quite a different problem: damaged relations between the European Commission (notably the Directorate-Generals for Environment and Agriculture) and the Dutch Ministry of Agriculture. But the price for solving that particular problem was to be paid by the farmers who tried to reground their process of production on ecological capital, in part by making good manure.

The issue of nitrogen excretion might also be approached in a different (more localized) way. As illustrated in Figure 8.2, *farm-specific* data on the amount of locally produced roughage and its composition (especially in terms of crude protein balance), the *farm-specific* amount of concentrates (and their nitrogen content) and, finally, data on the *farm-specific* characteristics of the conversion of feed and fodder into production might help to assess the real *farm-specific* excretion of nitrogen through slurry (Reijs et al, 2003; Eshuis and Stuiver, 2004; Reijs and Verhoeven, 2006; Reijs, 2007, p195). Such data can easily be obtained within the local situation; however, it is very difficult to manage in data sets used for global control, which is why the making of the Manure Law (in which applied science played a crucial role) did not pass through the local (see Figure 8.2) but followed, instead, the trajectory entailed in the global cow. This specific choice means that *real* excretion may be very different, both in a positive and a negative way, from that calculated using the global algorithm entailed in the global cow approach. Thus the 250kg of nitrogen per hectare level may, in practice, easily be exceeded. Ironically, this actual deviation is masked through the Manure Law.

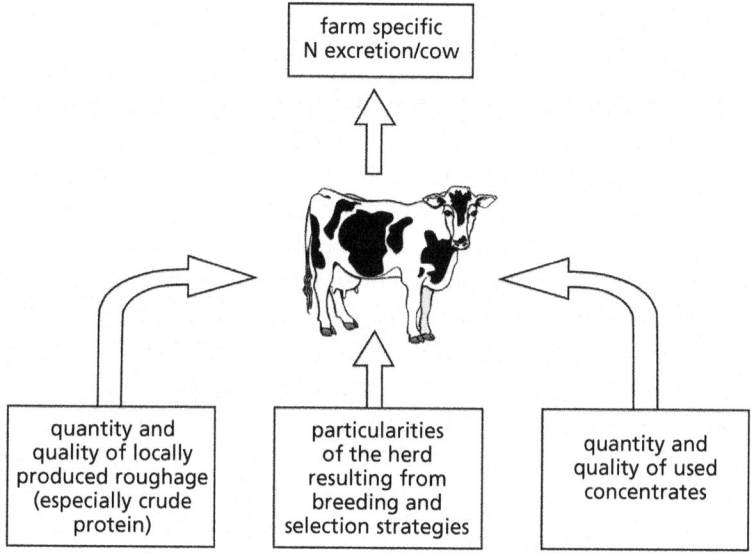

Figure 8.2 *Understanding the local groundings of manure production*
Source: Original material for this book

The imposition of standardized criteria such as those of the global cow produces a range of social and material effects, the combination of which implies a considerable distortion. First, such generic criteria are at odds with the local specificities that occur in a heterogeneous sector. This gives rise to frictions, which often translate into a coarsening or loss of practices that were previously the object of meticulous fine-tuning. Second, it negatively affects and sanctions all those farmers who actively created specific balances entailing high levels of sustainability (e.g. high levels of nitrogen efficiency). Third, the created system for accountability and control spurs an *increased* use of chemical fertilizers, industrial feed and fodder.[9] Thus, while a *virtual* sustainability is suggested, *real* sustainability deteriorates: ways of attaining the latter become lost, as well as the knowledge of how to achieve this. Fourth, any motivation to improve real sustainability is taken away, if not criminalized. The only thing that matters is whether farming is done according to the imposed global rules. Thus, the practice of farming is changed, in a way, into an institutionalized slow-down.[10] Novelty production and so-called disembodied technological change (which crucially depends upon local crafts and skills) are ruled out. It goes without saying that this again is highly detrimental to the creation of extra value added and the quality of work. The farmer is tailored into a Fordist system. Even when the farmer remains independent – in the formal sense of the word – the use of his resources is materially controlled in an Empire-like way. He or she is reduced to being just a passive receiver of the imposed generic sets of rules.

State apparatuses as important ingredients of Empire

In the case of hedgerows and manure, the state typically relates to the peasantry as an expression of Empire. From a range of interconnected 'centres', and through a widely extended and bureaucratically organized network that reaches down to all farms, simple (if not simplistic) norms, targets and yardsticks are introduced into a range of differentiated situations. The same network exerts a control that is defined according to criteria which suit the centre of the web. These aspects are all intrinsic to, and typical of, Empire. Yet, Empire adds little or nothing. It only renders a 'lookalike sustainability' (just as some dairy industries create lookalike milk). And here, also, Empire is not neutral to the practices affected. It produces a highly biased distribution of benefits and costs. Empire also distorts many practices. In several situations, farmers are forced to buy space elsewhere in order to get rid of their manure, even when its use on their own farms could be highly beneficial, including from an environmental point of view. To compensate for the reduced amount of manure, more fertilizer has to be applied. In other situations a careful extensification of production is barred – simply because reducing milk yields would imply a modest increase in the number of cattle, which would run counter to the imposed global norms. Although there are resourceful ways of dealing with the regime requirements while constructing elegant ways forward,[11] the transaction costs to do so turn out to be quite high. For the majority of farmers it is actually impossible to 'deviate'.

Degradation, one of the defining features of Empire, is omnipresent here as well. Alongside the elements already mentioned, degradation emerges from the legal requirement to inject slurry (more or less liquid manure) instead of applying it on the surface. True, surface application is associated with higher levels of ammonia emission; but farmers can easily control this danger, partly by using good manure and applying it only when skies are cloudy and rain is expected. Furthermore, the presence of hedgerows lessens the force of winds and this helps to reduce emission levels. Obviously, such an approach can only be grounded upon *localized* forms of control. The definition of good manure will vary slightly from place to place, and due to meteorological conditions, will change from year to year.[12] Quality control, therefore, can only be done at farm level (by checking the manure itself, its composition, its colour, its smell and through knowing the way in which the cattle have been fed). It cannot be assessed through binoculars from a car or helicopter. But localized control is at odds with Empire. It made the use of large machinery for slurry injection obligatory because it is controllable in a *global* way (Bouma and Sonneveld, 2004).

For most farmers this kind of machinery is far too expensive and they therefore hire contract workers to do the job. This gives rise to several problems. The first is that during the legally permitted periods (especially in early spring when the slurry cellars and basins are completely full), these contract workers operate continuously, regardless of weather conditions and preferably on large open fields. A second problem is that the heavy machinery causes damage to the soil structure and biology, especially in the wet circumstances of early spring. Another is the speed at which they need to work. It causes havoc to birdlife and nests in the meadows. Thus, multiple degradations are triggered. Behind the curtain of virtual sustainability, there is, in short, mainly regression and degradation.

The question of *hierarchical control* needs some specification here as well. Many peasants have been fined and brought to court for surface application of slurry. However, many of them convinced the judges that distributing improved manure, especially on cloudy days, was not harmful at all for local ecosystems or for the environment in general. It often reduced ammonia emissions (as was only much later affirmed by scientific research). Thus, many judges became convinced by the accused peasants. Although formally obliged to convict the defendants (who had, indeed, broken the law), the judges increasingly decided to refrain from prosecution. This was their way of protesting a law they considered to be unjust (at least in particular situations), unworkable and hard to enforce. However, after several years, this modest form of 'civil protest' was annihilated: the Ministry of Agriculture issued a formal instruction to all judges that ruled out any possibility of refraining from imposing sanctions on 'illegal' applications of slurry.

There are several connections between regulatory systems in agriculture such as manure policy, on the one hand, and the differentiated dynamics of peasant and entrepreneurial farming, on the other. In the entrepreneurial constellation, the business of farming has been disconnected, albeit in variable degrees, from available ecological capital. Milk production is disconnected to a considerable extent from feed and fodder production within the farm since much of it is bought in. Grassland production basically depends upon the use of fertilizer, which is again largely disconnected from manure produced on the farm. Thus, essential cycles that once moulded farming into one organic whole are segmented, and the resulting segments (milk production, fodder, manure use, etc.) are increasingly *standardized* on the basis of externally delivered artificial growth factors. Consequently, it is relatively easy to integrate *global* norms and *global* control methods into such standardized farming practices. They simply represent *further* standardization. In peasant farming, which is based far more on the use of ecological capital and which is therefore difficult to standardize, the imposition of such norms and methods not only creates more problems, it can also *paralyse* its intrinsic dynamics (i.e. the continuous search for improved use of available natural resources). This is especially clear when it comes to the most obvious response to imposed regulations. In entrepreneurial farms, this centres on a further increase in milk yields (by applying more industrial feed and through intensified use of fertilizer in grassland production – in short, by 'injecting' more nitrogen into the production system). As a consequence, the number of cattle can be reduced in order to create a fit with the new regulatory scheme. It is obvious that such a response is at odds with the rationale of peasant production. In Chapter 5, I discussed how, through the dynamics of the entrepreneurial mode of farming, the external squeeze on agriculture is transformed into an internal squeeze, which affects peasants, in particular, in a detrimental way. Here we are witnessing the same kind of mechanics. The evolving regulatory systems function as an administrative squeeze that is being tightened: indeed, they compose a 'regulatory treadmill' (Ward, 1993; Marsden, 1998) that particularly affects the peasant mode of farming.

Science as a Janus-faced phenomenon

Science relates in a double way to the creation of such a regulatory treadmill. For many domains of the natural and social worlds, science translates the regular and the similar into 'laws' that are assumed to explain and represent the behaviour of these worlds (the algorithm entailed in Box 8.1 is a perfect illustration of such a law). Knowing such laws allows one to intervene in and govern these domains. To do so the laws must be rebuilt into technologies and schemes of governance (Koningsveld, 1987; Ploeg, 1987a).

There is an almost 'natural' co-evolution of this side of science and Empire. Science constructs the patterns of regularity (the laws or sets of generalized rules) required for the latter to develop and materialize as a mode of ordering. At the same time Empire increasingly standardizes the world, among other things, through compressing the relevance of the local and especially its capacity to produce novelties (read 'deviations') of whatever kind. Thus, path dependency (North, 1990) emerges that feeds back as 'inverted path dependency' into science itself. Science mainly studies what is thought to be possible and relevant, while it simultaneously avoids enquiry into the 'impossible' and the 'irrelevant' (like 'good manure' and the other novelties discussed in Chapter 7). The 'horizon of relevance' used within science increasingly coincides with the order imposed by Empire. Thus, by studying mainly or exclusively the 'relevant' (and ignoring the 'irrelevant') science strongly contributes – even without knowing or acknowledging it – to the Empire-like patterning of the world.

Science is, of course, Janus faced in that alongside its focus on the regular and similar, it *also* focuses on the exceptional, the dissimilar and the seemingly impossible. In fact, it even *produces* these in its own 'locales'. This part of science is curiosity driven: it tries to identify and understand the *potentials* hidden in reality, to examine the exception and the novelty, and to uncover the extraordinary that is sometimes enclosed and concealed within it. While recognizing that there are regularities – sometimes so persistent that one tends to perceive them as 'laws' – this second side of science also claims that such regularities can 'move' and/or become the object of complex transitions that will re-pattern them (Ploeg, 2003a, pp145–224; Ploeg et al, 2004b). Regularities may change and rules be redefined precisely because, in many localities, local actors try to go beyond the 'platforms of commonalities' (Hofstee, 1985b). This, in short, is why and how deviations emerge that might potentially prove superior. Of course, local deviations do not necessarily entail potential superiority. That would perhaps be the exception. And even then it would require an adequate and convincing framework for assessing its relevance. It must then be proved to be superior in practice; finally, through and after its unpacking, it would need to be translated into new (and probably modified) general terms.

Thus, attention to deviations, the unexpected and the local is (or should be) an indispensable part of science. It is the daring and risky part that will always have a somewhat uncomfortable relation with the settled, slow and self-confident aspect of science that exists out of accumulated knowledge of regularities and similarities (see Rip, 2006, on contingent and rational repertoires in science). The second side of the Janus face will contest the first, while the latter will oblige the former to go beyond experiments only. Together they make science move, or not. The balance between the two sides is decisive, just as the struggles over what is accepted as a legitimate 'experiment' are decisive (Ploeg et al, 2006). If social domains are increasingly materially and symbolically ordered around generalized rules explained by science, then the questioning of such rules becomes a nasty and irritating, if not provocative and potentially dangerous, activity. The balance between the two opposite poles entailed in science might further shift when science becomes further 'embedded'[13] through mechanisms that make it dependent upon external funding and through university boards that are dominated by industrial and state interests.

Increasing parts of the social and natural world are *de facto* governed by models developed in expert systems: they are models that define particular domains as they are assumed to be and as they are assumed to function. These models might be loaded with empirical data; they might equally be theoretical constructions that are void of empirical references – that is, they are about things as they ought to be. Mostly, however, models used to govern particular parts of the world will be a combination of theoretical construction and empirical input. Nature for instance, and consequently the range of activities that produce (and reproduce) it, is defined, at least in The Netherlands, by means of 'nature-type objectives' (*natuurdoeltypen*). The definition of a hedgerow is just one of a long list. Meadows are defined, in part, through an assumed 'nitrogen delivery capacity' (see Figure 8.3, which indicates that sandy soils, for example, cannot deliver more than a *fixed* amount of 200kg of nitrogen per hectare per year).[14] Cows are, in turn, partly defined by the global cow that specifies the nitrogen content of the manure. And so on and so forth.

Regardless of what is being specified, these models always have a nomological structure (of the if–then type). Desired outcomes are related to a neatly delineated set of conditions. If the latter occur (or are created through specific interventions), then the desired results will emerge. These models establish, in short, uni-linear cause–effect relations between means and ends and might therefore be used to subordinate and control growing parts of the world.

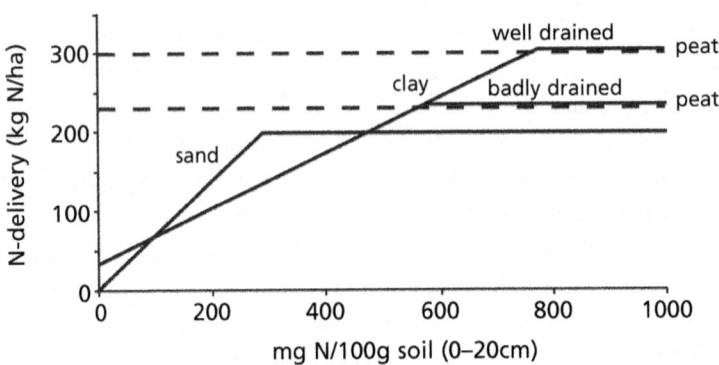

Figure 8.3 *Nitrogen delivery of sand, clay and peat soils*

Source: Hassink (1996)

Generally speaking, there are several problems related to the current generation of models. These problems are not necessarily intrinsic to the models as such, but can emerge when the models are applied to highly complex and dynamic domains that contain human and/or non-human actors (e.g. meadows, hedgerows and cows) that may counteract them in often unexpected ways.

A model might describe fairly well the *average* or the *desired* situation within a respective domain, but it will always be difficult to integrate within it all the relevant heterogeneity existing in the respective domain. For example, in reality, cows and meadows are the specific outcome of processes of co-production through which they have been made into what they are. Some meadows will have, due to the history of land use, an elevated nitrogen delivery capacity, others a far lower one (Figure 8.4 shows *empirical* levels of nitrogen delivery capacity in the NFW area; see also Sonneveld, 2004). The implication of real-life heterogeneity is that models that are correct in as far as the average situation is concerned might misrepresent specific local situations. Thus, the latter emerge as deviations from the rule. Compared to the richness of real life, a general model pretending to represent it might very well signify an impoverishment.

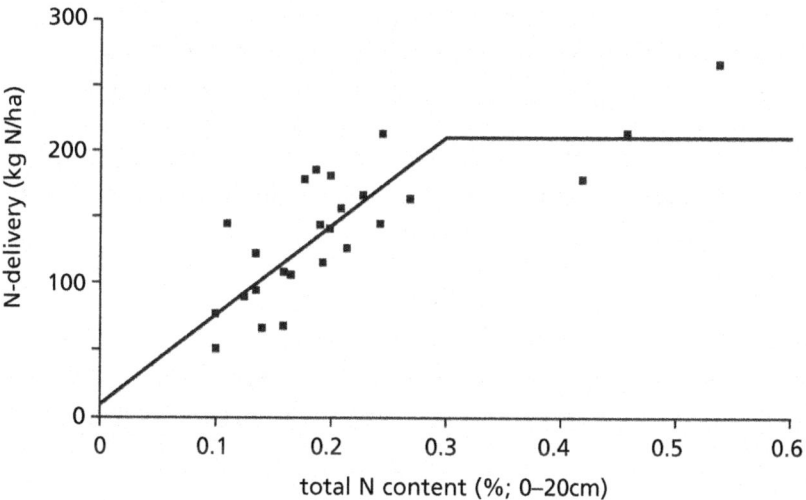

Figure 8.4 *Nitrogen delivery of the soil (empirical observations)*[15]
Source: Eshuis et al (2001, p90)

The second and probably decisive problem emerges when the general model is subsequently used to order and govern real life (whether meadows, cows, landscapes, etc.) and when deviations must then materially be avoided or redressed. The production and reproduction of the respective objects (e.g. the making and maintenance of a hedgerow) has to follow the formalized rules; the objects must contain and show the characteristics as defined in the model. This implies that the particular domain is standardized, regardless of whether this is intended or not. What initially was *assumed* to be the case increasingly *materializes* as a general feature of reality. Deviations are materially wiped out through rigid schemes for control[16] and thus the world becomes standardized.

The third problem derives from the particular relations between desired results and the means that are assumed to be effective in creating them. When the models are transferred to real-life domains that are to be controlled – in other words, when they are converted into a technology for control – the identified means automatically become central (if not exclusive). Theoretically, there might be other, equally effective or even superior, means; but if they are not included in the model, they will not be recognized as relevant: they are considered illegitimate (clearly an effect, mostly unintended, of the initial impoverishment – that is, the negation of heterogeneity). There may, for example, be many ways of distributing manure in a 'low-emission' way (this is the term used in the Legal Decision Regarding the Use of Manure); but only those ways or means specified in the model will be accepted.[17] The 'translation' goes further: the means must also be *controllable*. Take, for instance, the objective of having nitrate levels in the groundwater equal to or lower than 50 milligrams per litre of water. This is evidently impossible to measure everywhere. Thus, the focus shifts to the means of doing so: no more than 170kg (or, after derogation, no more than 250kg) of nitrogen entailed in animal excretion might be applied per hectare.[18] As discussed earlier, such levels are translated, through a complex model, into a maximum number of animals per hectare. Since this is basically a sophisticated way of 'counting tails', it is controllable. Through such an operation, animal husbandry as a whole is *materially reassembled*: it has to operate within the room defined and imposed by the new means.

Fourth, it is important to note that this reassembly goes far beyond what is indicated by the initial model. The model is built on, and for, an artificial segment cut out of a far wider reality. Yet, when reintroducing the model into the reality of real life, many unexpected consequences occur. Re-patterning the single relation between herd and acreage will affect the economy of many farms; it will create a new market for slurry and new brokers ('slurry traders') will enter the scene, and land will be rented or bought elsewhere in order to make for a new balance.[19] The same occurs in the case of the legally prescribed technology for slurry injection. As summarized in Figure 8.5, the use of the prescribed technology produces a widening range of additional effects that relate to the soil, the organization of the labour process, the coordination of weather and farming, energy use, meadow bird populations, and so forth. Thus, growing parts of the social and natural world are re-patterned.

A fifth problem results from the mechanics of control. In order to avoid high costs, control necessarily materializes as *control at a distance*. The prescribed practices are made controllable through digital registration that allows for centralized control or through monitoring by using helicopters. Controllability takes priority over effectiveness. The presence of a slurry injector in the field is what matters, like the visibility of the prescribed barbed wire

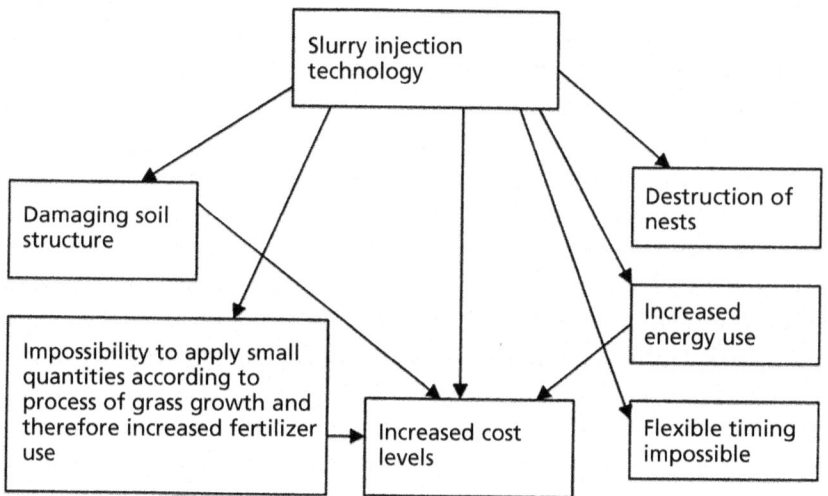

Figure 8.5 *The wider effects of legally prescribed slurry injection*
Source: Original material for this book

along the hedgerow. It is 'representation' in line with the rules that matters. However, behind a highly visible fence there could well be an ugly hedgerow. The highly visible injector machine might be operating with its knives raised and thus simply be distributing slurry on the surface (saving the operator costs in petrol and time); but this is not visible from a distance. And the required animal–land ratio might be constructed through administrative operations that add, on paper, additional land to the farm. All of this is precisely what occurs; it is triggered by the imposed models and control mechanisms. The irony is that the same models and mechanisms ban highly effective alternative solutions. Take, for example, the 'duo spray': an ingenious machine that produces, indeed, two sprays. A first flow consists of slurry and another spray, located above the first, sprays water. The effect is that the slurry is literally 'rained' into the soil. Technical measurements have shown that this device is highly effective. Nonetheless, it has not been accepted by the Ministry of Agriculture because 'the controlling civil servants cannot check, from a distance, whether the two flows are properly balanced'. *Controllability clearly dominates over compliance and effectiveness.* But in a lookalike world, this apparently no longer matters.

A sixth problem is that once the models or rule sets are introduced and made to work, it is very difficult and expensive to change them. This was already shown for the rules defining hedgerows. It equally applies to global cows and to manure application through injection. Apart from the already mentioned institutional interlocking, this rigidity is grounded in the fear that changes might trigger financial claims (one reason to stick to injection is that lifting the legal obligation will provoke a range of claims from contract workers who have invested heavily in the required machinery). It also resides in the boundaries that separate knowledge from ignorance and which have indirectly shifted. Currently available models are, as it were, a digitalized condensation of knowledge. This applies to the global cow, to models that describe emission reductions through injection, to grassland production models and to nature-type objectives that specify a hedgerow. The worrying aspect, however, is that *behind* this formalization two other processes are at work that for the moment go unnoticed. The first is the fact that in the domains mentioned, there are few actors remaining who are knowledgeable. There are only a few experienced grassland specialists, for instance, and there are no new specialists being trained because there is no need for them *for it is the model that now functions as the expert*. Second, material and social realities naturally change and evolve over time. The cattle of today are not identical to those of, say, the 1980s (and the same goes for meadows, for farmers, etc.). However, several of the interrelations on which current models are based are actually derived from empirical studies (e.g. conversion of feed and fodder into production) conducted during the 1980s. Thus, in the end, we have highly sophisticated models and control systems that are increasingly unrealistic. Indeed, following Bray (1986), it could be argued that these sophisticated models and related control systems constitute together a *mechanical* (and therefore poorly functioning) *technology* that tends to become central in the ordering and governing of the social and natural world as a whole. *Skill-oriented* solutions that are potentially far more effective are blocked.

Today, Empire as a mode of ordering and control does not only reside in large corporations of the Parmalat type. Nor is it only related to large flows of capital that browse the world for additional profits, nor with expanding and aggressive states. Expert systems and (applied) science equally constitute important sources of Empire in so far as they create the models, means and related control systems that are used to influence increasing segments of the world in which we live. They create the systems that are typical as well as strategic for Empire; they create systems that relate *as Empire* to agriculture, food production, nature and the countryside. Before elaborating upon this further, let me first touch on the other side of the Janus face of science.

The creation of a bug

When the first version of the Manure Law was debated in parliament, a large group of farmers signed a declaration which argued that the proposed law would exclude their own active (and skill-oriented) role in creating a sustainable agriculture, and that it was detrimental especially to those farmers who were improving the quality of manure (by simultaneously reducing its nitrogen contents). They argued that the overall effect could easily be negative. The declaration, an initiative of the NFW co-operative and widely supported by similar groups and several scientists from Wageningen Agricultural University, acknowledged the need for intervention, but argued that the *generic* or *global* approach of the proposed Manure Law was basically wrong. These arguments were shared by a majority in parliament and thus the minister was obliged to formulate an 'outstretched hand', as it was called: farmers who could *de facto* demonstrate that the level of nitrogen excretion of their herd was lower than assumed in the general formula would obtain the corresponding room to deviate from the Manure Law and the global algorithms built into it.

Thus, the NFW co-operative developed a method of calculating that would reflect real levels of nitrogen excretion and also allow farmers to actively reduce such levels. I will discuss this method at some length: first, because it entails a re-localized mode of ordering that strongly differs from the global one entailed in the Manure Law; second, because I believe that this new way of calculating represents a form of social struggle that will grow considerably in the years to come, especially since it is about 'bugs' created to neutralize menacing global threats; and, third, because the finally created bug may backfire strongly on the entrepreneurial type of farming that was to be protected by the Manure Law.

The bug that was created consisted of a clever combination of four pieces of software. The combination was already present in the particular trajectory towards sustainability developed in the NFW co-operative, but it lay dormant until the first versions of the Manure Law activated it.[20] The first piece of software was the agricultural cycle of soil–crop–animal–manure used in the co-operative to rebalance the process of dairy production in a more sustainable and more rewarding way (summarized in Figure 7.6 in Chapter 7). The second piece was the so-called mineral balance (nutrient accountancy system, or MINAS) previously used to reduce nitrogen losses (but deactivated with the arrival of the new Manure Law). The third was a model to deduce the nutritional needs of the herd from *farm-specific* data on milk yield, milk composition, composition of the herd, etc. Finally, and this was new, there was a translation of the nutritional needs *into the nitrogen contents of the silage produced on the farm*. The latter is always a 'black hole' in models that aim to represent the relevant flows in dairy farming (NRLO, 1997). The key to exploring this domain was the nitrogen:energy ratio (and the associated crude protein) that had played a major rule in the previous construction of 'good manure' and which was, therefore, well documented and well known within the realm of the NFW co-operative.

Thus, new software was developed that could assess the farm-specific level of nitrogen excretion with far more precision than could ever have been done with the general formula of the Manure Law. At the same time, this new mode of calculation resulted in a range of relatively easily managed and transparent guidelines on how to further improve performance (while the Manure Law only offers the difficult parameter of urea content of the milk and milk yields). Further improvements (and therefore a better environmental performance) also translate into material benefits, such as more room for development and growth and/or less costs or additional benefits associated with manure flows. Thus, this new calculation mode potentially becomes a new localized mode of ordering (Reijs and Verhoeven, 2006).

Taking all this together, we can conclude that through this new software *superiority* was created. To be more precise, it allows for a wider range of adequate actions, results in a better performance and is combined with a series of significant incentives. But what are the reasons for such superiority? One of these is, I believe, the fact that the new software centres on *local specificity* (see Figure 8.2) and *places the involved actors centre stage*. Whereas the Manure Law superimposes a global formula on the local situation, thereby eliminating the relevance of the local (as illustrated in Figure 8.1), this new software, instead, reintroduces the local as relevant space and place (in this respect, Figure 8.2 indeed represents the mirror image of Figure 8.1). Second, the Manure Law reduces farmers to passive receivers, whereas the software of the NFW co-operative empowers them. And, third, while the Manure Law further strengthens the 'artificialization' and industrialization of farming, the new software helps to reground farming on improved manure and improved soils.

It is, of course, not the software alone; rather, it is the software together with the socio-institutional relations and rules within which it is embedded that make for superiority. The NFW co-operative developed, in this respect, the following mechanics. In the first place, the NFW contracts farmers or their sons and daughters (people who are knowledgeable about manure, nitrogen and cows) to visit the farms for the collection of data. Second, the data are converted by applying the new software to each farm into an overview of all relevant cycles and their specific outcomes (among them, nitrogen excretion). This is done at the office of the co-operative, where all relevant documentation is deposited in order to facilitate eventual controls. Third, the reports are discussed in local study groups (where strange deviations would immediately be seen). Fourth, the NFW co-operative issues a certificate in which the level of nitrogen excretion is specified. With this certificate, farmers can demonstrate vis-à-vis the ministry that they farm in a way that complies with the general objectives, even if the means used represent a deviation. The certificate makes clear that this is not a deviation due to fraud, ignorance or mismanagement – *it is, instead, a deviation grounded on and therefore legitimized by superiority*.

Although the world is globalizing, the new software and the (local) procedures on which it is based function as devices that allow for a significantly different way of patterning the social and natural worlds. Without these devices (these 'smart technologies'), further artificialization and industrialization would have occurred (as unintended consequences of the Manure Law). But with these new devices, dairy farming can continue to reground itself on ecological capital, thus enlarging its 'peasantness'.[21]

The rest of the story is amusing. The experts of the ministry had to accept the new software (because it is technically correct and helps to induce better performance). It thus became the official 'outstretched hand'. However, instead of having a try-out in the NFW co-operative only, it was directly declared applicable to the country as a whole. Again, this reflects the preference for generic approaches. Thus, a potential bug was born. By extending the new calculation mode beyond the borders of the NFW co-operative, a wide range of unexpected consequences could well emerge – hence, the experts simultaneously tried to 'detonate' it by stripping its attractiveness. Using the *global* approach (as specified in the Manure Law), an overall 5 per cent reduction of nitrogen excretion was applied[22] because, as was finally admitted, the model applied to the average situation and therefore might be less accurate. When, according to the global formula, the nitrogen excretion per cow is, say, 100kg per year, one was allowed to assume that the 'real' level was 95kg per cow per year (the fact that inaccuracies happen on both sides of the average was ignored). The same experts also argued that due to its far higher degree of exactness, this downward correction had to be skipped wherever the *alternative approach* was applied. Thus, a two-level playing field was created that put those wanting to apply the new approach in a disadvantaged position from the beginning.

The issue was again raised in parliament and the motion requesting an equal playing field was supported by a large majority. The minister initially denied accepting and implementing the motion. Several members of parliament then proposed a vote of censure; but after considerable turmoil (not related to this issue), the minister finally accepted. Thus, the bug was installed and activated. I write here of a 'bug' when referring to the new software because it locally disables the global rules and helps to create ways of patterning the social and the natural that global rules exclude. Bugs (or 'viruses') of this type help to 'attack' the Empire-like forms of coercive governance that are emerging everywhere; bugs create holes that allow for experiments and dynamics (which in their turn may feed into processes of transition), which would otherwise have been impossible. Bugs are, just as novelties, change agents *in disguise*.[23]

There is an additional reason for talking about a bug. If the new software were to be widely applied to the Dutch countryside, the Manure Law and the related agreement with Brussels would probably explode. Given the calculation entailed in the global cow approach, the *lower* nitrogen excretion, as demonstrated by those using the new device, will mathematically translate into the conclusion that those farmers not using it are realizing levels of nitrogen excretion that are *higher* than those prognosticated by the general formula and therefore *exceeding* the 250kg limit (the total amount of nitrogen is known and fixed; therefore, if one group uses less, the other consequentially uses more). Whether this will literally blow up the Manure Law and the agreement with Brussels is uncertain.

This same episode dealing with the creation and introduction of a bug highlights another crucial feature: Empire aims to impose order, but it only partially succeeds in doing so. The order imposed by Empire is not a definitive one – it is not made from granite. In this respect, Empire is, above all, a 'project' (Holloway, 2002, p234) and a partly unintended outcome of 'interlocking worlds' (Rip, 2006). It seeks to impose an order; but that order is only partial. The same partial order provokes responses, just as it entails frictions and failures (some as big as the *Parmacrac*). I will come back to this in Chapter 10.

Postscript

At the beginning of 2007, results of empirical research became available (see Figure 8.6) that convincingly and directly demonstrated that, at a given level of milk yield, the nitrogen excretion per cow is, indeed, highly variable. At a production level of, for instance, 25kg of standardized milk (FPCM) per cow per day, it ranges from 250 grams to 475 grams of nitrogen excretion per cow per day. This validates the thesis that the 'global cow' may be 'globally' correct; but it may also turn out to be *locally* wrong – not a bit wrong, but very wrong. Figure 8.6 also indicates that the variation in nitrogen excretion is associated with, and indeed explained by, the type of roughage made by the farmers – exactly what the peasants of the NFW co-operative have been arguing for years.

I close this postscript with an additional observation. The data contained in Figure 8.6 were not generated in the expert system that developed the global cow and that informed (and prepared) the Manure Law. Nor could the expert system ever have developed it because it would have undermined its very 'expertise'. The data contained in Figure 8.6 stem from research realized by Joan Reijs in close cooperation with the NFW co-operative. The research was funded through channels other than conventional ones for agricultural research. Within the currently reigning organization of knowledge production, research on promising deviations is becoming itself an exception.

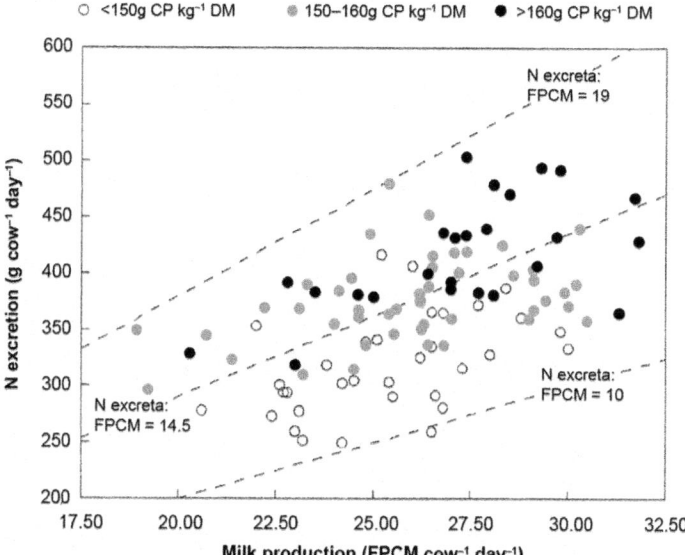

N excretion (g cow⁻¹ day⁻¹) in relation to the level of milk production (kg FPCM cow⁻¹ day⁻¹) for winter diets containing <150g CP kg⁻¹ DM (○), 150–160g CP kg⁻¹ DM (◐) and >160g CP kg⁻¹ DM (●) on 12 VEL and VANLA farms in the period 2001–2004. The dotted lines indicate constant N excretion per kg FPCM produced.

Figure 8.6 *Empirical levels of nitrogen excretion in relation to milk production per cow*[24]

Source: Reijs (2007)

9
Empire, Food and Farming: A Synthesis[1]

Introduction

Throughout the world we are witnessing the emergence of a new, powerful mode of ordering that implies a far-reaching re-patterning of both the social and natural worlds. Following Hardt and Negri (2001), Howe (2002), Stiglitz (2002, 2003), Chomsky (2005) and others, I refer to this new mode of ordering and associated forms of governance as Empire. In politico-economic terms, the emergence of Empire is strongly associated with increased mobility of enlarged flows of capital throughout the globe. Central to Empire as a form of governance is control and appropriation. According to Hardt and Negri (2000, pxii), Empire is 'a decentered and deterritorializing apparatus of rule that progressively incorporates the entire global realm within its open, expanding frontiers'. Consequently, controllability is central to Empire as a mode of ordering, and this often requires a far-reaching reordering of the social and the natural.

The constitution of Empire can be conceptualized as the specific way in which processes of globalization are currently manifesting themselves. Global movements of people, ideas, commodities and gifts are hardly new phenomena; but their intensity and speed have increased dramatically. Such acceleration and amplification, however, do not explain the qualitative shifts we are witnessing today in nearly all domains of society. The essence of the *current* stage of globalization is that it introduces literally everywhere *sets of generalized rules and parameters* that govern specific local practices. These sets of generalized rules represent the core of Empire. As a result, Empire materializes as an ongoing *conquest* that takes over once relatively autonomous and self-governed local constellations or, as Friedmann (2006, p464) puts it, 'self-organizing spaces', and reassembles them in a way that ensures controllability and exploitability. In so doing, it eliminates the local, transforming it into a 'non-place'. The only relevance of the local is as a set of coordinates – one of many other such sets – in which generalized rules are applied.

Imperial conquest similarly proceeds in the economy through the takeover of small independent enterprises by large holdings and through their comprehensive reordering – the former to nourish the needs of the latter. The dramatic cases of Ahold and Parmalat highlight this type of expansion and the disturbing combination of fragility and arrogance (Osmont, 2004) that results. Through conquest, takeovers and expansion, global requirements (e.g. to raise market shares, cash flow and profitability to centrally defined levels) are imposed everywhere – and then translated into ever-wider circles. Imperial conquest also impinges upon non-market institutions of whatever kind through the imposition of all-embracing procedures that prescribe, condition and sanction, in a meticulous way, whatever practices and processes exist. The resulting codification and formalization tend to rule out autonomy at the level of the shop floor – that is, they eliminate not only responsibility, but also tend towards annihilating agency. Non-agency is created because everything has to be carried out in accordance with pre-established and centrally defined rules. In this way an institutionalized slowdown is systematically introduced into many realms of social life (and, ironically, in nature as well through the protocols of nature conservation). Deviating from the rules in order to make things work better is judged an infringement.

As yet, it is very difficult to present an all-embracing and sufficiently distinctive definition of Empire as a mode of ordering – the more so as it crops up in domains as diverse as universities, public health, state apparatuses, private enterprises, non-governmental organizations (NGOs), agriculture, food processing and nature conservation. Simultaneously, it applies that Empire, as an empirical reality, represents a dazzling and confusing mix of old and new elements, which, of course, makes it extremely difficult to develop a clear, well-delineated and theoretically grounded representation of it.[2] In this chapter I discuss several such elements, as well as making a few historical comparisons. Then, in the final section of the chapter, I try to tie the different strands together.

Empire has not one single origin. It is, indeed, the outcome of a range of socio-technical worlds that increasingly interlock. It partly stems from big multinational corporations and their networks of transport, communication, assembly and control, and is partly rooted in the possibility of transferring enormous amounts of capital from one side of the globe to the other within seconds. But Empire also resides in state apparatuses and various supranational arrangements. Moreover, it is highly interwoven with new, centralized, but far-reaching, modes of organization (that build heavily on information and communication technology, or ICT) and with specific modes of knowledge production and their associated expert systems. It is the interweaving, steadily constructed coherence and mutual strengthening of these different ingredients that is currently rendering Empire powerful. Its multiple origins, its often highly confusing dynamics and its many-faceted and sometimes highly contradictory expressions clearly contribute to the difficulties of gaining a full understanding of Empire.[3]

In this chapter I focus on three particular domains in order to explore and comprehend the nature of Empire. These domains are farming, food production and consumption and, third, their regulatory schemes. Such a focus, I think, can offer a useful contribution to the rapidly expanding literature on Empire, especially since the latter tends to centre mainly on boundary shifts located at the outer borders and on the politico-military aspects that are associated with them. Instead, in the following analysis, attention will be concentrated on the *inward* expansion of Empire – that is, on how it penetrates, and materializes at the level of, fields, animals, food production, trade, people's livelihoods and the ways in which their practices are ordered. For Empire is not only to be conceptualized as external expansion; it is also, and simultaneously, a reordering of relations, practices, processes and identities within the very core of the imperial constellation.

From the Spanish to the current Empire

In a brilliant analysis, Henry Kamen (2003) convincingly shows that the 'Spanish Empire, the first globalized enterprise of modern times' cannot be explained by the often assumed power of Spain, and especially Castilla, itself: 'alone, the Spaniards never had sufficient resources'.[4]

'Conquest and power often resulted to be less important than *enterprise* – that is, the capacity to generate and to manage resources' (Kamen, 2003, p12).

The wars by which the Spanish Empire was created were fought mainly by soldiers from Germany, England, Switzerland and, in particular, Italy. Often the presence of Spanish soldiers was under 15 per cent, and the same applied to the commanders. In the *Conquista*, the conquest of large parts of Central and South America, the engagement of many native people who fought alongside the initially very small groups of Spanish warriors was indispensable. The Spanish navy was manned with sailors of the Basque Country and Portugal, while the art of navigation (especially the making and reading of nautical maps) was taken from the Dutch. The explorers, equally decisive for conquest and expansion, came from Portugal and Italy and the diplomats were recruited from the Low Countries. Cannons were constructed elsewhere since the necessary knowledge was not available in Spain. The money to finance the armies, the wars, the expeditions and the court was obtained from bankers from Genova and Antwerp. 'Promises' (often expressed in contracts) of a share in future wealth (in *El Dorado*) also played an important role. And when the debts were to be repaid (in order to obtain new credit) the gold and silver coming from 'las Indias' (especially from Chile and Peru) turned out to be decisive.

The centre of the Spanish Empire was almost void of resources. With its relatively small population, it could never have fought the many wars in which it eventually became engaged. Hence, it made *others* fight its imperial wars, just as it used the resources from *elsewhere* (credit, navigational capacities, diplomatic crafts, knowledge, etc.) to assemble the constellation that became known as the Spanish Empire: 'Empire was possible, not due to Spain only, but due to the combined resources of Asian, American and Western European nations that participated fully and legally in an enterprise that is mostly perceived, also by professional historians, as "Spanish"'(Kamen, 2003, pp13,14). Throughout his extensive analysis Kamen makes it abundantly clear that the representation of Spain as a powerful Empire was a myth; in retrospect, it was nothing but a legend. Empire was, above all, a (continuously expanding) *network* through which a range of resources from elsewhere were assembled together. It was only due to such a network that power was created. In this context Kamen talks of the 'joint use of resources' and typically refers to the network as 'a multiple enterprise'. It was an *actively constructed* network. The network, as Kamen (2003, p74) observes, was 'arranged in such a way that three basic needs could be met:

1   the possibility of obtaining money whenever and wherever it was needed;
2   the maintenance of safe ways of communication for the emission of orders and the circulation of correspondence; and
3   the possibility of disposing of armies.

Current food empires are as *void* as the Spanish Empire. They do not represent value (e.g. the debts of Parmalat and Ahold were as high as, if not higher than, their assets); neither do they produce any value of their own: they drain values produced by others. Food empires neither own, nor develop their own independent resources; they basically usurp and/or control the resources of others, as was demonstrated for the Chira and Piura valleys of Peru. Food empires need no direct ownership of a resource base, nor do they necessarily represent accumulated value. Their networks simply arrange the social and the natural world through the assembling of resources, processes, territories, people and images into specific constellations that channel wealth towards the centre. State apparatuses are likewise void in as far as they are part of Empire. They are, as I argued in an analysis of the Dutch Ministry of Agriculture and related expert system, 'ignorant and incapable' (Ploeg, 2003a). Where knowledge is needed to inform policy-making, it is instead virtual images that reign. And where capacities and capabilities are needed, it is mostly a certain 'non-agency' that dominates. There are similarities, just as there are dissimilarities that together define the historical specificity of *current* food empires. Following Colás (2007), who presents an extensive overview of historical and contemporary Empires, let me first discuss the similarities. I will do so by starting from three 'structural features'[5] that specify, according to Colás, the nature and dynamics of imperial networks. These are expansion, hierarchy and order (Colás, 2007, pp6–11).[6] I aim to show that these general features apply, perhaps more than ever, to contemporary expressions of Empire.

Expansion

'Empires are built on expansion' (Colás, 2007, p6). 'The imperial organization of political space has assumed the absence of permanent and exclusive borders' (Colás, 2007, p7). In general terms, Colás (2007, p31) argues that 'Empire reproduces itself through open[ing] and shifting frontiers'. Today's food empires are equally characterized by permanent and multiple boundary shifts, which redefine the very notion of food. Fresh milk was once a clearly delineated concept; but Parmalat (and several other industries as well) radically shifted this notion of freshness. It no longer refers to milk processed within 24 hours after milking and to be consumed in the following 48 hours. 'Freshness' today can stretch over weeks, if not months. This redefinition is the outcome of technological interventions within which microfiltration and repeated heating are central elements. Alongside this conceptual boundary shift, geographical and temporal frontiers have also moved radically, which potentially carries far-reaching politico-economic changes and impacts. As a matter of fact, some 80 per cent of research and development in food industries and related research centres is oriented towards the manufacturing of these kinds of boundary shifts. The tenderness and taste of chicken, for example, are no longer related to breed, feed or treatment. They can also be the result of *tumbling*: that is, the injection of water, additional proteins, softeners and flavours into any breed of chicken. Neither is colour any longer associated with breed, feeding, treatment, the absence of stress or proper storage and processing. Dark-coloured chicken meat (probably also emitting a bad smell and appearing of poor quality) is milled, mixed with water into a meat ooze, centrifuged and cooked, after which a whitish chicken filet (of the *simil* type) is obtained.[7] This 'upgrading' (as the official language puts it) of 'chicken separator meat' is just one out of many examples of shifting boundaries in respect to hygiene, healthiness and quality. These boundary shifts evidently relate to, and often allow for, associated shifts in primary agricultural production – the latter increasingly crossing borders once imposed by nature. Not long ago, meat, milk, vegetables and, consequently, land, buildings, animals and their management had to meet a range of conditions in order to produce tasty high-quality food. But once certain boundaries are moved, so can others.[8] This allows for massive scale increases in primary production and/or in complete reallocation of productive activities to other settings that offer conditions more favourable to food empires.

At the beginning of the third millennium, food empires literally materialize as ongoing expansion. This expansion proceeds as the *conquest* of nature, life, food and agriculture. This conquest affects patterns of consumption, health and consumer identity. It feeds aspartan (a substitute for sugar that easily provokes diabetes) to those who only want a soft drink. It feeds too many digestible fats (through the homogenization of milk) to those who simply want something healthy. Nature, food and agriculture, but also health and freshness, are redefined, materially reordered and reshaped, and, consequently, subjected to the specific rationale of various food empires. Inside the food empires, expansion is, indeed, perceived and consequently organized as *conquista*, as noted by González Chávez (1994) in his analysis of fruit and vegetable production in Mexico.

In Chapter 5, I referred to the enormous growth of value added in the Italian food industry compared to other industrial sectors, a growth also encountered in other countries. In The Netherlands, for instance, the gross value added (GVA) of the food industry grew from 22.5 billion Euros in 1985 to 33.0 billion Euros in 1997 (RLG, 2001): a growth of 46 per cent, comparable to the growth rate of Italian agribusiness reported in Figure 5.4. This exceptional growth is the outcome of a double movement: the squeeze exerted upon agriculture and the growing dependency of consumers upon supermarkets and the food industry. Together, agriculture and food consumption compose one of the *Dorados* of our times. Where once the mines of Chile and Peru funded the maintenance of the Spanish Empire, it is now, among others, the production and consumption of food that generates the enormous flows of wealth accumulated by the different food empires. In turn, this wealth triggers and spurs further expansion and converts it into ruthless conquest.

However, the expansion of food empires is not limited to food industries and supermarkets. The principles of Empire transmute into far wider segments of society. A telling example is the accelerated scale increase in primary production described in Chapter 5. Food empires, through the squeeze they exert on agriculture, provoke a search for lower costs of production, to be realized through a new and grim form of agricultural entrepreneurship: the usurping of space that belonged to others (whether land, quota, access, image, etc.). As one of the leading Dutch agricultural entrepreneurs expressed it: 'we can now start to *attack* vigorously' (Prins, 2006). Thus, another boundary is moved: that which separates an inner group of 'good entrepreneurs' from an outer circle of 'bad and failing farmers' who are to be removed from the sector.

Hierarchy

Imperial expansion is 'a hierarchical process' (Colás, 2007, p7). 'Expanding social formations [and empire-like networks] proclaim and generally enforce their political, cultural, [economic] and military superiority by codifying the subordination of subject peoples, and thereby leaving no doubt as to where power and authority reside' (Colás, 2007, p7).[9] Compared with current constellations, there are interesting differences as well, which at first sight seem to indicate discontinuities, but in the final analysis reaffirm the power, hierarchy and control exerted by Empire as a multiple enterprise, as Kamen (2003) would argue.

As argued earlier, current food empires do not own resources, nor are people directly engaged in stable dependency relations through, for example, long-term labour contracts. Current empires control *connections*. They are coercive networks that exert control over strategic connections, nodes and passage points, while blocking or eliminating alternative patterns: 'Inside the networks, new possibilities are relentlessly created. Outside the networks, survival is increasingly difficult' (Castells, 1996, p171). By specifying the rules that govern transactions and connections, current empires exist as monopolistic networks and, hence, in an *indirect* way, they control people and resources. By specifying, for example, that asparagus is to be shipped (or flown) from places of poverty to places of wealth, they generate a large range of detailed requirements (concerning quantity, quality, price, time and place of delivery, packing materials, paying time, mode of production, etc.). These requirements subsequently define the resources that are to be used, the way in which they are to be combined, and the different labour activities required. Empire operates, in short, as control at a distance. It is a kind of control exerted through the specification of the technical and economic requirements at every interface in the network. Through such (at first sight, nearly invisible) control, the *assembly* of specific sets of social and material resources is governed (see, for example, Figures 3.5, 4.3, 4.4, 5.6, 5.7, 8.1, 8.3 and 8.5). In other words, governance functions here by means of what at first sight seem to be neutral and value-free *technical specifications* that together compose regulatory schemes.[10] In synthesis: Empire is not only a network for patterning the world in a specific way, but, above all, represents *hierarchical control* over such networks.

The regulatory schemes imposed by state apparatuses on agriculture, the food industry and nature function in exactly the same way.[11] The state withdraws as a visible (and disputable) actor from several domains of public life, but re-enters as the omnipresent 'regulator' that imposes administrative and financial rules, procedures and agendas on all relevant social and natural elements. For instance, the 'global cow' represents a set of rules that specifies particular connections between global and local patterns, between fields and cattle, between pasturing and zero-grazing, and so forth. Empire rules, in a hierarchical way, through all-embracing and rapidly expanding regulatory schemes that specify required codes of conduct, and govern the (re)allocation and use of resources. These schemes emanate from both the state and large corporations. One could even argue that there is considerable, probably unintended, congruence and intertwining at precisely this point because large corporations can meet the state-driven regulatory machinery far better and far easier than small- and medium-sized enterprises that suffer high transaction costs because of these same regulatory schemes (see, for example, Marsden, 2003, on hygiene criteria). Furthermore, large corporations are often able to influence, through lobbying, the constitution of new regulatory schemes.[12]

The invention of printing technology was a decisive requirement for the establishment of the Spanish Empire: it allowed for sending messages, information and reports all over the world – and for subsequently storing and processing them. Contemporary information and communication technologies function in the same way for current empires:[13] they permit effective outsourcing and implementation of regulatory schemes. ICT allows for generalized control that functions like the circular set of windows of the panopticon (Foucault, 1975). It allows for connections between a myriad of places and one controlling centre, thus creating and extending visibility, controllability and, in the end, determinability. I will return to the crucial role of ICT later in the chapter.

Order

Empire is not only a hierarchical form of governance, but also a mode of ordering. It reshapes the social and the natural worlds in a particular way. One of the main particularities of the current order is that large segments of the social and the natural worlds are being reshaped into controllable phenomena. Control is of strategic importance because of the enormous capital flows invested by Empire in whatever domain that must be repaid in the near future. This implies that control over the conditions within which profitability is to be realized becomes crucial. Thus, Empire expresses itself also as a widely extended administrative ordering of nature and society. And although it is a foolish illusion to think that everything and everybody can be planned and controlled, the unavoidable deviations from the rule are represented, within this order, as ever so many infringements that are to be sanctioned.

The imperial order displays, at least at first sight, an overwhelming dynamism, although, paradoxically, it turns out to be strikingly slow and inefficient. A highly formalized world that functions according to strict rules generally finishes up being characterized by an institutionalized slowdown. Perceived from the centres of power (from the *cupolas*), it might appear well ordered and rational; but viewed from contrasting angles it reveals itself to be markedly chaotic, sometimes schizophrenic, and often highly contradictory. It is an order that fosters counterproductivity as much as it spurs productivity levels within particular sub-segments. Equally, it is an order characterized by the degradation of resources, food, work, environment and the quality of life.

Following Scott (1998), who describes mega-projects as 'tragic episodes of state-initiated social engineering', Empire can be depicted as the generalization of mega-projects[14] that specify how artefacts, activities and processes are to be connected to each other and how they are to be shaped and reshaped in order to remain connectable. The regulatory rule sets have a profound impact upon social practices: responsibility and agency are marginalized if not banned, at least for those who are to operate according to the rules designed and imposed by Empire.[15]

Empire introduces a contradictory order. On the one hand, it promises a beautiful, efficient, clean, sustainable and safe world, while on the other, it is generating a chaotic mess (Ziegler, 2006). There is a good deal of rhetoric about safe food; nonetheless, food scandals, often of major dimensions, are continually emerging. Potentially, food engineering carries considerable danger, now and in the future (Bussi, 2002). Commercials suggest beauty all around us, while ugly 'outcasts' are created all over the place (Bauman, 2004). Behind its virtual façade, Empire is both contradictory and schizophrenic. In practice, Empire comes down to *African* paprikas, contaminated with aflatoxins, being legally presented and sold as imported from Hungary, thus potentially destroying the livelihoods of many Hungarian producers.

Imperial patterns: Expansion, hierarchy and order

According to Colás (2007, p9), expansion, hierarchy and order are 'features common to most historical empires', and as I have shown in the previous sections, they also characterize current food empires. It is important to underline these characteristics. Food empires barely link to other domains such as farming and food consumption through interchange and cooperation. The interlinkages are structured through expansion and takeover. During this process hierarchical relations of governance are created. Empire does not co-ordinate ongoing processes and activities; it *imposes* blueprints, implying their remodelling (or reassemblage) in order to fit its interests, dynamics and requirements. There is, so to say, no subsidiarity whatsoever: everything is to proceed according to centrally defined rules and parameters. As a consequence, 'Empire involves [here] an expression of power that *aspires to* [exercise] *control over outcomes*' (Colás, 2007, p185).

Food empires are integral expressions of Empire as a mode of ordering and governance. Put somewhat differently, Empire unfolds not only in the battle for oil in Iran and Iraq and other hot spots of the globe such as Afghanistan (Chomsky, 2005), but also in farming, food processing and consumption, and nature conservation. Empire not only expresses its consequences in the slums and malnutrition of Latin America; it is equally present in the restructuring of European agriculture, in rates of obesity, or in the unknown risks associated with genetic engineering and the widespread inclusion of genetically modified organisms (GMOs) in food (Hansen et al, 2001).

Alongside the general features identified by Colás, some additional features can be formulated. These concern the specific nature of wealth creation and distribution, and the role of so-called 'extra-economic forces'.

A regressive mode of wealth creation and distribution

The dynamics of Empire is increasingly expressed in terms of two separate but interlinked spheres. There is the 'real economy' where things are made, changed, processed, repaired, grown, moved, shipped, packed, upgraded, designed, digitalized and so forth. And there is the 'virtual economy', composed of the imperial networks that control and manage the first sphere while simultaneously appropriating the value produced in it. The 'real economy' is everywhere; but owing to the ordering and reordering imposed by Empire it is continually on the move and made invisible.

Currently, this first sphere, that of the real economy, is being reshaped into a deregulated or informal economy – that is, one in which all kinds of civil and labour rights (the right to unionize, to job security, to decent remuneration) hardly apply,[16] which in the context of Empire makes it attractive as a place of poverty (see also Bové, 2003). Empire provokes and strengthens processes of deregulation in a twofold way. It does so by moving large parts of the real economy to East Asia (and, to a lesser degree, countries in Eastern Europe, Africa and Latin America), while fostering deregulation through far-reaching and as yet far from completed reforms in the economies of Western Europe, North America and Australia.

Deregulated economies might be conceptualized as large reservoirs of *freely available* resources and people awaiting the required *connections* in order to produce, trade and grow. People in these reservoirs urgently need such opportunities in order to make a living; and they are even willing to compete fiercely with each other (and other areas) in order to obtain the needed connections.

Alongside the accumulation of breathtaking wealth, Empire generates widespread poverty (also in the form of a gigantic environmental crisis that will be paid for in the future). Beyond that, it introduces a slowdown in the economy and a regressive moment in the creation of wealth. Compared to the possibilities theoretically available, the growth created by Empire is far less than might have been, partly because many of those who could be productive are condemned to idleness. It also often relates to inefficient use of resources and high levels of waste. Furthermore, it stems from the fact that new production in one place is associated with demise in other places. In particular, it occurs through Empire-triggered conquest in agriculture: the takeover of peasant farming through entrepreneurial farming tends to decrease the production of total value added (see Table 5.3). Empire represents, in synthesis, a regressive mode of wealth creation and distribution.[17]

Monopolization and extra-economic forces

Empire stimulates monopolization. It defends the 'entry points' to networks, such as credit and capital, and regulates who will have access. Empire also controls the 'selling points'. Outside of Empire, consumers are usually difficult to reach. Having them in its control represents the extra-economic power to monopolize markets. Food empires, for instance, do not simply function *within* markets; instead, they represent centralized control *over* the markets. Empire is market in disguise. It makes the world *look* like a market since there are many processes of buying and selling; but it monopolizes the *routing* of these processes and associated transactions: they can only be realized according to conditions imposed by Empire itself. For those who have to sell, the entry points of Empire are increasingly *obligatory passage points*, the more so since Empire actively seeks to eliminate all possible alternatives. And the same applies for those who want to buy. I return to this feature in the following section where I discuss earlier railway systems and corporations.

On railway systems and corporations

The widespread dissemination of corporations in the 19th century was closely associated with the construction of railway systems. They were needed to mobilize the enormous quantities of capital to finance the construction and rapidly extending railway systems. In turn, control over the railway systems gave these corporations enormous power (Bakan, 2004). Such corporations are not only an antecedent of Empire; they are also one of its sources. In both, networks play an important role.

The classical network of railways represented connections, points of entry and exit; it also had its nodes for transfer and conversion, and offered the opportunity to link territories that were separated from one another. The network allowed for flows of goods, services, people and ideas that would be very difficult, if not impossible, without them. It allowed (especially when railways became connected to steamship transport) for the imposition of conditions reigning in one territory to other territories located in other continents (as the first agrarian crisis of the 1880s made clear). Such networks represent power since the entry and exit points function as obligatory passage points. Whoever needed a connection had to accept the conditions formulated by the controlling corporation. Thus, corporations and the railway systems they controlled emerged as extra-economic power – just as extra-economic power had been a major requisite for their construction. The continous reproduction of their monopoly position thus became a strategic condition for their very existence.

These initial railway systems figure as a strong metaphor for Empire as it is now crystallizing. The construction of a railway system represents a gigantic investment that is to be revalorized. It must be used as fully as possible in order to quickly reach the break-even point. The use of the system must therefore be stimulated, if not outright planned, and, if necessary, imposed. To echo Colás: control over future outcomes has to be secured. Railway systems, once created, will exert an ever-widening impact upon time and space: they start to order the future according to their own specific needs and rationale.

It is remarkable that today many domains of society are organized in a manner similar to these 19th-century railway systems. In this respect, we are still (or again) facing a truly Fordist organization of society (Braverman, 1974). Take, for example, the supermarket. Different from what the word seems to suggest, it does not refer to a large, well-covered *market* where traders from many different places come together, having travelled different roads and using different means of transport to get there, bringing different products to be sold in the 'super'market. Instead, the supermarket as we know it today is rather like a railway system, both internally and externally. It recognizes fixed supply routes and the associated flows are directed and controlled from one central node. And the same applies internally: the workers inside the supermarket are far from being independent traders, having their own businesses and their own responsibilities. They are wage labourers (probably youngsters on temporary contracts) who carry out specified tasks according to fixed protocols and procedures defined by the central management.

Today's supermarkets determine how each shelf will be stacked, with what products and from where they come. And the turnover and required circulation time and sell-by dates are likewise pre-established. All this, and many more details, are calculated using highly complex models that indicate for all selling points what will be offered, in what quantities and in which combinations. The outcomes are differentiated by postal code area, which gives information on income levels, ethnicity and previous patterns of consumption within local branches.

The products circulating in and outside of these interlocked 'railway systems' must meet a range of criteria. Sufficient supply must be guaranteed in order to stock all outlets. Products that are produced only in minor quantities are therefore excluded. They cannot enter the 'system'. To deliver to only a few shops is too costly and troublesome. The product must render sufficient cash flow and profitability. According to supermarket criteria, if a product turns out to be successful, then the producers must guarantee that it can be supplied over long periods of time; otherwise, alternative suppliers will be sought. Exclusivity is often demanded such that producers or intermediaries are expected to deliver only to a specific retailer. Thus, intake or entry points start to function as 'filters' that at the same time order activities, processes, relations and prospects downwards in the food supply chain.

The third level

Currently, infrastructures (i.e. all material and institutional frameworks and systems that make the technical functioning of our societies possible) are conceptualized as composed of three levels (Twist and Veeneman, 1999). First are the physical patterns and means (e.g. the railway routes, the trains, the security systems). The second level consists of the flows that occur due to level one (in this case, persons and freight). And above this there is a third level, where the direction, control and, increasingly, the appropriation of value are located.

Level one, in contrast to the historical situation, is increasingly public property. Level two (i.e. the actual use of level-one facilities) is sometimes public, sometimes private by nature, and sometimes organized in public–private partnerships. Level three is again increasingly private. Take highways. The material infrastructure clearly represents level one, and the circulating cars, lorries, people and freight, level two. Sometimes access to level one is organized as a market controlled by private enterprise – such as is the case with toll systems, or new systems for direct payment for transport, etc. New digital navigation systems are increasingly shaping a third level, not only because it is at this level that considerable profit is realized (e.g. through a global positioning system, GPS), but because this emergent level three will probably unfold in a fascinating and impressive way – i.e. the movements of all vehicles together and of each single vehicle will be registered through radio frequency identification devices and GPS. Hence, each single user of the highway system will be guided along the most efficient routes, avoiding traffic jams and accidents and securing the shortest travel time. Were level three to materialize in this way (and technical preparations are advancing well), then direction and control of level two will undoubtedly be located at the 'top' (i.e. at level three).[18] Control will then, so it seems, be complete. At level two a major change also occurs: one has to follow the indications communicated by the 'system'. If not, havoc could easily arise. In other words, responsibility is shifted from individual participants in everyday traffic to a new locus of direction and control.

Empire is increasingly materializing as a level-three phenomenon. It has few resources or infrastructure itself. Empire is 'simply' the coercive ordering of flows the world over and the appropriation of any raised value. The constitution of Empire as level three implies a reconceptualization and, subsequently, a material reorganization of level one. Level one is, as it were, made fluid. There was time when land improvements (such as irrigation systems), buildings,[19] knowledge and skills were elements of level-one infrastructure that were tied to specific territories. Now, through the development of Empire, level-one production facilities might easily be reallocated (partly because sunk costs are no longer very relevant), just as flows can be re-routed whenever it is convenient.[20] In certain respects the creation of level-three constellations is attractive: especially if it helps to avoid traffic jams or makes exotic products from elsewhere easy to obtain. However, there are highly problematic aspects. Let me here discuss one worrying trend concerning food production – namely, the accelerated expansion of commoditization processes.

In the global economy as it is currently taking shape, it is no longer products such as milk, asparagus or cell phones that function as the main commodities. Far more important than being able to produce asparagus is the opportunity to access the connections that allow the product to be routed to areas of wealth. Thus, access becomes an important commodity. Paul and Jennifer Alexander (2004, p63) stress that 'the economies of ... former industrial powers [increasingly] concentrate on the *commodification* of transactions, services and knowledge rather than production'. Far more important than production as such is the direction and control over it, both of which are exerted through a third-level type of control over the many flows associated with production and consumption. Producers increasingly have to buy the right to have their 'raw material' processed and routed to the different entry points of the 'systems'. Thus, new markets emerge that tend to be far larger, and far more decisive than the ones that until recently governed primary production: 'by far the greatest amount of exchange now consists of "non-things"' (Alexander and Alexander, 2004). An indication of such 'amounts' might be obtained through a reconsideration of the system for milk quotas. Although the quota system was created during the 1980s for completely different reasons, it has functioned, in practice, as a market centred on the right (the opportunity) to produce milk and to have it processed and routed further afield by the dairy industry. Not only milk figured as a commodity but the *right* to produce milk also became a commodity that even took priority over the former. The right to produce milk could be sold, bought and leased; it could equally be shifted from one part of the country to another.

Together, the milk quotas in The Netherlands represent an astonishing value of 20 billion Euros (2006 data). This towers over a gross value of yearly milk production in the country of some 3 billion Euros. Until recently, some 400 million kilograms of milk were traded annually in the national quota market. This implies, per year, a value shift of 720 million to 780 million Euros (which equals, assuming an average interest level of 5 per cent, an additional cost of 36 million Euros per year). Knowing that such levels of commerce characterized a ten-year period and reckoning on the fact that there was fairly little repayment, the total cost of obtaining 'access' amounted, in the end, to probably some 360 million Euros per year for the sector as a whole. This is more or less 50 per cent of the total family income realized in the sector. Access payments also reside in admission fees and charges for auctions and co-operatives. They are entailed in the shares taken out by farmers in their co-operatives, as well as in the joint assets created during the years. All such forms, admittedly, are in an embryonic state and systematic knowledge of their magnitude and impact is still to be assessed. However, it goes without saying that such access payments will be one of the main future battlegrounds in and around agriculture and food production.

The central but contradictory role of information and communication technology

The rapid and massive dissemination of ICTs within the realm of agriculture, food processing, trading and regulatory schemes is basically explained by five factors, some of which are of a general nature, and others typically limited to agriculture, food production and the way in which agrarian policies currently operate.

First, worldwide outsourcing of the production of goods and delivery of services, and especially its interaction with a rapidly expanding spatial division of labour, requires a meticulous specification of all the different parts that must be assembled. Fluidly flowing processes of decomposition (e.g. the 'global chicken' described by Bonnano et al, 1994) and recomposition (ranging from the classical remaking of milk from butter oil and milk powder to the new and quickly evolving range of health and convenience foods) require an exact knowledge of all the *properties* of each and every ingredient and the exact *requirements* each must meet. The point is that, analytically, a new food item, composed of, say, ten ingredients (including stabilizers, additional flavours, colorants, etc.) represents, in as far as its 'construction' is concerned, 45 interfaces (at least mathematically). Particular specifications are detailed for each interface involved in the new food product and these define the characteristics that the ingredients must meet – that is, they constitute the requirements. Then every ingredient that enters the place of assembly will have particular properties that must match requirements; if not, sooner or later, an error indication will be generated.[21] In artisan processes of food production, such as that for Parmesan cheese (see Roest, 2000), it is the cheese-maker who sets the requirements and the properties to be met. This is achieved through long experience, continuous observation and an extensive knowledge of the primary producers and the way in which they operate. Evidently, this is impossible in industrialized processes, where food processing is automated. The centrality of craftsmanship is eliminated and requirements must therefore be defined in protocols that need detailed information on the properties of all ingredients. Finally, the fit between the former and the latter must constantly be checked. This implies huge and permanent flows of information that require constant analysis: hence the need for widespread ICT.

Second, the current need for precise timing of delivery (to avoid large stocks and associated costs) implies that the different flows associated with single ingredients must be planned, monitored and controlled. In turn, the planning of these flows through time and space requires that their production is likewise planned and controlled. This again implies widespread application of ICT over wide geographical distances and over a wide range of different producers.

A third reason for the need for such ICT-based 'chain management' is related to risk and legal responsibility. If something goes wrong, during or after the final assembly, the origins of the 'error' need to be traced so that legal responsibility can be delegated to others somewhere down the line. This implies, in turn, that all units of production that are somehow linked to, or are part of, the 'chain' must document their respective production processes according to specified formats. In synthesis, all steps within extended and complex lines of conversion and delivery must be accounted for in a documented way in order to make sure that, in the end, requirements are met. Given the current levels of internationalization of food production this implies a gigantic flow of data (and of data storage and analysis). Without ICT this would be impossible to handle.

A fourth reason for widespread ICT application relates to regulatory schemes imposed by state bodies. In essence, these function according to the same logic as the control schemes for food processing industries. In order to 'make for' landscape quality, good environmental records, hygiene, animal welfare, groundwater protection, prevention of erosion, reduction of ammonia emission and so forth, explicit and highly detailed protocols are defined that specify the associated requirements. In order to secure, for example, an acceptable level of ammonia emission, a specified storage facility for a specified volume of slurry and specified machinery for its distribution must be available; a specified calendar and set of ratios between milk yield, urea content of the milk, number of cattle and acreage must also be followed. This is one side of this particular interface. The other side regards the actual behaviour of the farmers. They have to show, mostly according to a pre-established format, that their practices meet requirements. Thus, the protocol translates into coercive procedures that order a wide range of agricultural activities. Typically, the once quite prominent extensionist (who, in a way, mediated between state policies and farmers) has disappeared, just as the *casaro* or cheese-maker has disappeared from the food industry. In both domains (i.e. state and food industry vis-à-vis farmers), it is now control at a distance that dominates. However, since there is a generalized fear that farmers might try to cheat 'the system', there is also an extended 'inspection service' to check whether actual practices coincide with electronically generated representations of reality. Thus, overhead costs rise again.[22]

A fifth reason relates to recent changes in agrarian and rural policies. From the McSharry reforms of 1990 onwards, economic support for agriculture has been gradually disconnected from the volume of production. It increasingly tends to be replaced by payments per hectare (a so-called 'flat rate'), paid only when farmers meet a range of legal requirements. The policy changes and the obligation to meet certain criteria are justified when public support is involved. The problem, though, is that again a huge control system is needed through which requirements and properties are checked to see if they match effectively.

So far, I have discussed five reasons that explain the widespread application of ICT in food processing and in state regulation of farming. Just as, historically, printing techniques turned out to be a technological prerequisite for the rise of the Spanish Empire, so the current food empires and the imperial behaviour of state apparatuses vis-à-vis the agricultural sector would be impossible without ICT. Or, to formulate this more positively, the huge development of food industries, supermarkets, international trading and 'chain management' would have been unthinkable without ICT. Yet, this same ICT application has another side, which, in several respects, turns it into a nightmare.

A first problem is that ICT cannot deal with *concepts*. When real life is full of concepts where each translates into a myriad of specific *expressions*, ICT can only operate at the level of expressions that distinctly differ from each other. One can talk to others, for example, about a 2 litre engine and nearly everybody will understand this. However, feeding such a notion into a computer system that commands the assembly of engines for automobiles triggers the danger that it will render you, say, an engine having outer dimensions of 20cm by 10cm by 10cm – that is, exactly 2 litres, which evidently is at odds with the very notion of a 2 litre engine (Thiel, 2006). Thus, ICT can only function for the regulation of complex processes if there is a meticulous, formalized and all-embracing specification of requirements and properties. A concept or metaphor is not sufficient; what is required is a full specification of all technical characteristics. Following the same line of illustration, a 'green car' is an inadequately specified requirement. It could be met by a car with green-painted windows, headlights and steelwork. The result would be a monstrosity.

Now, in the manufacture of artefacts such as cars this can be perfectly resolved. The real problems arise when dealing with concepts such as 'beautiful landscape', 'fertile and well-crafted land', 'good manure' and 'farming gently'. None of these concepts (and the same applies to good medical care, stimulating lectures and good academic research) has simply one specific and unambiguous interpretation. Hence, there will doubtless be many different, contrasting and probably mutually incompatible and debatable expressions. Just as the cheese-maker will ask whether this year's product 'really is a good cheese' (and, if not, both farmers and consumers will definitely pose the question), the inhabitants (and visiting tourists) of a particular area will always ask, debate and differ as to whether a landscape is 'beautiful' or not. Central to such a debate are nuances, shades and degrees, which, of course, go beyond digitalization. In the digitalized world one must work with 'yes' or 'no', and not 'maybe' (neither can 'fuzzy logic' resolve this problem). In the digitalized representation of the world, things (and constellations) *are what they are*. Within the framework of ICT it is difficult (if not impossible) to handle unfolding, debatable and, as yet, unclearly delineated entities. Hence, when ICT becomes dominant, the real world runs the risk of being reduced to simplicity and uniformity that is, in essence, alien to it.

This danger is the more immanent – and this is the second problem – when one takes into account that both in agro-industry and state bodies, two design principles apply. These can be summarized as 'simplicity of prescription' and 'simplicity of control'. This implies that objectives, protocols and procedures are not to be differentiated according to the specificities of local situations. That would signify their negotiability, which would increase transaction costs. Consequently, state bodies and food industries alike tend to link, as far as possible, with agriculture and the countryside in a global (non-differentiated) way. Yet, because the latter are highly differentiated and continue to produce differences, exceptions, deviations and novelties, such global approaches can only be but 'rough and ready' and therefore tend to generate frictions and distortions that result in an undesirable homogenization of the local – especially when they are coercively applied.

These problems interlock with a third one – namely, that once an ICT-based regulatory framework has been established, it becomes extremely complicated and costly to modify it (Beuken, 2006; Roos, 2006; Straten, 2006; Thiel, 2006). Even slight changes in some of the defined requirements imply that a long chain of associated interfaces will need to be rewritten. Building on a case described in Chapter 8 – the brambles growing in the hedgerows – one observes that in itself a small modification in specifications requires, among other things:

- a modification in control protocols;
- an adaptation of the modules that regulate payments; and
- repayments to farmers who have been sanctioned for 'deviations' that now, in retrospect, are redefined as having legitimate 'properties'.

But even more complicated is the fact that the initial programme and the modules implied by it have been formerly accepted by the European Commision in order to co-fund them with European resources. This implies that

- any change, in whatever detail, will also have to be renegotiated with 'Brussels': the respective interface located at supranational level will have to be 're-scripted'.

With the establishment of such widely extended command, control and funding schemes, made up of many formalized and ICT-based connections that entail numerous interfaces (and, consequently, sets of requirements and properties), the social and natural worlds indeed become extremely rigidified and subjected to institutionalized slowdown, if not an overall sclerosis. Within this structure it becomes too troublesome, too difficult and too costly to introduce improvements (let alone to really renovate the existing 'system').

Here we encounter many interesting parallels with what ICT experts refer to as 'embedded software' – that is, software built into complex machinery (e.g. machinery designed to construct new chips and medical machinery to monitor, control and, if need be, to correct the workings of an attached human body). The implied software simultaneously registers and controls many different functions, many of which imply a trajectory over time. Thus, many different interfaces emerge, several of which concern complex feedback and forward processes. Beyond that, the same software has to be able to control and correct itself, as well as its associated mechanical devices. If anything goes wrong, then there is immediate and automatic intervention and correction. This evidently implies another extended range of interfaces. 'Fine-tuning' of this kind of 'machinery' is, as practice shows (Straten, 2006), extremely difficult and costly. Since all interfaces potentially impact upon each other, the software *as a whole* has to be rewritten.

This short excursion into 'embedded software' throws up yet another contradiction. ICT (i.e. the languages and tools for programming, the division of labour between ICT experts and users, etc.) is hardly adequate for running complex *machinery*; yet it is used as an organizational device for *social processes and associated institutions* that are far more complicated, dynamic, contradictory, differentiated and flexible (at least potentially) than any mechanical artefact could possibly be. Nonetheless, many organizations such as universities, hospitals, nature conservation bodies, state organizations, security services and food industries, and many relations such as those between hospital administrations and patients, state bodies and farmers, and, consequently, many domains of the social and natural world are currently governed, ordered and controlled by ICT-based systems that relate as *mechanical technologies* to the subjected domains. They are basically unsatisfactory, crude, difficult to improve and often an insult to the qualities, responsibilities and skills of those on the ground. But, then, that seems to be the price that has to be paid for imperial control.

State, markets and institutions

The state and the market (understood as ordering principles) flow together and converge within Empire. In this respect, Empire emerges as the mutual co-penetration, interchange and symbiosis of state and markets. State bodies and their relations with 'clients' in public health, safety and education, for example, are increasingly structured, ordered and organized as a market, and state functions are transferred to market agencies. At the same time, markets increasingly cease to be governed by an 'invisible hand'; instead, they are subjected to new loci of control that exert different forms of extra-economic control. Imperial networks, with their obligatory points of entry, conversion and release, are concrete expressions of this newly emerging 'visible hand'. Through such networks, the economy tends to become subordinated to cycles of planning and control (as occurred in other times and places through state bureaucracies). Planning and control are evidently related to accelerated rates of expansion and conquest, which often occur by means of a massive mortgaging of available assets. Thus, future profitability and shareholder value become strategic for current operations; and the rationale and justification of any given activity no longer rest with that activity (and the specific place and time associated with it), but are, instead, linked to, and therefore dependent upon, their (assumed) contribution to the profitability and expansion of Empire. Precisely for this reason, tight cycles of planning and control are enforced.

The new symbiosis of state and market penetrates deeply into and reorders civil society, subjecting it to external controls, prescriptions and planning. Autonomy, responsibility and trust – three important vehicles of civil society – are increasingly eliminated and replaced by procedures, rules and protocols. As a mode of ordering, Empire tends to be *superimposed* on existing orders (state, market and civil society), aligning them and introducing new contradictions and development tendencies that, until now, have been unknown to mankind. This superimposition of Empire as ordering principle implies that market and state no longer counterbalance each other, even partially. In and through Empire they are increasingly aligned and fused into one comprehensive technology of regulation – a technology that exerts a disembodied but seemingly irresistible expression of power over nature and society.

In social science, hierarchical organizations are generally understood to be diametrically opposed to those of the market, though economic life might be organized within the former or through the latter. Nevertheless, theoretically, there is and remains a fundamental difference: within the enterprise there is no market, while within the market there is no hierarchy (Saccomandi, 1991, 1998).

It is precisely at this point that Empire emerges as a radical breakthrough. The essence of Empire is that it blurs the boundary line between market and enterprise. Empire is, in this respect, a double movement. It interlinks and orders markets in a hierarchical way, while at the same time introducing the market principle within institutions. The enormous power of Empire stems precisely from this mutual penetration and the subsequent intertwining of market and hierarchical organization. Empire is a new structure that actively interlinks markets (mostly in asymmetrical ways), while simultaneously subjecting them to centralized planning and control cycles.

Markets are actively interlinked and (re)ordered. Linking land and labour markets in Peru to the food market in Europe is a clear example of this and illustrates simultaneously the asymmetrical nature of these newly emerging patterns of production and trade. Parmalat and the drastic reshuffling of the dairy market via the *latte fresco blu* project is another example. What such illustrations highlight is that Empire is not governed simply by markets and the assumed 'invisible hand', but by the opposite since Empire is, to a considerable degree, able to govern the markets that it controls. At the same time, enterprises are dissolved by dividing them into a range of interlinked markets, each governed by price and cost levels established at the third level and which are related to each other through obligatory passage points that are equally controlled from this third level. In other words, Empire systematically introduces the market as an organizing principle into the enterprises that are controlled by it.

This also occurs within institutions: they are reconstituted into internal markets that are governed by sets of rules imposed by the respective '*cupolas*'. Universities, for instance, are conceptualized and then materially organized as markets in which the different departments 'offer' a range of courses, and students 'purchase' a set of courses, after which the supplying departments are financed according to the level of demand. To be able to give the courses, the departments must now 'rent' the required lecture halls from the university administration. In other words, the resources of a university (of which I have only mentioned a few) are converted into commodities.[23] However, the equilibrium which such a market is assumed to attain does not result from the 'free encounter of supply and demand', but from prior decisions made by the university authorities that assess 'price levels' and 'commodity flows'. Control over these flows and their requirements and properties are effected through the widespread application of ICT as the main organizational device. This provokes an all-embracing formalization of activities that negatively feeds into the university as an academic institution and thus threatens it as a place where ideas, information, debates, staff and students freely interact.

The role of science

The dream of *El Dorado*, a land full of gold on the other side of the boundary that delineated the known from the unknown world, was a crucial element in the creation and unfolding of the Spanish Empire (Kamen, 2003, p144).

Without such dreams or promises, there would have been no journeys of discovery, no colonization and no gold and silver flowing back to the imperial mainland. Current sciences equally excel in 'dreams' – that is, in the elaboration of the many promises that the life sciences, nano-technology, food engineering, biotechnology and so forth claim to make true. Scott (1998) refers, in this respect, to the 'imperial pretensions of agronomic science'. The promise of new *Dorados* composes one of the many links that tie, at this moment, considerable parts of science and Empire together. In its unbridled race towards new opportunities for realizing high returns on capital, and, especially, for the 'gold and silver' that is to pay for the many liabilities (just as the *latte fresco blu* was to pay for the huge debts accumulated by Parmalat), Empire continuously needs new resource fields to 'mine'. And it needs them desperately – hence, the importance of large research programmes (partly financed by big corporations) geared to exploring new technological possibilities and plotting road maps on how to reach them. The latter are as crucial for the current Empire as were the first generation of nautical maps for the Spanish Empire. Thus, the search for new forms of food engineering and the rebuilding of nature through biotechnology are converted into a series of nearly ruthless interventions into life itself – with all the risks entailed by such interventions (Hansen et al, 2001).

Through science, new resource fields are explored and materially constructed. Beyond that, science often legitimizes and defends their exploitation by affirming, for example, that 'food never has been as safe as it is today' and that risks related to genetic engineering are 'fully under control'. In this way the relation of science to Empire resembles that of the Catholic Church (and the Inquisition) to the Spanish Empire. The *Conquista* was literally blessed, from the beginning, by the Church, while opposition was met by the Inquisition. The current alignment of science and Empire is so strong that it hardly seems possible anymore to even think about it in contrasting terms. Yet, it is less than seven or eight decades ago that science stimulated economic growth through relations that were patterned in a completely different way. The introduction of chemical fertilizer, for instance, proceeded because the agrarian sciences of the time critically checked and assessed its quality, its performance and its impact measured in terms of yields and profit. Being transferred through an opportunistic market, farmers never knew (especially during the epoch in which fertilizers were introduced) whether the purchased bags really did contain fertilizer. It could, for example, be worthless sand or cement. Furthermore, they didn't know whether the recommended quantities, proposed application techniques and timing would prove effective in their own fields. Here it was science that bridged the gap – not through simply aligning with the fertilizer industry (as would occur later), but through a *critical stance* and an ongoing *scrutiny* of the many practices promoted by industry.

There are many other linkages between science and Empire. In Chapter 8, I outlined some of these, indicating how deterministic models developed by science are converted into technologies for governing specific social and natural domains. In line with all of this, large-scale expert systems and universities are increasingly being reorganized in accordance with the Empire mode of ordering and governing. Of the many reorganizational processes that are taking place, one tendency seems to be dominant: science, its products, its operators and its institutions are reshaped in such a way that their final properties meet the requirements of 'the system'.

Synthesis

Empire is a rigid and expanding framework composed of regulatory schemes of a political *and* economic nature that are imposed upon society and nature. In and by means of this framework the state and the market have become increasingly interlaced.[24] The one translates into the other and vice versa. Empire is not primarily about products, people, services, resources, places and so forth. Neither is it composed of such elements. Empire is, above all, a complex, multilayered, expanding and increasingly monopolistic set of connections (i.e. a coercive network) that ties processes, places, people and products together in a specific way. In the preceding chapters I attempted to elucidate the nature of these connections and their implications. Slightly rephrasing some of the previous arguments, it might be stated that the historical specificity of current food empires resides in the contradictory, but systematic, combination of two ordering principles: the global market and the assembly-line system. Taken together, these two principles combine, for example, into so-called food chains. Within food empires, production (farming included) has been segmented into an endless series of subtasks that are in themselves relatively simple and monotonous, and which perform operations that form part of a far longer assembly line. However, this assembly line is not located anymore in one large factory, within which hierarchy is the central ordering principle. The parts that compose this line are now distributed all over the world in the form of an archipelago that is continually changing. The *interrelations* between the different elements (the connections existing along the line) are constructed through the market. This allows for radical and quick shifts. Whenever a particular element can be obtained more cheaply in another reservoir, the overall pattern (of interrelated connections) will immediately be adapted. But then another problem arises: how to be sure that the elements will fit materially together? This is secured through the detailed prescription of all requirements and properties – a prescription that is communicated and controlled through ICT. With the dazzling expansion of the social, spatial (and even temporal) divisions of labour and the subsequent expansion of markets, control becomes absolutely crucial. Therefore, the assembly-line principle (effectuated through control at a distance) is reconstituted and combined with the market. Simultaneously, the borderline that once clearly delineated enterprises and the market has become obscured, if not, to a degree, liquidated. The 'enterprise' penetrates the market by prescribing what is to be done (and how, and when and by whom) even in the remotest of places; and the 'market' penetrates the enterprise by reshaping internal relations into market relations. The market is governed through a range of assembly lines (which emerge as extra-economic force that shapes and conditions the markets), and the assembly line, in its turn, is operated by continuously connecting different markets while it is simultaneously camouflaged and legitimized by the market principle (at least with neo-liberal discourse).

The general intermeshing of state and market, together with the widespread application of radically new technologies that allow new forms of conquest and control, relates to the specificity of Empire as it manifests itself at present. This general feature is reflected in food empires as the intertwining of market and assembly line into coercive networks that increasingly pattern agriculture, as well as the processing and consumption of food.

In her illuminating work on international food regimes, Friedmann (1980, 1993, 2006) distinguishes between two: the 'settler–colonial' (1870 to 1939) and the subsequent 'mercantile–industrial' food regime (1945 to 1990). The settler–colonial food regime basically centred on the free trade principle, and the mercantile–industrial regime followed the regulation principle. Following this periodization, I would argue that since 1990 there has been a shift to a third regime – the 'imperial food regime', which essentially embodies a complex combination and alignment of free trade and international food regulation. Forms of regulation that were once well established are de-moulded (such as agrarian policies at national and supranational levels, and institutions such as the peasant communities of Latin America), and new forms of hierarchy grounded in the converging interests of agribusiness and state bodies are emerging. At the same time, markets are drastically reordered. Food markets are globalized and aligned (through new forms of regulation) with global processes of accumulation, and new spaces are opened for large corporations, which operate as coercive networks within each of these spaces.

Citizens, who are assumed to move 'freely' in the newly created 'free markets', are subjected (especially if they want to produce or engage in service delivery) to asphyxiating protocols and procedures for planning and control that tend to exclude agency and responsibility. They are facing 'markets' that, in practice, turn out to be coercive structures that only allow for specific routines. There is a basic shift in hegemony as well. While the settler–colonial food regime was characterized by British hegemony, and the following one by a clear US hegemony, the imperial food regime has no clear political or territorial centre. Associated with this there is yet another distinctiveness concerning the nature of food itself and its consumption. The settler–colonial regime basically reduced initially rich and highly diversified diets to the centrality of meat and bread, while the mercantile industrial regime gradually added fats and sweeteners supplemented with starches, thickeners, proteins and synthesized flavours. The imperial food regime, in its turn, centres on the *artificialization* of food. *Latte fresco blu* is just one out of the many examples of this new trend in which genetic engineering is increasingly becoming dominant. This artificialization of food, and its expansion, is needed because the production, processing and distribution of food are, within the new imperial food regime, re-patterned into a worldwide vehicle for generating cash flows to meet the highly elevated levels of expected profitability.[25]

The major shifts that link and distinguish these three food regimes are clearly reflected in the three longitudinal studies presented in this book. The large cotton-producing *haciendas* in the north of Peru (typical expressions of the settler–colonial regime) were finally changed, through land reform, into state-controlled co-operatives that might be understood to exemplify the mercantile–industrial regime. The *imprenditori agricoli* from the north of Italy are typical exponents of the epoch of regulation and, thus, again the mercantile–industrial regime. With the slow but persistent demise of this regime and the unfolding of the new imperial food regime, these agrarian entrepreneurs increasingly deactivated their farms. It proved to be the *contadini*, the peasants of the 21st century, who were far more able to face the harsh conditions introduced by the new imperial constellations. The same applies to the North Frisian Woodlands where farmers decided to engage in the production of goods and services not directly, nor fully, controlled by Empire. They joined to create a new territorial co-operative that operates as a mechanism for relating directly with nature and society at large, while clearly functioning as a defence mechanism vis-à-vis new imperial constellations. While repeasantization has its historical roots in the processes of marginalization and exclusion related to previous regimes, the newly emerging imperial food regimes amplify marginalization and exclusion to levels hitherto unknown that stimulate and strengthen the process of repeasantization.

The agenda

The imperial food regime emerging since the 1990s – a process in which the World Trade Organization's (WTO's) 1995 Agreement on Agriculture represents a major landmark (Weis, 2007, p128) – introduces new challenges into both political and scientific agendas. I can be brief in so far as the political agenda is concerned. Many such issues have already been presented and discussed throughout this book. There is, however, one overarching question that synthesizes the other more specific issues. Will agriculture, in the decades to come, continue to feed the world's population in a healthy and sustainable way, or will it, to echo the title of a recent article by Harriet Friedmann (2004), increasingly be 'feeding the empire'? The consequences of such a reversal, summarized in several debates as 'profits versus people' (Bernstein et al, 1990) and more recently as the 'fuel versus food' dilemma, will probably be dreadful.

The nature of the imperial food regime also raises a range of issues that are challenging the main theoretical approaches and models used to understand the world. The fusion of market and assembly line is again pivotal. I will limit myself here to one, probably central, issue: the evaporation of the very notion of value. Due to Empire, value has, as it were, become footloose; it is increasingly becoming 'a ghost'. As amply explained in the first part of *Das Kapital* (Marx, 1867/1970), the double character of commodities resides in the enigmatic combination of use value and exchange value. From the industrial revolution onwards, both were specified *within* the factory (literally, the place where things were made) or farm (also interestingly referred to during the past in Italy as *fattoria*). The specific characteristics of products that made them useful (and were thus built into them) were specified *within* the factory itself according to routine or to explicit and evolving designs. Equally, exchange value was assessed *within* the factory: the labour time needed for its construction determined the exchange value that was to be realized through subsequent transactions. Within and by means of Empire, the factory and the farm now tend to be radically eliminated: 'Networks, not firms, have become the actual operating unit' (Castells, 1996, p171). Places of production are no longer the spaces where utility, usefulness and aesthetics are defined and constructed. At best, they are the temporary spots ('non-places') where particular elements are made according to the specifications formulated elsewhere – that is, at the third level. Subsequently, the different parts are assembled together, again according to the blueprints defined and imposed by Empire. Then the assembled product is made to flow to other spots where it is paid for and used. The second-level flows are likewise administered by Empire. Prices no longer bear any relation to labour time or, more generally, to production costs – let alone social and environmental costs. They are, above all, the expression of the imperial needs of having market shares as large and as rapidly expanding as possible, realizing the highest possible return on invested capital and, at the same time, enlarging shareholder value.

Once, the factory and the farm were places where use value and exchange value were created. They were relatively autonomous units of production (self-organizing spaces), with a well-functioning resource base that was employed and developed in a skilful way. Furthermore, both factory and farm embodied a certain *agency*: they could strive to do better than others, to make a difference. Currently, however, farm and factory are primarily appendices of a worldwide Empire that has taken over the symbolic definition as well as the material organization of use and exchange value. As a consequence, the different units of production might formally be autonomous; but substantially they are completely dependent because it is now impossible to function outside the assembly lines (and command lines) of Empire. Concomitantly, the notion of use value is blurred. The main utility of products such as *latte fresco blu* (or any other lookalike product) is that they allow for accumulation and enrichment at the third level, just as the main use of imagery or discourse (e.g. of sustainability) is that they further strengthen control and help to solve typical level-three problems. The same happens with the notion of exchange value: it now mainly resides in, and is created through, the combination and reproduction of places of wealth and places of poverty.

In short, rethinking value for a context in which it no longer seems to matter is probably one of the most important contributions that science can deliver to our present world. This applies especially to post-modern peasant studies.

10
The Peasant Principle[1]

Introduction

The peasant condition is composed of a set of dialectical relations between the environment in which peasants have to operate and their actively constructed responses aimed at creating degrees of autonomy (Gouldner, 1978) in order to deal with the patterns of dependency, deprivation and marginalization entailed in this environment. Responses and environment mutually define and shape each other; it is impossible to understand one without the other. There is no 'external' relation between them: the two are linked by internal relations through which the responses shape the environment as much as the environment generates the responses. Such mutual articulation unfolds dynamically over time as each side of the equation impacts upon the other. Typical of the peasant condition is that the responses unfold by means of constructing a resource base that allows for the co-production of man and nature.

Although wider society strongly puts its imprint on every time- and place-bound expression of the peasantry (Pearse, 1975; Shanin, 1990), there are also autonomous moments, histories and collective memories within the peasantry itself. The latter contribute to the way in which the peasantry constitutes itself within society. Hence, any form of uni-linear determinism that assumes the peasantry to be a straightforward *derivative* of its structural context must be rejected (Long, 2001). The same applies to any *a priori* scheme that separates and hierarchically orders 'dominant institutional frameworks and discourses' and 'subordinate actors' (Long, 2007, p66). The two are far more interwoven, as Long (2007) has recently argued in a convincing and theoretically underpinned way. According to him: 'we need to probe more closely into the dialectics of "dominant" and "subordinate" social forms'.

The peasantry represents a historical subject, as demonstrated, for example, by the 'peasant wars of the 20th century' (Wolf, 1969; Huizer, 1973), by the many micro-situations that tell the story of the sturdy construction of progress (Ontita, 2007) and by the rich morphology of farming styles: several farming styles can be seen as critical responses to the rationale imposed by current regimes. Nonetheless, the specification of context matters, precisely because the particularities of context impact upon the ways in which peasant responses are articulated (see Paige, 1975, on collective struggles). The point I wish to emphasize in this final chapter is that the growing hegemony of Empire as an ordering principle[2] implies a far-reaching redefinition of the 'hostile environment'. Consequently, Empire reconstitutes the peasantry as a new phenomenon that crosses, in several respects, the borders of the peasantry as we knew it in the past. Empire increasingly changes the context in which today's peasantries are embedded: Empire articulates with these peasantries as a radical negation of their very existence. Thus, Empire equally provokes new forms of resistance, struggle and response. Through the many contradictions and confrontations between Empire and the peasantry, the 'peasant principle' is strengthened and extended. The peasant principle is an emancipatory notion. It outlines the potentials entailed within the peasantry – potentials that are currently being blocked by Empire but at the same time are (re)activated by it. Thus, the peasant principle embraces the countervailing powers generated within the peasantry itself.

Empire and the peasantry

As I have discussed throughout this book, Empire is an ordering principle that is expressed in, and by means of, many different entities and relations. It has many different drivers and sources and also takes many different forms that build on a wide range of contrasting mechanisms. For example, Empire is expressed as Parmalat, as the usurpation of water in Bajo Piura and as the tightening of the squeeze exerted on European agriculture. It is present in the Dutch Manure Law and in the associated 'global cow'. But Empire is also expressed in many forms not discussed in this book, such as the reduction of the world's grain reserves, genetic manipulation, and in the curricula and research pursued in many agricultural schools and universities. Empire is, in short, multicentred.

Through all of its manifold expressions, Empire impacts in a characteristic way upon the peasantry. Wherever located, the peasantry currently faces three highly destructive tendencies. These tendencies – no matter how they are articulated – relate to and reside in Empire.

First, the resource base on which peasant modes of farming are grounded is the object of considerable distortions, if not of abrupt processes of disintegration. Normally a well-balanced and well-crafted constellation, the resource base becomes disassembled (mostly by interrupting some of its strategic connections) while the associated process of co-production is slowed down. This occurs by means of various mechanisms, of which some have been closely analysed in this book. Blocking access to lines of credit, the usurpation of water, the abrupt elimination of market outlets through the introduction of lookalike products, the destruction of important institutions such as communal landownership and tenancy arrangements, and the imposition of regulatory schemes that prevent the further unfolding of well-balanced forms of co-production – all of these processes erode historically created resource bases. By taking away or negatively affecting one or more of the strategic connections, the resource base will, in the end, dissolve into a dispersed set of unused (and therefore useless) assets.

Second, due to extraction processes (or what I termed 'drainage') to which large segments of agriculture are subjected, Empire tends to introduce a generalized precariousness into the farming sector. As signalled before, some 40 per cent of Dutch farming families derive less than the legal minimum wage/income from farming, and the effects of globalization and liberalization may increase this significantly. In Italy, similar or even higher levels of 'poverty in the countryside', as the title of an official publication of the Italian Ministry of Agriculture puts it (MPAF, 2003), are encountered. There are, of course, sharp contrasts in living conditions and rural livelihoods between European and developing countries where some 800 million of the 850 million who suffer chronic under-nourishment are located – many of them peasants. Nonetheless, according to local social (and legal) conventions, precariousness and deprivation are experienced in both the centre and the periphery.

Third, through the takeover of strategic resources – land, genetic material, water, market outlets – Empire often creates new parallel circuits for the production of specific commodities. This often implies that large numbers of peasant producers (and many other producers linked to them through local networks) are *de facto* sentenced to redundancy.

Admittedly, features such as disintegration, precariousness and redundancy are hardly specific to Empire.[3] Yet, what is unique to Empire (as a mode of ordering) is that it converts the disassembling of local forms of production, the draining of wealth, and the associated induction of precariousness and redundancy into unprecedented phenomena. This becomes clear when the scale and intensity of these phenomena are taken into account and their effects scrutinized. Due to the centrality of control – enabled by new planning and monitoring technologies – Empire as a mode of ordering tends to be ubiquitous and all embracing: it extends over many domains of social and natural life, and introduces forms of control that hardly leave any spot unaffected. It penetrates into the smallest details of the social and natural world (even affecting how brambles grow). Through Empire, a widening range of connections, processes and outcomes are specified in a strict non-negotiable way. This tends to reshape the world into a totalitarian constellation in which 're-feudalization' is becoming a new and worrying feature (Ziegler, 2006; see also Benvenuti, 1975b, for one of the first discussions of this phenomenon). Empire extends control beyond the boundaries of imagination (and, for that matter, beyond the boundaries of historical precedent as well). The emergence of 'level three' is a clear expression of the unprecedented nature of this extension.

In equal manner, the commoditization of increasing parts of productive infrastructures introduces levels of drainage that are hardly yet known, the more so since such drainage is quickly generalized through the mechanics of global markets. Low prices paid for vegetables in Senegal or Kenya (or for asparagus in Peru or China) will translate immediately into downward pressures in Europe. The world market is not primarily a mechanism that makes the best products and services generally available; it tends, instead, to generalize the worst conditions of production on a world scale (Bové, 2003).

Finally, there is the element of redundancy. Beyond the already strong tendency to produce a relative surplus population (Ploeg, 1977, 2006d), there is now the widely distributed practice of outsourcing by agribusiness groups. This implies that many (if not potentially all) production areas and groups of producers can, overnight, become superfluous.

According to José Carlos Mariátegui (1925), a rural sociologist *avant la lettre* from Peru, there is no intrinsic problem with social change or transformation. The only thing that matters is that the new is better than that which it replaces. Evidently, Empire is a *new* mode of ordering, currently being imposed upon large segments of the social and the natural world. However, whereas increases in productive employment and in the value added produced are what is needed, Empire results in the abrupt reduction of both. Where development is necessary, Empire creates places whose main quality and *raison d'être* is enduring poverty – and if value added is produced, Empire siphons it off. The same applies to sustainability and the quality of food, life and work. Empire produces only virtual sustainability and virtual qualities. By prescribing and controlling the work of millions of people (through the allocation of resources and, especially, through the authorization of their use by means of strict cycles of planning and control), production is, as it were, frozen. Dynamics, innovation and heterogeneity are ruled out. To slightly paraphrase Knorr-Cetina (1981, p7), one could argue that 'social order is no longer that which comes about in the mundane but relentless transactions of individual wills – [under Empire] social order increasingly [tends to be] the outcome of a monolithic system which regulates individual action and controls individual wills'. Associated with this is a second, deeply troubling, feature: Empire creates dependency; yet it simultaneously introduces turbulence and insecurity. Thus, on the one hand, 'the network of interdependence caused by the growth of specialization widens' (North, 1990), which makes 'institutional reliability essential'. On the other hand, though, Empire destroys institutional reliability as much as it assumes it.

Resistance

In its relation to Empire, the peasantry increasingly represents resistance. This is a multiple resistance that expresses itself at many different levels, unfolds along different dimensions and involves a wide range of different actors. Peasant resistance (as we witness it at the beginning of the 21st century) is not only, or primarily, articulated through overt struggles (demonstrations, marches, occupations, road blocks), though such expressions are never absent. Neither is peasant resistance limited to the everyday acts of defiance summarized by Scott (1985) as 'weapons of the weak' (see also Torres, 1994, for interesting examples). Building on Long's (2007) perceptive reformulation of the issue of resistance, I think we have to recognize that there is a far wider and probably far more important field of action through which resistance materializes. Resistance is encountered in a wide range of heterogeneous and increasingly interlinked *practices* through which the peasantry constitutes itself as *distinctively different*. Resistance resides in the fields in the ways in which 'good manure' is made, 'noble cows' are bred and 'beautiful farms' constructed. As ancient and irrelevant as such practices may seem if considered in isolation, in the context of Empire they are increasingly vehicles through which resistance is expressed and organized. Resistance equally resides in the creation of new units of production and consumption in fields meant to lay barren or to be used for large-scale production of export crops only. In synthesis, the resistance of the peasantry resides, first and foremost, in the *multitude of responses* continued and/or created anew in order to confront Empire as the principal mode of ordering.[4] Through, and with, the help of such responses, they are able to go against the tide.

At first sight, the old and lonely Dutch peasant in Figure 10.1 seems to be just another 'potato' out of the well-known sack of Karl Marx: isolated and lost in a seemingly incomprehensible activity that must be, according to the primitive kind of instrument he is using, a very traditional routine. However, as I underlined in Chapter 2, immediacies are highly treacherous, especially in the world of peasants. As a matter of fact, this particular peasant is engaged in a highly meaningful, recently developed (or rediscovered) activity. The image reflects one of the many responses to imposed schemes, procedures and scripts. He is actively distancing grassland management from the dominant patterns of herbicide use. Instead of spraying the thistles (that every now and then reappear in the meadows) with toxic chemicals, he carefully pulls them out of the soil using a large pair of wooden scissors. The use of this artefact represents a 'retro-innovation', as Stuiver (2006) would argue. It is applied with considerable skill, at the right time, and handled in such a way that no particles of the roots remain in the subsoil. But the picture reveals far more. It shows that the standard routine of having just one 'entrepreneur' on the farm is overcome here: it is more than likely

Figure 10.1 *Lost or carving out new pathways?*

Source: Ploeg et al (1992, p48)

Figure 10.2 *Humility or pride?*

Source: Photograph by Jan Douwe van der Ploeg

that this old father works alongside his son (or daughter). There is a shift in economy as well. Cost reduction is realized here through the elimination of external inputs (herbicides and perhaps the mobilization of a contract worker to apply them). These are replaced by the improved use of internal resources. Simultaneously, mechanical technologies are replaced by a skill-oriented technology. It is even possible that this practice is (re)generated in order to contest the strict regulation schemes associated with weed and pest control.

As Figure 10.1 indirectly shows, the newly emerging responses are related to narratives that specify (or obscure) their social meaning, relevance and embedding. These narratives might well compete with each other. It may also be argued that the response entailed in Figure 10.1 is accidental – it is there simply because an old father cannot accommodate himself to the luxury of rest. It is also possible that government agencies 'expropriate' the response ('black boxing', in its original meaning) and claim that farmers 'are finally taking seriously the politics of sustainability'. Although, it is equally possible that state agencies will make this response illegal because it is not easily controlled.

Figure 10.2 indicates another response. It shows a peasant family from Catacaos in Peru. They are settled, as so many others, in *tierra de lucha campesina*: in the area that once was the place of massive and repeated invasions and where later one of the first communal units of production (San Pablo Sur) was created. To some the image may refer to being lost in messiness and poverty (just as the previous Dutch peasant seemed to be lost in emptiness, solitude and lack of meaning). There is some fodder, a corral with a few sheep, the wall of a temporary shelter basically made out of straw and, of course, there is a man and his wife. For them, however, the story is different: it is about a resource base they have constructed throughout the years, a resource base that grounds the expectation that with their own labour they might, *si Dios quisiera*, forge some progress.⁵ It is about relative autonomy and, above all, about the actively constructed response to an ugly regime that condemns many people to despair, to 'sell your strength for a few *centavos* and then even be obliged to be grateful for that'. In both images there is also pride and dignity. These are important elements because 'countervailing power resides in the dignity of everyday life' (Holloway, 2002, p217).

The mainstream of peasant resistance flows through the multitude of responses that are actively created to face and counter, as Long (2007, p64) argues:

> ... the inequities in international trade agreements, unacceptable levels of labour exploitation, controversies concerning the role of science, GM crops, and methods of controlling environmental pollution, as well as the implementation of bureaucratized systems for the measurement and regulation of product quality and food

> *safety ... [in short] the battlefields [in which] food fights are taking place.*

Alongside this mainstream of materialized responses there are others: the overt struggles, and the food-dragging, ear-cutting, irony and well-camouflaged sabotage. The three are always present, although the relevance, visibility and force of each component are continuously changing. However, the mainstream (that often feeds the others) should never be left out of the analysis: it is the bedrock of peasant resistance. Analytically, the construction of the many, often actively interconnected, responses goes back to the following mechanisms.

As discussed earlier, Empire tends to disassemble existing constellations by eliminating, taking over and/or redefining strategically important connections. The new peasantries tackle such disassembling through a rich spectrum of *re-patterning* techniques. This occurs in the context of peasant-driven processes of rural development described in Chapter 6 – just as it occurred in the case of Catacaos. Wherever connections with consumers are disrupted by Empire (through paying extremely low prices to producers, or by condemning the latter to redundancy as would have been the fate of the Italian dairy farmers in the *latte fresco blu* case), peasants search for and actively construct new connections through direct selling (Milone and Ventura, 2000; Schuite, 2000), farmers' markets (Knickel and Hof, 2002), the creation of new alternative agro-food chains (Marsden et al, 2000a, 2000b), and public procurement schemes (Morgan and Sonnino, 2006). That which has been the object of disassembling (in order to be assembled anew in accordance with the ordering principle of Empire) is actively reconnected and re-patterned by peasants. For instance, lack of credit is increasingly countered by the mobilization of resources through trans-local livelihoods (Sivini, 2007). What is denied on one side of the border is mobilized on the other. This same type of response is also encountered in current peasant-driven forms of land reform (Borras, 1997; UNRISD, 1998).

A second mechanism relates to the precariousness introduced by Empire. Many responses aim at, and succeed in, constructing new ways of creating (and protecting) new and more elevated levels of value added – precisely at those points where Empire drains value added away. The emergence of new peasant technologies (see Chapter 6) is an outstanding example of this type of response. Third, several responses aim at enlarging autonomy. The development of the North Frisian Woodlands (NFW) co-operative described in Chapter 7 and the construction of what finally became a *bug* are clear expressions of this. Fourth, in as much as Empire tends to make parts of the peasantry redundant, so will new peasantries start to reposition themselves (both symbolically and materially) as *citoyens* whose rights cannot be neglected.[6] Examples of such responses are the 'return' of Dutch peasants to nature reserves,[7] the sturdy and resilient development of farming in areas considered by experts to be unsuitable for agriculture (as the Northern Frisian Woodlands or the mountains of Abruzzo; see Milone, 2004) and the remaking of good manure.

A fifth and equally important set of responses centres on the reassessment of visibility. Empire tends to create invisibility (see also Holloway, 2002, p214) since production is moved to 'non-places' where the origin of food (or of its many ingredients) is hidden behind the façade of lookalike products, and primary producers are made anonymous and interchangeable. They tend to be converted into 'non-persons' whose identities and skills do not matter. For example, as long as Hazard Analysis and Critical Control Point (HACCP)[8] criteria are met, it does not matter *who* produces the asparagus, cheese, milk or tomatoes. In this respect, there is a strong resemblance to the position of *indios* within the *hacienda*-dominated system that formerly characterized many Latin American countries. The fear of *invisibility* turned out to be a strong element in the cultural repertoire of the indigenous population (reflected, for example, in Scorza, 1974; see also Montoya, 1986) – a fear that was clearly nurtured by the created redundancy, precariousness and often brutal marginalization caused through the takeover of land, water and market access. Currently, Empire reintroduces such a threat; but at the same time it triggers a range of responses. Peasant communities such as Catacaos proudly claimed their existence, making themselves, indeed, *visible* through the declaration of 'shared values' (see Box 3.1 in Chapter 3). The same applies to the territorial co-operative of the Northern Frisian Woodlands (see Box 7.1 in Chapter 7). It also applies at the micro level when, through endogenous processes of rural development (described in Chapter 6), farms regain a name and unique identity. The internet is an important medium for this recovery of their distinctiveness and associated visibility. In a brief survey (carried out in 2005), Henk Oostindie encountered more than 2000 Dutch farms with their own website.

A sixth category of responses relates to the presence and use of *conversion mechanisms* that differ from those imposed by Empire. In a world ordered by Empire, conversions occur through monetary transactions aimed at maximizing profit. Through Empire, exchange value and profitability are made dominant over other kinds of use value (Holloway, 2002, p262) or, to paraphrase Burawoy (2007, p4), 'the mode of exchange oppresses the mode of production'. Consequently, resources, labour force, knowledge, products, services, etc. are all converted into commodities, and markets function as the exclusive domain through which all connections, transformations and translations are organized. Hence, many connections become impossible, many resources lay idle, many lives are wasted, and many conversions are blocked. The factory and farm (having no internal market) and/or socially regulated exchange are thought to be irrelevant and impossible materially. However, the interesting feature of agriculture is, of course, that alternative conversions (that do not imply *all-purpose money* or assume markets) are omnipresent.

The same applies to reciprocity as an important connection between farms that convert the idle time of both oxen and peasants into productive labour – without engaging in any monetary transaction whatsoever. Such non-market conversion mechanisms are everywhere in agriculture. They range from exotic examples in which a shortage of labour is matched by the breeding of piglets which, in turn, are converted into food and gin, and subsequently converted into a labour group that steps in where labour is lacking (Ploeg, 1990b), via other well-known (but often misunderstood) examples, such as the conversion of available cattle into improved (more productive) cattle (instead of buying them), to forms of socially regulated exchange that assume widely extended networks. An example of the latter is the mobilization of labour in the Italian hills for the olive harvest. Normally, farmers from the plains are invited to join the work. In exchange they receive bottled olive oil. Labour is mobilized without the need for any monetary transaction so long as farmers from the plains obtain their needed olive oil.

In this way new resources are created, forms of social security are established (Nooteboom, 2003) and activities, practices and development trajectories are constructed that would have been impossible if effectuated through the markets. Thus, the rebelliousness and reformism of the peasantry reside partly in its capacity to operate on the boundary that separates commodity from non-commodity circuits. The same occurs where new products are channelled through new circuits that link producers and consumers: transactions are, here, strongly embedded in new non-commodity relations that dominate, specify and legitimize particular commodity relations. These new constellations, it seems, often appeal to and attract young people (from agrarian as well as non-agrarian backgrounds), just as they are strongly supported by new social movements such as the Slow Food Movement. At the same time, they extend these new hybrid combinations of commodity and non-commodity patterns into a wide range of sometimes dazzling phenomena, such as we witness, for example, in 'community-supported agriculture'. In short, as a reaction to the current dominance of markets, 'old' non-monetary transactions are increasingly being disseminated in substantial ways, while at the same time new forms are being developed. Their importance should not be underestimated. They function as a symbolic critique of, as well as an alternative to, the type of market-governed conversions that are central to Empire. Indeed, they may be viewed as acts of *insubordination* to Empire.

Such alternative conversion processes allow room for manoeuvre (or 'space') that otherwise would be absent and they demonstrate that things can, indeed, be done better. In his recent discussion of the 'fate of society',

Burawoy (2007, p7) refers to the 'utopian dimension'. Following his line of thinking, the preceding discussion of peasant responses might be synthesized by stressing that together they represent an attempt to produce 'actually existing utopias', as opposed to 'imaginary utopias' (Burawoy, 2007, p7). Even when they are only partly aware of it, the peasants of Figure 10.1 and 10.2 attempt to create – together with millions of others – their own 'actually existing utopias' that present themselves as a concrete critique of Empire.[9] Hence, paradoxically, Empire provokes and triggers responses that create resources, connections, processes of conversion and additional wealth (as minimal as they might be) that Empire could never have assembled or provided.

Reconstituting the peasantry

In Chapter 2, I outlined the choreography of the peasantry – that is, the way in which it moves through time. It should by now be clear that this choreography is far from circular or repetitive. It unfolds in a dynamic and heterogeneous way. Its script is constantly being rewritten according to the difficulties and challenges that emerge along the road. At the same time, it is clear that the current conjuncture is as much characterized by processes that tend to regiment farming populations (aligning them to Empire as a mode of ordering), as by peasant responses that aim to go beyond the limits of the imposed regimes. As a matter of fact, what we are witnessing now is a completely new kind of resistance. It is not the resistance of frontal confrontation, enduring industrial strikes, occupations and painfully disciplined class organizations. Neither is it merely defiance. There are, of course, every now and then eruptions, overt fights and occasional forms of sabotage. But these dissipate, just as they emerge, into a multitude of responses. Equally important in this new form of resistance is that it basically searches for, and constructs, *local* solutions to global problems. Blueprints are avoided.[10] This results in a rich repertoire – the heterogeneity of the many responses thus becoming one of the propelling forces that induces new learning processes (Pernet, 1982; Reinhardt and Barlett, 1990).

This pattern reflects the new relations imposed by Empire: frontal confrontations are increasingly impossible, if not counterproductive, and claims over global solutions are deeply distrusted. Instead, responses now follow a different road:

> *Resistance is no longer a form of reaction but a form of production and action. ... Resistance is no longer one of factory workers; it is a completely new resistance based on innovativeness ... and on*

> *autonomous co-operation between producing [and consuming] subjects. It is the capacity to develop new, constitutive potentialities that go beyond reigning forms of domination.* (Negri, 2006, p54)

I think this describes fairly well the multitude of responses discussed. It is resistance that is difficult to capture. It is everywhere: it is multiple, it is attractive and mobilizing, it re-links people, activities and prospects. It comprises a constant flux of often unexpected expressions that time and again flows over the limitations imposed by Empire. Every single form is an expression of critique and rebellion. It is a deviation and it simultaneously articulates superiority. Yet, these expressions are innocent and harmless. However, when combined, they become powerful: they change the panorama. When bundled into one flow, they reconstitute peasantries (once again) as 'uncaptured entities' that play an important role within the complex interrelations that define power.

In these ways, Empire reconstitutes the different peasantries of the world. The triggering of a multitude of responses is one of the important dimensions along which this reshaping is currently taking place. A second, equally important if not decisive, dimension concerns autonomy. Empire creates, wherever it operates and penetrates, an all-encompassing regime that excludes autonomy – at whatever level and in whatever form – simply because centralized control and appropriation are its main features and mechanism of development. It is along these same lines that we have to rethink the peasantry – not the farming population as a whole, but precisely the part that is (re)constituted as *peasantry*. Empire induces in this newly emerging peasantry a deep distrust in large commodity markets and the ways in which they are governed. It likewise induces or strengthens the already existing distrust in state apparatuses and the regulatory schemes that they impose. By doing so, Empire makes autonomy an overarching need that is explicitly articulated by the new peasantries.

A third dimension, along which the peasantry is currently being reshaped, is related to the fact that for the population involved *there is hardly any alternative but farming*. This introduces a new doggedness into the construction of responses and alternatives. While, in the past, the cities and the associated urban economies appeared to represent alternative opportunities, the operation of Empire implies (through the changing international division of labour, as well as through the restructuring of urban economies) that such opportunities are now generally lacking. Thus, Empire is backfiring in a contradictory way. It tends to strongly decrease levels of rural employment, while simultaneously eliminating needed alternatives (Ploeg, 2006d, Annex 1). As a consequence, and despite the high rates of urbanization, the ranks of those working in and dependent upon agriculture (see Table 10.1) have remained strikingly constant over the past four decades.[11]

Such phenomena are not limited to Latin America since Central and Eastern European countries are witnessing the same tendencies towards repeasantization (Burawoy, 2007, p2). And if the qualitative dimension is taken into account (which relates, among other things, to the attractiveness of working in agriculture), the argument can be extended to cover large parts of Western Europe and the US as well.

Table 10.1 *Evolution of the agricultural labour force in Latin America (1970–2000) (thousands)*

Country	1970	1980	1990	2000
Argentina	1495	1384	1482	1464
Bolivia	872	1064	1249	1497
Brazil	16,066	17,480	15,232	13,211
Chile	715	800	938	980
Colombia	3080	3776	3696	3719
Costa Rica	243	290	307	324
Ecuador	997	1013	1201	1249
El Salvador	673	697	709	775
Guatemala	1106	1257	1569	1916
Honduras	580	684	693	769
Mexico	6541	7995	8531	8551
Nicaragua	350	393	392	396
Panama	211	197	245	251
Paraguay	409	514	595	706
Peru	1915	2183	2654	2965
Uruguay	207	192	193	190
Venezuela	829	751	874	805
Total	36,289	40,670	40,560	39,768

Source: adapted from Long and Roberts (2005, p63)

Together, these three dimensions (new forms of resistance, the search for autonomy and doggedness) feed into two highly important new features. These are the emergence of the 'peasant principle' and a redefinition of the 'agrarian question'. I discuss these mutually related features in the next two sections.

The 'peasant principle'

As indicated earlier, the 'peasant principle' is an emancipatory notion. It implies that being engaged in the peasant condition needs to be understood as a flow through time that entails the promise of offering some way forward. The realization of such a promise depends as much upon the availability of space (Halamska, 2004) or room for manoeuvre (Long, 1985, 2001) as it does upon engagement, involvement and dedication. In this respect the peasant principle tends to converge with social biographies; and since it is always about shared prospects and shared conditions, the peasant principle equally coincides, at least partly, with what is generally referred to as a mode of livelihood. However, as much as the peasant principle is rooted in the peasant condition, it also goes beyond it. Even when the direct circumstances imply deprivation and despair, the peasant principle contains hope: the hope that through hard work, co-operation, joint actions and/or overt struggles, progress might be wrought. The peasant principle allows the actors involved to reach beyond the immediacies of context. It also feeds into peasant resistance, into the proverbial resilience of the peasantry and into *bodily* struggles in the fields, cowsheds and corrals – struggles that aim to forge some progress. In short, the peasant principle is about facing and surmounting difficulties in order to construct the conditions that allow for agency. It may also be seen as the peasant condition *projected on the future* – that is, it synthesizes the script that projects peasants through time. It links the past, present and future, it attributes sense and significance to the many feedback and feed-forward mechanisms that relate the different stages to each other, and it embeds the many different activities and relationships within a meaningful whole. In short, the peasant principle carves pathways into the future. It is also about subjectivity – the peasant principle implies that particular worldviews and associated courses of action matter. It stresses the value and satisfaction of working with living nature, of being relatively independent, of craftsmanship and pride in what one has constructed. It also centres on confidence in one's own strengths and insights (see Box 10.1 for an illustration).

There is, of course, also a mirror image to this peasant principle. When things go wrong and/or when one has to quit farming, the emphasis on self-confidence (which is needed to be able to face a hostile environment) might translate into highly negative self-perceptions – that is, one may stress that one has failed where success had been assumed, which – as often postulated – exposes fragility and personal flaws (see Frouws and Ploeg, 1974, for an extensive inventory). Furthermore, the peasant principal might equally translate into asking far too much of women and children – then the *padre padrone* becomes dominant. However, the peasant principle normally entails various 'counterpoints' (Wertheim, 1971) that allow for a critique of such deviations.

I believe, therefore, that, apart from the concept of peasant condition outlined in Chapter 2, we also need that of peasant principle. The peasant condition assumes agency in order to realize the choreography synthesized in Figure 2.2. It is only through active and goal-oriented involvement that the peasant condition will progressively unfold. It is precisely this goal orientation and associated dedication that I wish to emphasize since it is by means of the peasant principle that the peasantry makes its mark on agrarian and rural history – and on the making of future trajectories.

Box 10.1 *An expression of the peasant principle*

'The beauty of being engaged in farming': A column by Monique van der Laan, dairy farmer, in *Agrarisch Dagblad* [*Agrarian Journal*], Wednesday, 4 October 2006

Prices are still decreasing, more and more rules are [imposed upon] us, we are dependent upon big markets (the world market, the supermarkets), controls are being augmented, our work is increasingly prescribed, financial burdens are rising and one has to make very long working days.

However, I definitely would not like to work under a boss. The freedom I have as a farmer, the possibility [of organizing] my own work and my own schedule – it is all very important to me. We are working in the open air, have a lot of alternation and variation in the work, both mentally and physically. We are working with nature, with animals. Thus, you are confronted every day with values that refer to life. We are proud of our animals, our products: they are fresh and tasty. All of this compensates for the negative elements with which we are confronted – the more so if you can escape asphyxiating rules by processing and selling your own meat and cheese, while you simultaneously obtain a better price.

Farming is also about anticipating the new needs that emerge in society at large. People are searching for quietness, space, active involvement in outdoor activities, authenticity and genuine products. As farmers, we can respond to all of this by opening a camp site or offering services for canoeing. Farmers invest in such activities in order to strengthen their farm. Risks are in this way spread and possibly profitability will be far better than when you're investing in milk only.

I know there are colleagues saying that those with additional activities are no longer farmers. But I think such a statement reflects a too limited worldview. Compare this with agrarian management of nature and landscape. When some ten years ago some farmers took the lead with such management, many colleagues replied that 'they wanted to be farmers, not nature managers'. Nowadays, however, it has become an integral part of most farms. An additional branch can render a lot of value to your farm. For us, the visitors to our farm are representatives of our consumers. Here, visitors taste our products; they take them home and give us their feedback. For us, that is a nice and cheap form of market research. People sometimes confront us with ideas we would never have dreamed of. With such information we try to improve our farm and our products in order to obtain a better position in the market.

There is no better form of public relations than direct contacts and word-of-mouth information. It makes you proud as a farmer when visitors appreciate your products and pay you a good price. This positive energy impacts upon the job as a whole. Thus, negative news moves into the background, precisely because you are going your own way.

Historically, the peasant principle was articulated and simultaneously operated as a line of defence against the many threats, dangers and temptations that surrounded the peasantry. Horse gambling, the cities, the neighbour's wife, the charms of alcohol, laziness and card playing, cheating consumers (my grandfather was very keen in this respect), hay trading, and accepting seemingly irresistible offers from bankers and traders – the pitfalls have been many. At the same time, the cultural repertoires – summarized here under the category of the peasant principle – contained many responses and, in practice, every incident and mistake led to discussions that in one way or another reaffirmed the norms.

In the current circumstances, though, it is different. The peasant principle is now both triggered by and operates as a response to Empire. This new relation turns the peasant principle into a many-sided negation of Empire. It inspires and informs resilience and multiple resistance. It allows people to communicate – even when significant differences in culture and language must be bridged. It is also the vehicle by which people actively engage in and further develop the peasant condition. The peasant principle appeals to growing sections of the farming population. Whereas the script of entrepreneurial farming increasingly fails to outline a convincing trajectory for development and survival, the peasant principle, with its focus on the construction of an autonomous and self-governed resource base, clearly specifies ways forward. This is especially so under current highly adverse circumstances.

The peasant principle is increasingly linked to the superior performance that it makes possible. Thus, the principle becomes a symbol that operates in wider society as a positive indication of what farming *might* be. Performances that are highly important, in this respect, are the superior energetic efficiency levels entailed in peasant farming (Netting, 1993),[12] the efficiency of water use (Dries, 2002) and the specific interrelations of nature, animals, landscape and people created in the context of co-production (Gerritsen, 2002). If developed along such lines, peasant farming might provide a major *liaison* between society and nature – a liaison that is positively valued and actively defended and supported by wider society.

In a recently translated essay, the Norwegian rural sociologist Ottar Brox (2006) refers to the fact that Norway, now a major oil-exporting country, combines the wealth derived from its natural resources with democracy and a relatively equal income distribution. Compared with many other major oil-exporting countries, this is quite exceptional (Ploeg, 2006a). According to Brox (2006), this situation can be explained mainly by the presence of a strong and autonomous peasant population. Being a poor country at the beginning of the 20th century, Norway had a large and independent peasant population that was also very much involved in fishing. When industrialization started, the peasants who became occupied in the first factories maintained their peasant properties. This allowed them to literally return to farm activities as soon as working conditions in the newly emerging urban enterprises turned out to be unattractive. Thus, the peasant principle induced, right from the beginning, a strong democratic tradition and a well-balanced income distribution in Norwegian society. In short, the peasant principle might have several direct positive effects; but it may also impact indirectly in a positive way upon society at large.

Throughout history the peasant principle has travelled. The massive postwar movement of Italian *mezzadri* to the urban economy introduced the concepts and practices of autonomy, networking, flexibility and novelty production into the small- and medium-sized enterprises established by former peasants and which, in the end, composed the basic nuclei of many of the vibrant economic districts that emerged as the cornerstones of the Italian economy (Beccatini, 1987; Ottati, 1995; Camagni, 2002). A similar process pertained to the growth of 'informal economies' in the developing world, which grew especially from the 1950s onwards, as well as to forms of 'urban agriculture' currently emerging in many of the world's large metropolises. Both embody elements of the peasant principle.

The peasant principle also travels to, and thus mobilizes, people from non-agrarian places. A key example is the Brazilian movement of landless people (*Movimento dos Sem Terra*, or MST), which played an important role in mobilizing people whose life in the urban *favelas* was reduced to *lixo humano* (human dirt) (Athias, 1999) to migrate to new rural settlement areas to regain dignity and the prospect of a better life, especially for their children – they move to the countryside and become peasants. Here, the peasant principle functions as the link between the construction of some new autonomy and the promise that things can improve through one's own dedication and willingness to engage in everyday struggles. Thus, the peasant principle points to ways of life that are impossible in the *favelas* of the big cities. Equally telling, in this respect, is the fact that the initial waves of organic producers in agriculture were also characterized by a strong urban background. These first organic producers not only constituted themselves as peasants – they simultaneously reorganized farming into a set of practices that was far more peasant like than agriculture was at the time.

In these ways, the peasantry might emerge as a strong adversary, if not antithesis, of Empire – especially since the peasant principle is diametrically opposed to Empire as an ordering principle and also because the new mode of resistance embodied in the peasantry makes the latter intangible (both Bakker, 2001, and Schnabel, 2001, highlight this feature in their analyses of the Dutch peasantry). Of course, the peasantry is far from being the only antithesis of Empire; there are many countervailing powers, as well as pockets of resistance within which a critique is being articulated and from which counter-tendencies have emerged. However, it is, I think, valid to argue that the peasantry especially represents a *continuous, multiple, massive, unavoidable, intangible* and probably *convincing* negation of the many expressions of Empire. The new peasantries represent insubordination; they are ever so many irritating bugs that might even provoke a *renewed travelling* of the peasant principle, which crosses the boundaries of agriculture and rural society, to inspire many other emancipatory movements within today's world – just as happened in the past.

The peasant principle and agrarian crisis

In addition to the above arguments, it can be argued that the relevance of the peasant principle also resides in the fact that it represents a powerful way out of the increasingly global and multidimensional agrarian crisis we are currently witnessing (see Figure 1.4 in Chapter 1). This centres on the sturdy reconstruction of ecological, social and cultural capital as the main resources upon which peasant farming is increasingly founded. With regard to the agrarian crisis, the peasant principle potentially entails three major reversals that together allow for a multiple reduction of dependency upon the state and financial and industrial capital.

Over the ages farming has been identified with the conversion of living nature (or ecological capital) into food, drinks and a broad range of materials. Through this process, the necessary resources were reproduced and increasingly reshaped in order to allow for more productive forms of conversion. During the modernization epoch, this deeply institutionalized pattern, which coincided with the creation and enlargement of autonomy, was interrupted: the centrality of nature was significantly reduced and farming became increasingly dependent (albeit in highly differentiated ways) upon artificial growth factors and thus upon industrial and financial capital. Currently, a strong counter-tendency has emerged that is rooted in the strategy of 'farming economically' aimed at reducing the use of external resources, while simultaneously improving and reusing internally available resources. This counter-tendency is further reinforced by organic farming. 'Farming economically', or what has been termed 'low external-input agriculture', is primarily a response to the squeeze on agriculture. At the same time, it goes beyond this in so far as it simultaneously represents a rediscovery of *ecological capital* as the main foundation for farming (Smeding, 2001). This reversal puts co-production and *art de la localité* again at centre stage, while a new scientific approach, agro-ecology, has begun to reflect and inform these newly emerging practices.

The return to nature is intrinsic to repeasantization. At the same time, it is a major response to Empire.

This also applies to a second major reversal. This concerns the development of local and regional self-regulation as an alternative to the currently dominant regulatory schemes promoted by agro-industries, supermarkets and the state that emphasize control at a distance. The struggle for self-regulation (embodied, for example, in *comunidades campesinas*, territorial co-operatives and *campamentos* of the MST) is strongly rooted in (and further strengthens) the *social capital* of new peasantries. The main ingredients of this social capital are available and effective networks, shared values, accumulated experiences and knowledge, the combination of trust and distrust, and the capacity to resolve internal conflicts, engage in learning processes, and acquire a clear vision of one's own role in today's society. Together, these varied components make it possible to re-conquer control over the organization and development of farming as a complex social practice. Whereas the classical state–market dichotomy has proved ineffective in resolving many tensions, frictions and contradictions (especially with respect to issues of sustainability; see Hagedorn, 2002), the new institutional solutions rooted in social capital are expected to develop and put their imprint on agriculture and the countryside (OECD, 1996; Rooij, 2005). In a democratic context, such a change is, in the end, unavoidable – all the more so since it is the only way to reduce the currently high level of transaction costs.

A third major reversal entailed in the peasant principle concerns the interrelations of the producers and consumers of food. Over recent decades these interrelations have increasingly been reduced to those controlled by food processing industries and large retailers (Wrigley and Lowe, 1996; Goodman and Watts, 1997). Together these industries and retailers have reorganized trade in agricultural products by establishing a completely anonymous market in which the origins and destinies of products no longer matter (Ritzer, 1993). At the same time, new lookalike identities have been added to the final food products. It is precisely this new contradiction that has created the space for a third reversal that centres on the creation and use of *cultural capital*. Origin, quality, authenticity, freshness and specificity of products, and of associated ways of producing, processing and marketing, are clearly articulated in order to attract consumers and to communicate the *distinction* embodied in food – a distinction that 'passes' to the consumers themselves (and to the act of consumption). The latter enrich their lives through acquiring, preparing, consuming and sharing distinctive food products.

The construction of cultural capital is also rooted in local and therefore knowable practices. The more that local production and processing are well crafted, visible, sustainable and ethical (e.g. with respect to issues of animal welfare), the higher is their cultural capital (Cork Declaration, 1996; Fischler, 1996; Countryside Council, 1997; IATP, 1998; Goodman, 1999; Benvenuti, 2005; Commissione Internazionale, 2006b). This, of course, does not imply an *adieu* to anonymous global markets. Cultural capital implies the emergence of *circuits* that link specific producers, and specific places of production, with specific consumers. Within these circuits, *social definitions* of quality (and fairness and sustainability) are a decisive feature. Food transactions take place in socially regulated (and therefore also differentiated) circuits (see Meulen, 2000, Ventura, 2001, and Miele, 2001 for ample descriptions of such circuits and the added value they render to those actors participating in them). Thus, 'nested markets', embedded in new institutional arrangements, are emerging that link the new peasantries in firm ways to consumers looking for distinction (Depoele, 1996).

Regrounding farming in ecological, social and cultural capital is intrinsic to the peasant principle. Under current conditions, the peasant principle can be expressed as constituting a threefold movement, as summarized in Figure 10.3. This way out of the agrarian crisis has been emerging over a number of years, but will reach its aegis in the decade ahead (Ikerd, 2000a, 2000b, 2000c).

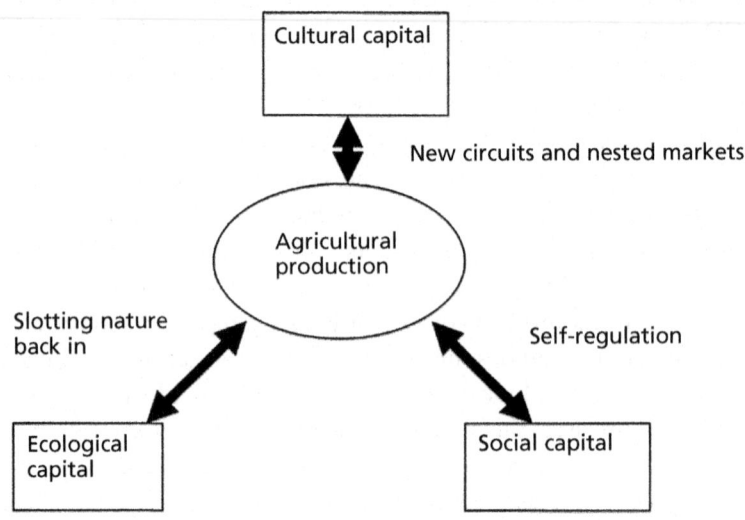

Figure 10.3 *Going beyond the agrarian crisis*
Source: Based on Ploeg (2006a, p268)

In a recent contribution, Henry Bernstein develops a reconceptualization of the agrarian question that significantly differs from the agrarian crisis (and its solution) that I have outlined here (see also Ploeg, 2006a). According to Bernstein (2007a), it is no longer 'useful to regard today's small farmers as "peasants" in any inherited historical sense'. Together with this goes the argument that the once classical agrarian question (that 'of capital') has lost its significance.

According to Bernstein, the agrarian question of the 21st century is primarily that 'of increasingly fragmented classes of labour'. However, the associated struggles over land 'do not have the same systemic (or world historical) significance as the agrarian question of capital once had' (Bernstein, 2007a, p449).

Let me now briefly discuss this position, especially since it allows me to specify some key points of my own approach. First, it is both theoretically and historically incorrect, I think, to oppose rigidly the agrarian question of 'capital' to that of 'labour'. In Bernstein's view, the latter is assumed to reflect 'fragments' of the labour class that are struggling for land in order to create a living.[13] This opposition of the agrarian question of capital and that of labour is theoretically inadequate in so far as it rigidly separates two elements that can only be understood through the internal relations that generate them. Capital defines and creates labour, just as labour produces and reproduces capital. Capital is, in the end, objectified labour, just as labour struggles provoke the further development of capital (Holloway, 2002, pp226–227). That which applies to the dialectical relations between a hostile environment and the peasantry – between Empire and the new peasantries – also applies more generally to capital and labour. Historically, the sequence proposed by Bernstein (first the centrality of the agrarian question of capital and then, from the late 20th century onwards, an agrarian question of labour) is also misguided. Agrarian questions have always had a multiple articulation. The Latin American *problema de la tierra* translated simultaneously into *el problema del indio* (Mariátegui, 1925). The Italian *questione agraria* simultaneously represented a problem for the bourgeoisie and for labour. And the essentially *double* character of the agrarian question again became very clear in Latin America during the 1950s and 1960s. On the one hand, peasants witnessed and suffered the lunacy of 'land without men and men without land'; on the other hand, the process of industrialization was seriously blocked while massive peasant movements and incipient rural guerrillas likewise created considerable threats to capital (Ploeg, 1977, p203, 2006d, p201). This double character explains why land reform is constantly a heavily contested process of change. In synthesis, domination and struggle must both be taken into account. They compose two organically related points of departure for analysis. They coincide, intertwine and mutually generate each other in the agrarian questions that arise.

Second, Bernstein stresses that 'predatory landed property had largely vanished as a significant economic and political force by the end of the 1970s'. This argument is needed, indeed, to sustain his point about 'the end of the agrarian question of capital on a world scale' (Bernstein, 2007a, p452). This is, I believe, likewise problematic. Since the 1970s (and especially during the 1990s), many new 'landed properties' were created worldwide, some of which I described in Chapter 3. These are probably far more predatory (if not vampire like) than previous forms: they waste land, water, energy and labour;

they threaten biodiversity and global sustainability; they destroy employment on both sides of the globe; and they block the production of badly needed social wealth. Analytically important is the fact that these new 'food empires' do not only backfire on labour; they also create many new problems and contradictions for capital (as highlighted, for example, by the Parmalat case).[14]

Third, although the state figures in the different descriptions offered by Bernstein, in theoretical terms its role tends to be minimal. Yet, policy can mediate between capital and labour, and the state might regulate *for* as well as *against* the market. These issues are discussed in the next section.

Some notes on rural and agrarian policies

From 21 to 23 September 2003, the European ministers of agriculture gathered for the informal meeting organized every six months by the member state chairing the European Union (EU) during that period.[15] In 2003, it took place in Taormina on the isle of Sicily. The Italian presidency had prepared a policy document centred on the 'sovereign right of every country to have and defend its own agriculture' (in this respect it resembles Bové's position on 'food sovereignty': Bové, 2003, pp208–209). The document recognizes the unequal levels of development at the global level and proposes a search for new forms of cooperation to redress these inequalities. As a whole, the document reflects the widespread concern that, within a framework of globalization and liberalization, European agriculture might be out-competed by cheap imports from Asia, Latin America and Africa, while the newly emerging productive systems in other continents are unlikely to trigger any sustainable development. This probably explains why the US delegation was not amused by the Taormina document.

In order to present convincing proposals for new forms of agrarian and rural policy,[16] the document referred to seven commonly shared values, indicating their validity and relevance in both developing world countries and in Europe (see Box 10.2).

Box 10.2 *A fragment from the Taormina policy document*

Convincing policies for sustainable development in rural areas that aim to guarantee the rural population a satisfactory quality of life may be reached only if grounded on a solid and widely shared set of values that underlie the different objectives and goals that govern agrarian policies and accompanying measures in both EU [European Union] states and developing countries. These values are:

- Public responsibility to protect public health through securing adequate and safe supplies of food and water.

- The necessity and duty of every country to create, maintain and defend its own agriculture and the possibility for citizens engaged in agriculture to earn a decent livelihood. This right is underpinned by the need for food security, especially in adverse times.[17]
- The necessity and duty to offer emancipatory prospects to marginalized people (the 'have-nots', the hungry and the poor), especially through access to, and possession of, land, which offers the possibility of food security and autonomy. This right reflects the long and worldwide history of land reform.
- The necessity and duty to protect agricultural activities in areas (especially those with complex and fragile ecosystems) that would otherwise be marginalized and subjected to ecological and/or social desertification. In this context it should be remembered that some 54 per cent of the total green area of the EU is classified and treated as less favoured areas (LFAs).
- The national (and sometimes supranational) responsibility to create the conditions required for the generation of acceptable income levels from agriculture and for ongoing agricultural growth and development. This entails the need to construct adequate and efficient institutional support structures.
- The public task of organizing and implementing rural development (RD) policies that promote and sustain liveable countryside. These policies especially concern the interfaces and linkages between agriculture and other sectors and aim, as far as agriculture is concerned, to create multifunctional enterprises – especially since the latter have highly positive multiplier effects on the rest of the rural economy. In LFAs, these multifunctional enterprises are often promising foci for new development trajectories.
- The necessity to develop education, training and research.

In historical and recent times, these values have given rise to public interventions in markets through, for example, different forms of price and income subsidies. However, as the world swings towards liberalization and free trade, it is increasingly the accompanying policies that become the strategic means for translating these values into practice. Recognition of this shift and the need to (re-)operationalize the shared values into adequate accompanying policies necessitates a critical examination of heterogeneity in agricultural production and processing, as well as at the level of marketing food and agricultural products.

As a whole, the Taormina document reflects society's present concerns: public health, food safety, employment, income levels, prospects for marginal people, ecological and social balances and the attractiveness of the rural. In synthesis, the document claims that agriculture represents an important *liaison* between society and living nature, and cannot be reduced, therefore, to providing food products only. Hence, it is the right and duty of every country to defend this. There is probably no need to underline that this thesis runs diametrically counter to Empire as a mode of ordering.

At the same time, however, the document makes clear that not all forms of farming or food processing and trading contribute to the positive role of agriculture in today's societies:

> *Agriculture is characterized, worldwide, by an impressive diversity. Such differences directly affect the levels of employment and income that are generated through agricultural activities. They also have important effects on ecosystems, sustainability and the development potential of industries and services related to agriculture.* (Presidency, 2003, p7)

And, in this vein, the document continues:

> *As specified in the discussions on the European Agricultural Model,* the EU needs particular types of agriculture. *Exactly the same applies to developing countries. That is, the range that goes from large-scale export-oriented agricultural enterprises, on the one hand, and the peasant economy, on the other, is not irrelevant as far as the creation and distribution of wealth are concerned. Choices have to be made.* (Presidency, 2003, p8)[18]

The recognition of agriculture as important for society as a whole and the simultaneous acknowledgment that agriculture entails different trajectories for development – trajectories that are not indifferent to society as a whole – are not limited to the circles of farmers and/or scientists specialized in the matter. They form part and parcel, as the Taormina document shows, of political debates at top levels within the EU;[19] at that level they are equally objects of dispute.[20]

The coming decade will, it seems, reveal five tendencies that, together, will result in a far-reaching reshuffling of global agriculture and an associated redefinition of farming as an integral, indispensable and non-negotiable part of our societies.

First, volatility will increase considerably. Relative overproduction will be combined with or followed by new scarcities. Consequently, output prices will exhibit fluctuations far larger than we have seen to date. Alongside promising opportunities, there will be large market segments that enter into decline. As I stressed earlier, as an institution the peasant mode of farming will be far more capable of dealing with volatility and insecurity than entrepreneurial and corporate farming, which assume and critically need long-term stability – especially since their reproduction is essentially future dependent.

Second, there will undoubtedly be a process of re-regionalization of agricultural production and consumption. This will partly stem from increased volatility (and associated new scarcities); but it could also be triggered by increased energy prices and transport costs. Changing consumer tastes and their preference for freshness, quality, authenticity and transparency will also favour re-regionalization. It is important here to emphasize that such a process will not be limited to the rich developed countries. Large exporters, such as Brazil, will also face the need to first secure the nutritional needs of their own populations, instead of dedicating large and increasing parts of their agricultural (and virgin) lands to export production.[21]

Third, it is highly probable that in the next planning period of the EU (2014 to 2020), agricultural policy will be abolished. Simultaneously, agriculture will reappear as a series of 'chapters' within food policy, regional policy, energy policy, cohesion policy and the defence of biodiversity. And farming will only receive support[22] in as far as it contributes to the different sets of objectives implied by each of these policies. Thus, a new framework will emerge that will definitely impact upon the divide between peasant and entrepreneurial farming. While peasant farming easily unfolds as multifunctional agriculture (being able to deliver simultaneous contributions to several or even all of the indicated 'chapters'), entrepreneurial farming will find it far more difficult to do so, although bitter clashes between competing claims will certainly accompany the required adaptations and transitions. Especially important within this context is the policy aiming at cohesion (and, thus, avoidance of high socio-economic inequalities between and within regions). With the recent growth in the number of member states and especially with the unavoidable entrance of yet another cohort of new member states in the decades to come, there will be an overarching need to create high employment and adequate remuneration levels in these new rural areas of the enlarged European Union. This definitively requires a reconceptualization of farming that goes beyond entrepreneurial and corporate models that tend to reduce employment levels and value added. Repeasantization will occur as a material need (if it is not already one).

Fourth, the ongoing and partly interrelated processes of peasant-driven rural development (see Chapter 6) and deactivation of entrepreneurial farming (see Chapter 5) will certainly be strengthened, partly because of increased volatility, articulated re-regionalization and the demise of agrarian policies. The latter process, especially, will cause the 'entrepreneurial condition' to literally erode: entrepreneurial farming will lose its very foundations as well as its safety net. Wherever the latter is shrinking, new spaces for peasant farming will be opened up. Of course, peasant farming will not flow automatically into such new spaces. A decisive factor will be the attitude of young people wanting to put value on the specificities of the local (including their own skills, tacit knowledge and new patterns for cooperation), while building new and steadily evolving resource bases that allow for autonomy.[23] New institutional arrangements (that partly unfold along the lines discussed in Chapter 7) will probably strengthen this process (for specific proposals, see Ploeg, 2005a, 2005b).

Fifth, it is more than probable that, in the medium term (that is, in the 2008 to 2013 period), important elements of current agrarian policies will be reconsidered and adapted. This applies especially to so-called 'income support' now being received by a proportion of European farmers. Two trajectories are discernible. It is not unlikely that income support will be redefined as a kind of 'insurance premium', paid by European taxpayers and channelled towards farmers (not to a part but to all of them) in order to keep the productive potential of the land and associated resources intact in order to allow for a quick regeneration of production whenever volatility on the international food markets makes this urgent. The other possibility, politically less likely, but intellectually more attractive, would be to use a considerable part of the funds now available to stimulate and support newly emerging small- and medium-sized enterprises that are actively involved in food processing. One of the main barriers to the further unfolding of peasant agriculture is, of course, the extremely high degree of monopolization of the food processing industry throughout Europe, especially in the north-west. New, small and medium food processing enterprises could help to redress this distortion (or 'visible hand') of the food market, while providing peasant agriculture with new forms of access to the market (for further details on these two trajectories, see Ploeg, 2005b).[24] Public procurement might also turn out to be strategic in this respect.

Nevertheless, the interaction of the trends I have indicated will trigger considerable landslides in the agricultural landscape of Europe (and elsewhere). Crucial to this is the fact that Empire as an ordering principle will be inadequate in addressing, shaping and coordinating the multitude of new responses that will be set off by such trends. There will be tough confrontations between those developing new responses and those involved in shaping the world according to Empire. But in the end the latter will lose control. Mediocrity cannot block superiority for long, especially not when the general public is interested and watching. The peasant principle – constructing autonomy in order to mould new ways forward – will orient and inspire many grassroots initiatives and drive forward new processes of repeasantization in both developing and developed countries. These processes will tend to be all the stronger when two sets of conditions are met. In the first place, the more peasants of the 21st century represent *citoyennité, lien local, autosufficience* and *autonomie* (Jollivet, 2001),[25] the more they will be able to push forward processes of repeasantization, resisting, at the same time, the different forms and guises of Empire. Second, the peasant principle might be actively combined with, and partly translated into, other important ordering principles that are present in our societies. I referred to one of these above – namely, the principle of *sovereignty* in relation to the essential needs for food security, food quality, the protection of valuable landscapes and the defence of biodiversity: needs that are increasingly defined as *non-importables*, thus delineating another line of demarcation vis-à-vis Empire. The peasant principle may also combine well with the principle of *subsidiarity*, ensuring the connection between society and living nature – a connection that is also central to many of the large non-governmental organizations (NGOs) of civil society. In the end, these NGOs will probably engage with and actively defend peasant farming. Thus, the peasant principle is likely to converge with the principle of *solidarity* at local, regional, national and international levels. And, finally, it may well coincide with the longing for *superiority* within a world where mediocrity seems to dominate – especially in relation to the quality of food and the sustainable use of natural resources.

Notes

Preface

1 I am not referring to 'post-modernism' as it currently figures in the social sciences. 'Post-modern' refers here, in the first place, to the fact that the studies on which this new approach is based were realized in the aftermath of the big agricultural modernization projects of the 1960 to 1990 period, which affected the countryside nearly everywhere in the world. Second, 'post-modern' implies a critical analysis of these modernization projects, as well as an attempt to go beyond their practical and theoretical limitations.
2 For a convincing argument, see Held et al (1999, p2); Aldridge (2005, p144) and Bourdieu (2005, pp223–232); .
3 As argued by Colin Tudge (2004, p3): 'We need again to see farming as a major employer – indeed, to perceive that to employ people is one of its principal functions, second only to the need to produce good food and maintain the landscape. Yet modern policies are designed expressly to cut farm labour to the bone and then cut it again.' See also Saraceno (1996) and Griffin et al (2002).

Chapter 1
Setting the Scene

1 I am very grateful to the universities of Leiden, Wageningen and Perugia, to the European Commission and to the Italian Ministry of Agriculture for the many research opportunities, debates and conferences on the themes outlined in this chapter. I am also much indebted to the following individuals for engaging in discussions and providing helpful critique: David Baldock, Bruno Benvenuti, Henry Bernstein, Rutgerd Boelens, Gianluca Brunori, Frederick Buttel, Ezio Castiglione, Paolo di Castro, Ada Cavazzani, Janet Dwyer, Harriet Friedmann, Benno Galjart, Pieter Gooren, John Harriss, Markus Holzer, Karlheinz Knickel, Catherine Laurent, Ann and Norman Long, Jo Mannion, Terry Marsden, Pierluigi Milone, Laurent van de Poele, José Portela, James Scott, Flaminia Ventura and Harm Evert Waalkens.
2 This is a nice neologism (compare with amphibian – creatures that move in and out of water) that is used by Kearney (1996, p141) to describe the circumstances of the many rural people 'who move in and out of multiple niches'.
3 I slightly extended the original definition in Rip and Kemp (1998). This extension anticipates the conclusions elaborated upon in Chapter 9.
4 Of the total world production of rice, only 6 per cent is traded across borders. In the case of wheat, which is the world's largest export crop among the cereals, only 17 per cent of the world's production is exported while the remaining 83 per cent

is consumed in the producing countries themselves. Meat is exported in growing quantities, facilitated by global cooling chains, which allow long-distance trading. Nevertheless, meat exports still represent less than 10 per cent of total world production. This minor share, however, does not exclude that the total value of global food exports (in 2000) was estimated by the World Trade Organization (WTO) at US$442.3 billion, representing 9 per cent of world merchandise trade and 40.7 per cent of world exports in primary products. Over the last 15 years exports of food products have been growing more quickly than total world production (Oosterveer, 2005, pp14–16; see also EC, 2006, which also gives the long-term tendencies).

5   Johnson (2004, p64) rightly observes that repeasantization implies a 'redefinition':

> *Today's peasantry is a population struggling for survival, clinging to control over the means of production that are increasingly unable to meet their subsistence needs, and excluded from the system that used to offer hope of development. Not aiming for an accumulation of profit, the peasants of today are instead in search of a sustainable livelihood that will ensure their survival ... into the 21st century.*

Johnson (2004) adds that this repeasantization 'may become the dominant process as agriculture becomes increasingly non-viable'.

6   Deactivation implies that agricultural *production* is contained or reduced. De-peasantization implies that *peasants* leave agriculture. This might occur without levels of production going down. But the two may also go together. When the 'space' left by disappearing peasants is not used by others to reinitiate agriculture production, there is, likewise, deactivation.

7   I am explicitly talking here of 'mode of farming' and *not* about 'mode of production'. Although the real problems do not reside in the words we use, but in the relations between *les mots et les choses* as Foucault (1972) phrased it, I nevertheless want to distance myself from the sterile debates of the 1970s on 'modes of production' (and their 'articulation'). See Chapter 2 for further discussion.

8   The following paragraph leans on a contribution I wrote for the *Handbook of Rural Studies* (see Cloke et al, 2006, Chapter 18).

9   Bernstein (2004, 2006) offers an alternative reading of the agrarian question. I will come back specifically to his interpretation in Chapter 10 of this book.

10 In addition, I could use my experiences in several parts of Africa (notably Guinea Bissau, Rwanda, Mozambique and the Republic of South Africa), Eastern Europe, Brazil and Mexico and also the often very rich material gathered by my masters and doctoral students.

11 This symmetry explains why peasants easily make the shift from the rural to the urban economy: from agriculture to the 'informal economy' of the cities (see, for example, Bagnasco, 1988).

Chapter 2
What, Then, Is the Peasantry?

1   This chapter is strongly based on PhD courses I have given in Porto Alegre (Brazil), Catania and Cosenza (Italy), Sevilla (Spain) and at Yale University (US). Parts have been published in Ploeg, (2006b, 2006c). For the development of my ideas, the congresses organized by the Insitituto Cervi in Reggio Emilia (Italy) and

by Nitra University (Slovakia) have been extremely helpful (see Ploeg, 2003b, and Ploeg and de Rooij, 1999). The discussions with, and help received from, Encarnación Aguilar, Rudolf van Broekhuizen, Lola Dominguez, Benno Galjart, Paul Hebinck, Norman Long, Terry Marsden, Raúl Paz, Henk Renting, Sergio Schneider, Eduardo Sevilla Guzman and Marta Soler have been very stimulating.

2   From Shanin's (1972) classic study *The Awkward Class* onwards, the term awkward has been normally associated with the notion of the peasantry. I use the term here on purpose in connection with the sciences that have studied the peasantry. I will come back to this issue in the final chapters of this book.

3   In the Marxist type of peasant studies (as well as in mainstream development economics) the peasantry is understood and represented as a major obstacle to the process of capital accumulation and to a full unfolding of markets (Bernstein, 1977, 1986; Byres, 1991). The very telling rural histories of, for example, France, The Netherlands and Norway, are completely neglected (or highly distorted) within these theoretical approaches. These rural histories show that the development of capitalism and the presence of a strong peasantry go well together, the latter often being the driver of the former. See Jollivet (2001), Brenner (2001) and Brox (2006).

4   Following Marx, it is possible to understand the peasantry as analytically representing 'petty commodity production' (PCP) since it is based on a *partial* integration within and orientation towards markets (see Ellis, 1988; Bernstein et al, 1990, p72; and Ploeg, 1990a). Agricultural entrepreneurs, then, emerge within the same framework as representing 'simple commodity production' (SCP) in which all resources (labour apart) enter the process of production as commodities. In capitalist commodity production (CCP), labour, too, enters as a commodity. In Ploeg (2006c) I have synthesized this approach, which might introduce powerful tools of analysis into empirical research (see Long et al, 1986). Within Marxist peasant studies such a line of enquiry became blocked due to the theoretical confusion introduced by Gibbon and Neocosmos (1985) that was consequently followed by Bernstein (1986) (see also Bernstein and Woodhouse, 2000). Gibbon and Neocosmos (1985) basically argued that there are just two degrees of commoditization: full commoditization or no commoditization whatsoever. This is completely at odds with empirical reality (see, for example, Benvenuti and Ploeg, 1985; Long et al, 1986; Saccomandi, 1991).

5   Relevant, but hardly tackled, research questions concern the following issues:

- To what extent has the change from peasants to entrepreneurs materialized?
- What have been the social, economic and ecological *benefits* and *costs* of this partial realization?
- To what degree are there 'pockets' that allow for contrasting development trajectories?
- In the era of liberalization and globalization, what are the prospects for the modernized entrepreneurial agriculture, and the 'pockets' of peasant-managed agriculture?
- What kinds of 'rurality' are produced by the contrasting development models?
- How are the different constellations to be coordinated politically?

6   This question reflects the title of the introductory chapter of Mendras (1976): '*Qu'est-ce qu'un paysan?*'.

7   This, of course, is partly due to the social division of labour. Peasant studies were firmly rooted in anthropology, non-Western sociology (!), development sociology

and development economics, while the study of Western European or US agriculture was rooted in other disciplines (and often even in other institutes, such as the Land Grant Colleges). In Wageningen there was, likewise, a sharp divide between 'Western' and 'non-Western departments'. Only from the 1970s onwards were research groups set up that were engaged in both Europe and the developing world and which actively discussed, compared and tried to integrate the different processes and outcomes.

8 Underlying this point of view is the wide range of rural sociological, anthropological and economic studies, carried out especially in Western Europe (but supported by a range of empirical, albeit somewhat neglected, studies in the US, Canada and Australia) that show that many (if not the majority of) farmers in the centre of highly developed capitalist countries are far from the textbook image of the 'agricultural entrepreneur' (a summary is included in Ploeg, 2003a). They are, instead, far more 'peasant like' than might, or ever will be, admitted in official political and theoretical discourse. Echoing Latour's (1994) finding that 'we never have been modern', it might be said that 'farmers never stopped being peasants'.

9 There is, of course, a far wider array of studies that *specify* how peasants practise agriculture, from, for example, French research traditions and also from Wageningen social scientists. But these have remained largely within the domain of (social) agronomy and have not been translated or integrated into peasant studies. The Anglo-Saxon tradition is relatively poor on the analysis of peasants' farming practices. It is very telling that in the heyday of peasant studies in the UK, there was a preoccupation with peasants in the developing world, and if peasant-type farming was acknowledged to exist in Europe, then it was usually located in the northerly Shetland isles or in relatively isolated parts of Wales, the Pennines and definitely Ireland.

10 Here, again, there is a specific background, which once more links with the internal division of labour within science. Until recently, and in most places, beta and gamma disciplines were neatly separated. Anthropologists are not trained agronomists. Neither have experts in peasant studies been skilled in understanding the technicalities of dairy farming or potato production. At best they were simply seen as technicalities, related to a domain that was assumed to be stagnant. However, yields, soil fertility levels, longevity, culling ratios, pruning techniques, etc. are not just data given since Genesis: they are outcomes of co-production. They are sociomaterial constructions that inform us about ongoing interaction and mutual transformation of man and nature. Hence, it is these very 'technicalities' that inform about peasant struggle in the arena defined by the process of production.

On the other side of the disciplinary divide a similar myopia is applied – agronomists trained since the 1960s, in particular, are barely able to relate technical outcomes to the specific and strategically inspired organization of the labour process. Technical data are perceived in isolation. Since the beginning of the 1990s there has been, internationally and likewise in Wageningen University, a strong tendency to overcome the divide between the 'social' and the 'technical' perceptions of farming. At the international level, the newly emerging agro-ecological tradition has been very influential in this respect (Toledo, 1981, 1994, 2000; Altieri, 1990, 1999, 2002; Sevilla Guzman and Gonzalez de Molina, 1990; Martinez-Alier, 2002; Sevilla Guzman, 2007). In Wageningen, outstanding work that goes beyond the indicated divide has been developed by soil scientists (Sonneveld, 2004) agronomists (Steenhuijsen Piters, 1995; Groot, et al, 2003, 2004, 2007a, 2007b) and animal scientists (Groen et al, 1993; Reijs, 2007).

11 In Ploeg (2003a, Chapter 3) I have given an overview and a synthesis of the major studies of farming styles. Later studies have consolidated and enriched the theory

of styles of farming. See Bakker (2001); Schnabel (2001); Wielenga (2001); Boonstra (2002); Flören (2002); Lauwere et al (2002); Mourik (2004); Schmitzberger et al (2005); Averbeke and Mohamed (2006); Slee et al (2006); Wartena (2006).

12 In *Types of Latin American Peasantry*, Wolf (1955, pp453–454) established three basic criteria of peasants as a social type:

1 primary involvement in agricultural production;
2 effective control of land and making autonomous decisions about cultivation (see also Wolf's *Peasant Wars of the Twentieth Century*, 1969, pxiv); and
3 a primary orientation towards subsistence rather than reinvestment.

These elements are included and simultaneously specified in my definition of the peasant condition. What is missing in Wolf's definition is the dialectical relation between dependency and the struggle for autonomy. But apart from this is the need to go beyond the assumed dichotomy of 'subsistence' and 'reinvestment'. Peasants are continuously 'reinvesting' (through their labour, among other things) in order to improve 'subsistence', and improved levels of subsistence allow for more 'reinvestment' and so on.

13 This can perhaps be partly explained by the fact that peasant studies grew out of the social and political sciences and not from agronomy, animal sciences, etc. Furthermore, peasant studies are typical of the Anglo-Saxon tradition and stem originally from the UK and the US, where peasants were seemingly less evident at the moment when the field of peasant studies was born; nor was there any tradition for studying production relations at the level of detailed farming practice. It seems that it has taken high-tech science and serious food scares to promote research of this kind.

14 Land evidently was, and still is, an outstanding indicator and metaphor of autonomy. But it is not only land that is an important vehicle for autonomy. Depending upon the specific context, control over marketing channels, having other economic means at one's disposal, and/or being able to express one's own identity in an unequivocal way might be equally important. It also follows that ownership of land is not a necessary condition. The aspiration and/or struggle for land might already be an important defining momentum for being and acting as a peasant.

15 It is telling that nearly all empirical research indicates that farmers refer to this encounter with living nature as one of the most attractive and rewarding aspects of their work. Equally significant is that craftsmanship (i.e. 'mastering' the relations with living nature) nearly always translates into pride. However, at the level of theory, these aspects have been reduced to just marginal phenomena, if not to mere 'romanticism'. This marginalization is rooted, at least partly, in the far-reaching division of labour within the agricultural sciences between technical and social disciplines. In the former, the process of production has been understood, since the 1930s, as the unfolding of scientific laws, principles and insights. Thus, the labour process is just viewed as a more or less imperfect version of the ideal specified by and within science. It definitely is not understood as a creative process of construction, implying among other things, an ongoing transformation of nature. For a long time, the social sciences also espoused this view.

16 In mathematical terms, recursive relations dominate over direct relations – that is, the situation at $t = n$ can only be understood and known if $t = n - 1$ (etc.) is known. In this respect, it is interesting to note that neo-classical economics reduces the 'farm as process through time' to just one moment in which only direct relations

are available. Building on this same line of reasoning, it could be argued that entrepreneurial farming increasingly is to be understood using 'inverted recursive relations'. From the future (t = n + x) is derived what is to be done now (at t = n).

17 As an Italian saying from the peasant world goes: *Moglia e buoi, paesi tuoi* (Your wife and your oxen had better come from your own village).

18 In a nice analogy to the 'polibians' of Kearney (1996), Paz (2006a) refers to the 'hostile environment' in which the peasantry operates as an anaerobic environment. He continues by specifying that 'the peasant is an anaerobic bacteria that learned to survive in a setting in which no capital is available, and in which capitalist enterprises ("aerobic" *par excellence*) are dying. To identify the mechanisms used in the reproduction of these peasants is a main challenge that is central in many new peasant studies.'

19 The study of Ventura and Milone (2005a) on peasant innovativeness carries a telling subtitle: 'The rediscovery of the peasant model: Regaining control over resource use'. It underlines the relevance of autonomy, its neat interweaving with the peasant mode of farming and simultaneously stresses the re-emergence of these issues within Europe.

20 In *Reconceptualizing the Peasantry*, Kearney (1996, p2) argues that the notion of 'peasant ... belongs to discourses that are being superseded'. His book, indeed, is based on the 'proposition that the category peasant, whatever validity it may once have had, has been outdistanced by contemporary history' (Kearney, 1996, p1). And further on he argues that 'some people may point to the presence of contemporary peasants and I would agree that some pockets [sic] of them remain in Latin America, Asia and elsewhere. But the point is that peasants are mostly gone and that global conditions do not favour the perpetuation of those who remain.' Kearney is also wrong when he argues that migrants/peasants involved in high-tech company production – wherever that takes place (in the US or Mexico) – are reduced to being the same type of worker (i.e. they are subjected to the same type of subordination). This denies the crucial importance of locality and culture in the lives of workers (see Arce and Long, 2000).

21 According to Pearse (1975), 'livelihood' has two elements: 'a way of living' that involves both the idea of 'life course' (implying a temporal or historical dimension to human endeavour) and 'shared' conditions and 'groupness' (i.e. collectivity). Hence, the concepts of 'livelihood' and 'neighbourhood' (like all other words ending in 'hood' in Old English) have in common the implication of acting within a community.

22 Critical discussions of the repeasantization perspective are entailed in Djurfeldt (1999); Goodman (2004); Gorlach and Mooney (2004); and Dupuis and Goodman (2005).

23 The data are calculated according to Figure 2.3 (relation a). Each cell gives the part of a particular resource that is mobilized on the market as percentage of the totally available amount of that particular resource. The reader interested in the technicalities of the approach is referred to Ploeg (1990a).

24 It is important to underline that the presented ratios are independent of technological levels and ecological conditions. In both The Netherlands and Peru it is possible to run a farm with family labour only (then dependency upon the labour market is 0 per cent), while the opposite is equally possible (wage labour only and 100 per cent dependency). The high dependency upon the labour market in potato production in the Andes in Peru, for instance, is not a reflection of a relatively low level of mechanization. It reflects the fact that socially regulated labour exchange has been replaced, to a degree, by wage labour relations. This is an effect of both

credit programmes (see Ploeg, 1990a, and Chapter 3 in this volume) and the dissemination of new Christian sects such as the 7th Day Adventist (see Long, 1977).
25 If the main conditions are equal, the peasant mode of farming results in yields that are superior to the outcomes of contrasting modes. For Latin America this has been abundantly documented in the CIDA studies of the 1960s (CIDA, 1966, 1973). However, the *ceteris paribus* condition is increasingly invalid: capitalist and/or entrepreneurial farms have access to technologies that are inaccessible to peasant producers. Beyond that, in capitalist and entrepreneurial farming time and space are often organized in such a way that very high yields seem to be, at first sight, their main feature. In feedlots, for example, an extremely high production per hectare is produced – but evidently this is due to imports of feed and fodder produced elsewhere. The same applies, for example, to the reorganization of time in breeding. Cows may produce a very high milk yield *per year*; but their longevity (the *total* number of years that a cow is lactating) is sharply reduced.
26 Within the cultural repertoires of the peasantry, 'consuming one's farm' is always considered to be a major mistake, if not a downright sin.
27 It is *relatively* abundant compared to the availability of land, animals and other physical resources. At a particular moment (e.g. land preparation or harvest) there very well might be an *absolute* shortage that is to be remedied through reciprocity and/or the labour market.
28 Italian sharecroppers (*mezzadri*) feared that the distribution of the harvest would be redefined in an unfavourable way (e.g. from 50:50 to 40:60) if yields improved. Thus, through the application of specific ploughing techniques (basically, from the centre to the margins) they created, on purpose, badly yielding fields.
29 Bernstein (2004) formulates a critique on what he interprets as an expression of 'neo-classical populism'. This critique is, I think, seriously flawed. It focuses on small property as such and not on the dynamics of labour-driven intensification.
30 The peasantry emerged as a conglomerate of free producers out of the chains imposed by feudalism. This occurred first in the periphery of the great feudal empires (e.g. in Friesland and Groningen: see Hofstee, 1985a), and during and after the French Revolution all over Europe. The decisive difference associated with these shifts is that the peasant is owner of his own land and labour – which is not the case for the feudal serf. However, the conversion of formal ownership into real ownership, possession and control turned out to be a long trajectory.
31 A classical example is to be found in Mediterranean dairy farms that are too small to produce the required feed and fodder. Available land is therefore used for tomato production. The harvest, when sold, is then converted into the needed feed and fodder. The advantage obtained is that the costs associated with this feed and fodder no longer 'pressure' the coming cycle of dairy production: they have already been paid for by the tomatoes and the work invested in it.
32 The low rent does not imply that the lessors are acting irrationally. It is true that by selling the land and investing the money in stock certificates, they could obtain, in the short term, probably far higher benefits. They would lose, however, the long-term security related to landownership.

Chapter 3
Catacaos: Repeasantization in Latin America

1 In 1973 and 1974, I lived and worked for 18 months in Catacaos and in 1977 I published a book based on my experiences there (see Revesz et al, 1997, p546). In

2006 the Instituto de Estudios Peruanos (IEP) in Lima published an enlarged edition that included the results of my last visit in 2004. In between, I had visited Catacaos in 1976, 1978, 1982, 1983 and 1987. Many people helped me over all those years. I want to express my gratitude to Rutgerd Boelens, Ruffo Carcamo Ladines, José del Carmen Vilchez Lachira, Julio More and Julia Yepes. I presented and discussed the main theses of this chapter in the international Water Law and Indigenous Rights (WALIR) conference on Legal Pluralism, Water Policies and Indigenous Rights held in Cuzco from 28 to 30 November 2006.

2   In local language, *pequeños propetarios* is normally abbreviated to *pequeños*, just as *trabajadores estables* is reduced to *estables*.
3   Slicher van Bath (1960) used the yield–seed ratio as a major indicator of progress throughout agrarian history.
4   Scorza (1974) described the notion of invisibility and the fear of it as entailed in the Andean peasantry in a convincing and beautiful way. For a theoretical elaboration, see Montoya (1986).
5   This is, of course, a general feature for Latin America and many other lesser developed parts of the world. See Long and Roberts (2005) for an overview of demographic trends in rural areas in Latin America.
6   In so far as Europe is concerned, one might refer to the relatively recent episode of repeasantization in Tras-os-Montes (compare Dries, 2002). Sevilla Guzman and Martinez-Alier (2006) analyse new social movements in Andalucia in terms of repeasantization. The Scottish Office (1998) does the same for Scotland. For processes that occurred in Eastern Europe, see Hann (2003) and Burawoy (2007). Contrasting expressions of a similar process of repeasantization elsewhere in Europe are described in Ploeg et al (2000) and Ploeg et al (2002c). A wide range of empirical cases is described in Coldiretti (1990) and Scettri (2001). For the US, see Joannides et al (2001). As far as Latin America is concerned, the most visible, massive and best-known example is, of course, the Brazilian *Movimento dos Sem Terra*; see Hammond (1999), Branford and Rocha (2002); Souza Martins (2003) and Cabello Norder (2004). Other episodes have been described by Zamosc (1990), Gates (1993), Vaeren (2000) and Enriquez (2003), while Schüren (2003) pays attention to the complexities of combined processes of de-peasantization and repeasantization at the micro level. Typical 'rural development' expressions that are omnipresent in Europe are neither absent in Latin America nor in other continents (see Gerritsen et al, 2005; Kop et al, 2006; and Gerritsen and Morales, 2007). Indirectly, the debate between the so-called '*campesinistas*' and '*descampesinistas*' that emerged during nearly all the great agrarian transformations of the last decades (in Chile, Nicaragua, Mozambique, Angola, Guinea Bissau and so on) is a theoretical expression of the 'space' that (at least hypothetically) allows for repeasantization. A solid theoretical overview and discussion is that of Feder (1977, 1978). A recent description and analysis of the same 'space' is given, for Mozambique, by Hanlon (2004).
7   Internationally this is known as the rise of 'urban and peri-urban agriculture' (Veenhuizen, 2006). In Catacaos, this comes down to most families having some cattle or pigs, while vegetable production for self-consumption and fodder production also plays a significant role. It is telling that currently some 35 per cent of people in Catacaos live in so-called *asentamientos humanos,* an administrative understatement for slums.
8   Their condition changed in this period from being wage labourers to being small farmers.
9   The rest of the area is composed of the districts of Chalaco, la Matanza, Salitral,

Santa Catalina, San Domingo, las Lomas, Rinco Llenada, Tambo Grande, Piura and Castilla.
10 Normally, the expansion of large irrigation schemes is associated with a decrease in the man–land ratio (see Ploeg, 2006d, especially Chapter 5).
11 Currently, 300 to 500 Peruvian soles (US$100 to US$165) are paid to rent 1ha of land for a year (this allows for planting two crops per year). The owner will normally use this rent to pay for the fertilizer needed for his own plot.
12 The process of production in these *haciendas* was, overall, relatively extensive and large scale. This extensive type of agricultural production was abundantly documented and analysed in the CIDA studies of the 1960s (for Peru, see CIDA, 1966; and for a general overview, see Feder, 1973). The point here is that the Iowa Mission missed the opportunity to introduce, right from the beginning, the far more intensive mode of production practised in the peasant economy. This would have allowed for far higher employment levels. See Ploeg (1990a, Chapter 4) for a description of similar struggles aimed at the realization of more intensive levels of production.
13 These principles were elaborated upon by a small group of Peruvian agronomists, lawyers and social scientists who worked closely together with the Catacaos community. They derived the principles from the study of the history of the community and, especially, from extensive consultations within the many villages and UCPs of the community. They made explicit what was widely felt among *comuneros*. The political situation of that time really required a clear articulation of commonly shared principles in order to challenge the land reform imposed by the state (see Ploeg, 2006d). The principles were officially accepted in a large meeting of more than 6000 *comuneros* in Cruz Verde in Catacaos during early 1973. The principles are used frequently in internal meetings, in solving internal conflicts and, especially, in response to the state – that is, they are *guidelines* for a rich multi-level *practice*.
14 Halamska (2004) relates expansion and/or reduction to 'the will of the political centre'. Typical for the Peruvian situation is that the strength of the peasant movement has been strongly reduced, especially since the middle of the 1990s; and that internal stability is no longer dependent upon national food production by the peasantry (cheap food can be imported from Asia). Thus, the 'political centre' could shift its agrarian policy from focusing on the peasantry towards favouring the newly emerging 'agro-export economy'.
15 The much discussed 'yield gap' is here, as in most other places, not a problem to be resolved through more technical assistance, knowledge and inputs. Low yields are mostly due to lack of 'space' – that is, to the way in which socio-technical networks are organized.
16 This is an important difference from farming organized on an individual basis. The need to engage in complex transactions or the risk of illness, for example, quite often cause delays in the timely realization of specific tasks. This has negative effects on yields.
17 There have been and are many situations in which these non-agricultural revenues are used to finance, albeit partly, the next cycle of production. Due to very high input costs compared to the very low salary levels (on average, 10 Peruvian soles per day), this is now fairly impossible.
18 This phenomenon has been perceived (see Kearney, 1996) as demonstrating the definitive disappearance of the peasantry. This, of course, is a completely wrong interpretation. What we are seeing here is, in the first place, a defence mechanism to survive, *as peasants*, under harsh conditions. Second, during the centuries and

over all continents peasant economies have demonstrated this phenomenon. It is not the beginning of a definitive *adieu*, but rather a periodically returning feature that tells us something about the interrelations between the peasant sector and the economy as a whole (and, consequently, about the levels of poverty that the peasantry suffers).

19 Both in the sierra and in the coastal area a considerable de-mechanization has occurred over the last decades. Compared to deteriorating (and highly fluctuating) prices, mechanization has become too expensive and too risky. The turnover of co-operatives into small peasant units has also contributed considerably to the re-emergence of animal traction.

20 All rice produced is exported, while rice consumption in Peru is largely based on imports of cheap rice from China, the Philippines, Korea and Ecuador.

21 Apart from capital, organizational capacities and access to international trading channels are also crucial here.

22 A considerable risk of current rice production technologies is salinization of the land.

23 In fact, this land could only be sold by the Ministry of Agriculture since it had neglected the claims of the surrounding communities that the land was theirs. The ministry argued that it was not tilled and therefore could not be the property of the communities. The latter argued that they could not till it because there was no water to do so. This is the terrible kind of confusion that is omnipresent at every level in Peru. The same applies to the use of irrigation water for the production of river crabs within artificial lakes. Formally this was forbidden; but the enterprise argues that the water is basically for crop irrigation and that crabs are harvested only between times.

24 This expression of Empire is controlled by an investment group composed of eight members, of whom the biggest is known locally as *Mustafa*. He has an Arab background.

25 *Gamonal* is a highly negative term used in the countryside to refer to large landowners and those associated with them.

26 Comparable cases are reported in Feder (1977), Llambi (1994) and Barros Nock (1997).

27 Peru became the largest exporter of asparagus at world level. It has replaced sugar and cotton as the main export products. Recently, however, China took over the position of largest global exporter.

28 From an economic point of view, this is very favourable since far less so-called land improvements (e.g. levelling) are needed. Hence, sunk costs are low because few long-term investments are made that link the enterprise to the particular location. The enterprise can easily move elsewhere.

29 The large production area specializing in the cultivation of asparagus in Navarra, Spain, suffered a sharp and far-reaching setback as a consequence of imports from Peru and China. The same applies to comparable regions elsewhere in Europe.

30 All major counter-reforms attack precisely this neuralgic point. A well-known and abundantly documented case is the peasant co-operatives in the Algarve that were created after the Carnation Revolution in Portugal.

31 Such considerations indicate that many of the changes introduced during the period of 'liberalization' (during which the Washington Consensus reigned) are, to a large extent, irreversible. This implies that new political projects must go beyond the schemes that were more or less successfully applied before this liberalization epoch.

32 This argument on the dialectal relationship between institutions and their organi-

zational expressions has been eloquently developed by Anton Zijderveld, among others, in his 1999 publication.

Chapter 4
Parmalat: A European Example of a Food Empire

1   This chapter relies heavily on the book entitled *Latte Vivo*, which I co-authored (Ploeg et al, 2004a) with some of the best-informed insiders of the Italian dairy industry, among them Corrado Pignagnoli, former secretary general of Coldiretti and former secretary general of the Italian Ministry of Agriculture during the Marcora period; Enrico Bussi, former head of the Centro di Ricerche Produzione Animale (CRPA) (an outstanding research institute that specializes in issues regarding dairy production), and later head of the Unalat organization that administered the quota system in Italy; Bruno Benvenuti, previously professor of rural sociology in Trento, Wageningen and Orvieto; Giuseppe Losi, the foremost expert in Italy on processing technologies of milk; and Cees de Roest, researcher of the CRPA and adviser to the Consortium for Parmesan Cheese. An introduction to our Italian book was written by Carlin Petrini, the president of the Italian Slow Food Movement. A *postfazione* was written by Sergio Nasi, president of the National Association of Agricultural Co-operatives.
2   It is telling that the subtitle of Gabriele Franzini's book on Parmalat (2004) directly refers to Empire: '*Il Crac Parmalat: Storia del crollo dell'impero del latte*'. On page 13 there is another provocative reference: 'It was the Empire of Calixto Tanzi, the milk empire that knew no sunset', for Parmalat had establishments 'from Canada to Australia, from Europe to South Africa and from Russia to Latin America'.
3   In the available body of literature, Empire is often identified with a *global* expansion of the market (i.e. with 'spatial moves' that, in the end, embrace even the most remote places of the world). This is, I believe, a somewhat misleading notion. Today, the globe hardly contains any places that are, as yet, 'unconquered'. Global expansion is not following the 'classical' pattern of imperialism. What we are facing is, instead, *a new wave of 'marketization'* (Burawoy, 2007): non-market areas (e.g. nature, environment, health) are converted into new markets and non-commodities are converted into new commodities. At the same time, classical barriers that contained and disciplined the market (state regulation and the many institutional arrangements entailed in civil society) are increasingly being eliminated in order to create further space for commoditization.
4   An interesting element is that this value can be somewhat modified within the framework of specific social relations. A co-operative dairy, for instance, could aim to be a long-term guarantee for dairy farmers to have an outlet for their milk (Dijk, 2005). Within the context of development aid, a processing unit could operate as a central trigger for the rise of small dairy farmers. For current consumer groups, the main value of a co-operative dairy unit might be that it can supply locally produced high-quality dairy products. This does not deny the unit's need to reproduce as enterprise; the point, however, is that the value of the enterprise was, to a degree, negotiable and mouldable.
5   For the Italian Cirio company a price was paid that equalled 'the weight of Cirio in gold' (Franzini, 2004, p64), especially because Parmalat had to also take over the debts of Cirio (these debts were superior to the value of the Cirio company). At the same time, the acquisition was related to the way in which politics, industry and banks intertwine. Tanzi, the owner of Parmalat, was urged by the government to

take over Cirio (to get rid of a nasty problem implied by massive dismissals and political unrest) and also by the Central Bank (in order to avoid big losses by the involved banks and the probable crash of at least some of them).

6   In retrospect, much of this institutionalized trust turned out to be virtual. As only recently became clear, the big Italian banks decreased their share in Parmalat during 2003 from 229 million Euros to just 31 million Euros. The Deutsche Bank openly 'fooled' the market by announcing a 5 per cent participation in Parmalat while, at exactly that moment, it completely ended its involvement (*La Repubblica*, 2007).

7   Also in quantitative terms: while Parmalat as enterprise had, according to the official denunciation of the Milan procurators, an accumulated but camouflaged loss of 7 billion Euros, Tanzi had, as a person, 'only' extracted 800 million Euros.

8   When stock value decreased in the last months of 2002, many shareholders thought this was due to a brilliant move of Tanzi to bring most stock back into the family.

9   There are evidently other conditions – for example, an already widespread change in food consumption patterns. Without the 'fast food nation' (Schlosser, 2001) – that is, the presence of people caring less and less about what they eat – food empires would be less prominent. The same applies for highly unequal power relations between nation states and the increased interweaving of large corporations and the state (Korten, 2001).

10 Central to the project of Tanzi (the owner of Parmalat) was the assumption that the positive connotation of Parma (obtained through the *Parmigiano-Reggiano* cheese and the Parma hams) could be attached to his dairy products.

11 Through the global financial markets and the private equity and hedge funds operating in it, enterprises and institutions *as a whole* are converted into commodities. Outside these new markets it is hard to conceptualize physical infrastructure, networks, enterprises, market shares, expected rise of profitability, etc. as commodities.

12 It is increasingly recognized (due to the activities of hedge funds) that in particular cases the 'third level' (or holding) not only renders no added value whatsoever, but *costs* a considerable amount of money – in fact, it *drains value*. Consequently, an enterprise might function better, and represent more value, if the 'headquarters' are eliminated (*Volkskrant*, 2006).

13 This remarkable contrast is due to the differential impact of a multinational (in this case Parmalat) compared to the co-operative structures that govern the processing of Parmesan cheese. Here the telling title of van Dijk's (2005) handbook on co-operatives in agriculture – *Where the Market Fails* – applies very well.

14 The differences between Tables 4.1 and 4.2 are due to yearly fluctuations and to inclusion versus exclusion of value-added tax.

15 These imports are associated with the history of the dairy deficit in Italy, which since the middle of the 1980s has been institutionalized through the European quota system: Italy was not allowed to raise its national milk production. However, it is important to note that milk imports are not necessarily limited to these kinds of deficit situations. In The Netherlands, having huge surpluses of milk makes the country into a large net exporter; there are, nonetheless, also considerable imports. With the growing liberalization of the milk market such imports will increasingly become a part of daily life, especially since outsourcing is highly profitable.

16 This was also due to the solidarity of the Italian public who wanted to support the workers. Thus, the sales of Parmalat products rose after the crash.

17 Peppelenbos also highlights the price to be paid for such an organization: 'blurred

task divisions, high transaction costs, institutional distrust and low innovation capacity'. Consequently 'Tomatio' 'entered [in the end] into deep crisis'.
18 These debts and the losses that occurred every year were, however, well hidden from the outside world. Even falsified credit letters (suggesting, in one case, a US$3.8 billion credit line from the Bank of America) were used to construct the required virtual images of a solid and powerful enterprise. Adding virtual turnover to the balance was another important instrument (just as happened in Ahold with the famous side letter affair). Franzini (2004, p113) notes that central for these companies is 'the need to create the image of cash flows that in reality do not exist'.
19 There are also resemblances. In the course of 2004 and the beginning of 2005, Albert Heijn, one of the main units of the Ahold group, developed an extended package of so-called 'house' brands that in nearly all external aspects (packing, presentation, colours, accompanying slogans) resembled already existing benchmark products of different food industries (such as Becel, Bertoli, Lipton and Calve). These 'lookalikes' took over considerable market share. Although I am not able to judge the content of the claim, Unilever sued Albert Heyn by stating that its 'lookalike' healthy margarine (evidently a copy of the well-known Unilever product Becel) did *not* contain the type of fats, etc. needed to sustain the health claims. This strongly resembles *latte fresco blu*, which evidently was just a lookalike of *real* fresh milk. I will come back to this issue further on in this text.
20 Alongside these new initiatives, nearly always farmer driven and supported by particular groups of consumers, there are some historical 'pockets' of direct delivery of raw milk. In, for example, the Basque country in Spain, there are widespread networks built around the production and delivery of raw milk (Broek, 1988). Current food empires are exerting heavy pressure on the European Commission to forbid these 'pockets', as well as newly emerging initiatives that equally aim at the creation of new short circuits.
21 A parallel line is to be encountered in high-quality raw milk for cheese-making. Several high-quality cheeses in Europe (such as *Parmigiano-Reggiano*, or Parmesan cheese, the French *Gruyere* and Dutch *boerenkaas*) are made of raw milk. It is partly the microbic flora that renders its special taste. The artisan process of processing is also crucial. To meet the requirements of cheese-making, the raw milk must be of an exceptionally high quality. In this respect it is telling that already in the 1960s a leading Dutch expert observed that 'raw milk from the average Dutch farm no longer allows cheese-making'. In the cheese industries of today, milk has to be treated in a complex way before it can be converted into cheese.
22 See Saccomandi (1998) for an excellent discussion of this kind of phenomena. Apart from being a distinguished economist, Vito Saccomandi knew the industry very well through consultations and a multiple involvement as commissioner. He was also for several years Italy's minister of agriculture.
23 A tragic case was the DES hormone, which affected the daughters of those pregnant mothers who used it. When it was banned as human medicine, it was still used for decades in the feed industry (especially for pig feed).

Chapter 5
Peasants and Entrepreneurs (Parma Revisited)

1 Since 1979, I have been involved in research programmes in this area. I have written previously about differentiated patterns of farm development, processes of commoditization and, more recently, forms of repeasantization located in this area

(Ploeg and Bolhuis, 1983; Bolhuis and Ploeg, 1985; Benvenuti and Ploeg, 1985; Ploeg, 1987b, 1990a, 2003b). I am grateful to the following friends and colleagues: Bruno Benvenuti, Eppo Bolhuis, Enrico Bussi, Corrado Pignagnoli, Andrea Pezzani, Bruno Riva, Cees de Roest, members of the Gruppo Bizzozzero and the researchers of the CRPA in Reggio Emilia. I owe much to Paul Hebinck and our students of Wageningen University with whom I was able, over time, to develop my ideas on the many differences between peasants and entrepreneurs. I am very indebted to the many researchers who joined the farming styles research group. They unravelled in a competent way the heterogeneity of today's agriculture.

2   It is important to note that entrepreneurs link the notion of the margin to the farm as a whole or they express it per 100kg of milk. They are not overly interested in, for example, the margin per cow. The concept of scale refers to the total amount of milk produced on the farm.

3   In 1971 entrepreneurs dedicated large amounts of their land to the then highly profitable production of tomatoes and onions, which were manually harvested. Thus, employment levels were somewhat 'inflated' through the high numbers of daily workers employed in the harvest. Later, the harvest was mechanized, and during the 1980s these crops disappeared completely from the farms considered here.

4   Technological level is assumed to be constant here. Hence, I am referring to small moves *along* the iso-curve that links the tilled acreage with the labour input per acre.

5   Socio-geographical patterns are, in this respect, very telling. The introduction of cubicle sheds (a highly visible expression of modernization in Dutch dairy farming) started in the south of the country during the 1960s, moved slowly to the east during the 1970s and was only applied massively in the north during the 1980s. Farming pride functioned, in the north, as an evident obstacle. Farmers would not accept the thesis that so far they had done it wrong.

6   It is interesting that this relation between huge state intervention in agricultural markets and the success of the modernization project has especially been observed by social scientists from the developing world (see, for example, Abramovay, 1992). They could relatively easily see this relation between the 'water and the fish' since the 'water' was quite often lacking under conditions at the periphery, thus hampering modernization or even making a burlesque of it. For most European observers the presence of the 'water', though, was all too evident. They tried to explain modernization and the associated rise of 'entrepreneurs' by reference to other variables.

7   *Ganaderizacion* often implies far-reaching spatial movements that interact with specific and highly unequal social divisions of labour. Small farmers open up parts of the forest, sell the timber to international companies, cultivate crops for a few years and, then, when soil exhaustion sets in, use the land for grazing and eventually sell it to large cattle-breeding companies. In Costa Rica, these farmers, who often live under very precarious conditions, are referred to as *precaristas* – a telling expression. It is expected that in a few more decades they, and behind them the large breeding companies, will have reached the shores of the Atlantic Ocean.

8   Throughout history, there have been different types of ranching. In South and Central America one encounters extremely large *haciendas* in which land was used mainly for extensive breeding. From the 1960s onwards, new (basically capitalist) breeding companies emerged alongside them. This was partly triggered by the global impact of McDonald's and other chains that occurred first in the US and later in Europe. Typically, however, the *feedlot* (i.e. the concentration of fattening

animals on a small area in order to be fed with feed and fodder obtained elsewhere) hardly emerged here since the extensive use of grazing lands (that often resulted in considerable ecological degradation) remained a central feature.

9   Cattle breeding farms may expand by purchasing land belonging to peasants. The *ganadero* farms described by Gerritsen own, on average, 100ha, while the average peasant farm controls only 3.5ha. The expansion of the former, though, often implies another mechanism called *pastura* in Mexico, where a peasant rents out, on an annual base, the right to pasture his lands. The land is therefore no longer available, or only partially so, for his own productive activities.

10 Here we encounter in the centre of the most modernized agricultural system of Europe one of the most traditional expressions of moral economy – namely, the image of the limited good.

11 The Dutch Farmers' Union (LTO) redefined itself during the late 1990s as an organization aimed at bargaining for those 'who have a future'.

12 At first sight this might be confusing. I remind the reader that the difference between peasants and entrepreneurs does not reside in magnitude as such. What is essential is scale (i.e. the relation between labour objects and labour input). Since peasant farms in Parma and Reggio Emilia dedicate decisively more family labour to the farm, they are, in absolute terms, larger, but at the same time smaller, in scale than entrepreneurial farms.

13 The irony, then, is that those referred to as 'good entrepreneurs' are 'good' due to their *peasant* way of farming, while those referred to as 'bad and failing entrepreneurs' are 'failing' precisely due to their *entrepreneurial* mode of farming. This evidently relates to the fact that reigning scientific and popular perceptions of 'entrepreneurship' mainly focus on *ex-post* results (while being unable to decipher the *ex-ante* foundations of it). Nobody can or will tell you what the entrepreneur has to do exactly in order *to become* a good entrepreneur. Many prescriptions and recommendations are given; but if they fail to function this is always attributed to 'bad entrepreneurship'. Entrepreneurship is about 'making money' (and about 'making more money out of money'); but it is difficult to know *ex ante* which combination of, for example, land, cattle, fodder and concentrates, which type of cattle and/or which alternative (continue farming or selling the land and reinvesting elsewhere) renders the best results.

14 This concept differs from the gross value added that is presented in Table 5.3. The difference between the two is located, in part, in depreciations and salaries paid.

15 This was already predicted at the end of Chapter 5 of the Dutch version of the *Virtual Farmer* (Ploeg, 1999, p244–245). Another early warning is given in Buckwell et al (1997).

16 This is an apt title that also suggests a somewhat hidden difference vis-à-vis the LEI (the state accountancy bureau). The latter produces figures that do not say much and are rather confusing.

17 The differential reactions of peasants and entrepreneurs to crises have been described in several empirical studies. For US agriculture, see Barlett (1984); Salamon (1985); Strange (1985). The assessment is that 'the financial crisis of the 1980s has affected big farms more than has been widely recognized', while smaller farming operations 'do not always prosper, but they persist, despite public policies that would wish them away ... they are resilient and difficult to crush' (Strange, 1985, pp6, 7). Salamon tends towards similar conclusions (1985, p338). Barlett (1984, p841, Table 3) demonstrates the same. However, such observations never entered theory. Dominant theories remain centred around the *invincibility* of large-scale entrepreneurial farming; indications of alternatives, doubts, etc. are

systematically omitted. Policy and science opt here for a self-imposed ignorance. As expert systems they are 'adrift' (Jacobs, 1999). However, as we will probably move from temporary crises towards a structural one (see Chapters 1 and 10 of this book), it will become difficult to ignore the demise of entrepreneurial farming any longer.

Chapter 6
Rural Development: European Expressions of Repeasantization

1 I have been involved, at the European level, in the preparation of the two big European conferences on rural development: in Cork (1996) and Salzburg (2003). In The Netherlands, I was closely involved in the practice of rural development through several non-governmental organizations (NGOs) such as the Wadden Group, Regional Products Netherlands (SPN), and Stichting Beheer Natuur en Landschap (SBNL; a conglomeration of farmers and landowners actively managing landscape and biodiversity). Our research group in Wageningen published the first Dutch documents on rural development in The Netherlands (see Bruin et al, 1992; Roex and Ploeg, 1993; Ploeg, 1994, 1995; Ettema et al, 1995; Broekhuizen and Ploeg, 1997; Broekhuizen et al, 1997).

2 In a discussion of the persistence of the peasantry in the developing world, Johnson (2004) rightly criticizes Bernstein (2001, p45), who claimed that 'the peasantries ... that inhabited "the world of the past" ... are indeed destroyed by capitalism and imperialism'. Johnson's point is that capitalism and imperialism have not resulted 'in the disappearance of the peasantry *but in its redefinition*' (Johnson, p64, emphasis added). This applies also to the 'developed' world. In Europe, 'today's peasantry [equally] is a population struggling for survival, clinging to control over the means of production ... and excluded from the system that used to offer hope of development' (Johnson, 2004, p64).

3 Regarding the Dutch situation, see Ploeg (2002); LEI (2005); LNV (2005); Schoorlemmer et al (2006); Venema et al (2006); and for other European countries, see DVL (1998); Stassart and Engelen (1999); Scettri (2001); Ploeg et al (2002c); O'Connor et al (2006).

4 See Oostindie and Parrott (2001), who give the 'average' biography of different rural development practices. Most of these developed long before talking about rural development became *en vogue*. The distance between the 'birth' of most practical initiatives and the *implementation* of rural development policies is even larger. See also Oostindie et al (2002, p225, Table 13).

5 I do not imply any 'structuralist' notion here. The point is that once certain shifts are made, they tend to become enduring. Beyond that, they start to mutually reinforce each other. And when cooperation emerges, between renewing farmers, the shifts also become, in a spatial sense, a stable phenomenon.

6 It is important to note that these tendencies are increasingly wrought together and translated to higher levels of aggregation. This occurs in the environmental (or territorial) co-operatives that have been constructed in The Netherlands (Renting and Ploeg, 2001), in the Italian wine routes (Brunori and Rossi, 2000), in farmers' markets in Germany (Knickel and Hof, 2002) and England (Banks, 2002), and in the French 'chestnut economy' (Willis and Campbell, 2004). The same reconstruction and strengthening of autonomy can be supported by cleverly designed

regional programmes, such as the Spanish Proder and the German RegionAktiv (Domínguez Garcia et al, 2006; Knickel, 2006).

7   Ireland, United Kingdom, The Netherlands, Germany, Spain and Italy (see Ploeg et al, 2002b).
8   Derived from Oostindie et al (2002). More detailed information is provided in Oostindie and Parrott (2001).
9   The survey was limited to professional farmers only (i.e. those farmers obtaining at least 25 per cent of their earnings from agricultural activities).
10 This percentage refers to professional farmers only. When part-time farmers and hobby farmers are also taken into account (as is normally the case in national and EU statistics), the percentage is much higher.
11 By now this will be far higher. There is, however, no statistical registration of the underlying flows or adequate registration of the implied activities. On the contrary, when the contribution of new activities to income goes beyond 50 per cent, these units are eliminated from agricultural statistics.
12 This is reflected in the fact that products resulting from processes of deepening entail a relatively high value added per unit of product compared to conventional products (see especially Roep, 2002). Thus, the worrying trend signalled in Chapter 5 (see Table 5.3 and the associated discussion) is again countered.
13 The concept of social capital is generally used to explain effective governance, a smoothly functioning civil society, properly functioning markets (elevated levels of social capital might imply strongly reduced transaction costs), etc. It is hypothesized here that higher levels of social capital will induce a higher quality of life.
14 See www.worldbank.org/poverty/capital/index.htm; see also Galjart (2003). Harriss (2002) contains a solid critique.
15 They were asked, in the first place, whether specific networks (out of a total of ten) performed better in the countryside than in the cities and in the second place to give a personal judgement of each network on a scale from 1 (very bad) to 10 (excellent).
16 Hence, within one and the same community some may dispose of a high level of social capital and others with little. This relates to Bourdieu's (1986) view on social capital and its associated 'closure'. At the same time, it applies that for a specific area as a whole (or for a combination of a specific area with a specific social category), the summed scores refer to the strength and openness of the networks as such. In this way a well-known controversy regarding the definition and measurement of social capital (see Lin, 1999, p36) might be resolved.
17 The assumed interrelation implies that social capital cannot be created or maintained without state-backed institutions, as Harriss (2001, pp1–92) strongly argued. It is not only about horizontal voluntary organizations. The role of the state and politics cannot be disregarded: 'Local or grass roots social organizations have to be viewed in the context of the overall structure of social relations and of power' (Harriss, 2001, p18).
18 An interesting detail emerging out of this particular research programme is that many (non-agrarian) rural dwellers consider multifunctional farming an interesting source of new and attractive employment in the countryside. This strongly contrasts with entrepreneurial farming that represents, above all, a 'closed shop'.
19 The interested reader is referred to Ventura et al (2007).
20 These interrelations again underline that social capital is not a simple and disembodied asset. It is only there when it is actively mobilized through multiple relations, encounters and confrontations.
21 Notorious in this respect was the introduction and step-by-step growth of Tiv

cultivation. Tiv is a grain of Ethiopian origin that is gluten free. Since an increasing number of consumers are allergic to gluten (e.g. in bread), this promised to be an interesting niche. Within a few years, with a lot of experimentation – and learning how to deal with setbacks and insecurities – they were able to develop 12ha of Tiv to a level of some 1100ha. This experience turned out to be one of the important resources in the following episode.

22 Several years later (at the beginning of 2007), the first of the required legal permissions had almost been granted.

23 Although I did not elaborate on it, there is clearly craftsmanship in the Zwiggelte design, especially in the 'feeding' of the fermentation process with different types of continuously changing slurry and carbon. This requires ongoing monitoring and continuous fine-tuning. One reason for the dissemination of bio-energy production among farmers is that it resembles cattle feeding: both processes require craft and associated skills.

24 Although the same objective is aimed for, skill operates in a way that is diametrically opposed to 'embedded software' (see Chapter 9). Skill puts highly qualified labour and the agency and responsibility contained in it centre stage. Embedded software excludes agency and, especially, responsibility. It therefore implies a degradation of work.

25 The fresh milk of northern Italy will definitely be different from fresh milk produced in the Basque country in Spain (see Broek, 1998).

26 This not only occurs at the level of the involved units of production, but also within the wider networks in which they are embedded. A beautiful example (the production, distribution and consumption of high-quality Chianina meat) has been abundantly documented in Ventura and Meulen (1994); Ventura (1995); Meulen (2000); Ventura (2001). These studies show how the concepts of quality circulate, mutually inform and adapt to each other, while together they create trust and shared interests and prospects – that is, they explore and describe the *social construction of quality*. In addition, they explore and describe the social construction of the market (in this case for Chianina meat, which has characteristics that are decidedly different from those of anonymous meat). For a similar reasoning, see Bagnasco (1988) and Miele (2001).

27 This also means that the involved unit of production needs to control the provision of inputs. The larger the scope and reach of the involved unit (the more its multifunctional character is developed), the more successful the skill-oriented technology will be.

28 These are nowadays often organized by provinces and farmer organizations for the integration of landscape and nature within farming.

29 I will limit myself to two references. The rules imposed after the explosion of cattle diseases, such as foot and mouth disease, require a strict registration of all 'contacts'. This is evidently impossible when walking routes cross the farm or when having an on-farm shop. Consequently, farmers are in constant infringement. The same occurs due to spatial planning that delineates specialized agricultural areas from recreational areas, nature areas and areas for housing, etc. Farmers crossing these boundaries (e.g. by developing agro-tourist facilities) will likewise be offending the rules. The case of the 'farmers' hotels' around Amsterdam is a well known case. With regard to such 'semi-illegality', banks will not offer loans. Thus, only farmers with sufficient resources are able to enter into such new, highly valuable trajectories. In the end, these kinds of contradictions are unnecessary. They arise only because the exclusive starting point for designing rules and procedures is made up by the specialized agricultural enterprises. Within policy design the *multi-*

functional farm evidently is a 'monstrosity' (this concept is frequently used in transition studies): it represents, indeed, an *uncaptured* peasantry.
30 And, in a way, as *cover* that was to allow the continuation of the same CAP. However, it is also the case that rural development policies were understood by others as establishing a new *bridge* between Europe and the mainly peasant-like populations that flocked to the European countryside. Nevertheless, the new rural development policies (or the so-called 'second pillar') constituted, from the very beginning, an extended battleground.
31 What was accepted as 'rural development activity' was subjected to very detailed and formalized prescriptions, and only those farms and those actors meeting these *a priori* defined requirements were to be supported.

Chapter 7
Striving for Autonomy at Higher Levels of Aggregation: Territorial Co-operatives

1 The content of this chapter has been discussed at farmers' meetings, at several national and international forums and in some, somewhat dispersed, publications. Especially important were the Società Italiana Di Economia Agraria (SIDEA) congress held in Pisa in 2005, the Seljord meeting in Norway in the summer of 2006, and the Regional Congress on Agriculture and the Countryside of Tuscany, held in Florence in December 2006. I owe many thanks to the following people: Jozias van Aartsen, Folkert Algra, Anita Andriesen, Geale Atsma, Joop Atsma, Fokke and Ella Benedictus, Johan Bouma, Jaap van Bruchem, Lijbert Brussaard, Jaap Dijkstra, Nico and Conny van Eijden, Jeroen Groot, Taeke, Dictus and Douwe Hoeksma, Hugo Hoofwijk, Douwe Hoogland, Pieter de Jong, Foppe and Boukje Nijboer, Albert van der Ploeg, Kobus Walsma and Bert Wijnsma.
2 Initially they were often referred to as 'environmental co-operatives' (the environment being their main object). Later this was widened to, and simultaneously specified as, territorial co-operative.
3 In The Netherlands, the notion of 'environment' refers to the physical environment (i.e. to the quality of air, soil and water).
4 These segments might easily conflict. Slurry injection in the spring, for instance, kills many young birds and destroys many nests. It can also be detrimental for soil biology and thus reduce available food for birds and chicks. Operating the huge equipment needed for injection can be difficult in small fields with surrounding hedgerows. Thus, pressure emerges to remove hedgerows and enlarge fields.
5 In more formal terms this might be defined as 'legally conditioned self-regulation'. The theoretically important element is that through such a framework *responsibility* is not eliminated, but, instead, clearly delineated and transferred to the level where it should reside. Such 'legally conditioned self-regulation' sharply contrasts with the hierarchical type of control at a distance exerted by Empire-like types of regulation.
6 The minister's intervention was decisive. During and especially after the signing of the contract, the bureaucracy of the ministry actively tried to demolish this new and apparently threatening process (see Ploeg, 2003a, for details). A later minister, Brinkhorst, tried to formally stop the 'experiment'. The co-operatives, however, had at that time gained such momentum that he failed.
7 Later on, this became one of the ingredients of the national programme for nature management by farmers (*Programma Beheer*).

8   In the NFW this is called 'farmer-guided research'. This programme was started in order to monitor, test and develop the range of novelties proposed by the co-operatives. Over time, it evolved into a multidisciplinary research programme to co-design new solutions.
9   In the case of Catacaos: 'We do not allow the exploitation of our resources and production by external elements' (see Box 3.1 in Chapter 3).
10 Potentially this change entails a rupture in 'protein dependency', or, to put it another way, in the ongoing draining of proteins from developing world agriculture through (among other things) the production and export of soya.
11 I do not imply here a general rejection of modelling. What I am criticizing are generic models that negate, per definition, the specificities of the local. Further on in this chapter and also in Chapter 8 I will present examples of alternative modelling that depart, instead, from the local and help to strengthen it.
12 The difference is that in the state programme, quality is only defined in a negative way (i.e. as a range of potential infringements, apart from which it includes no reference to regional differences). According to the national programme, a hedgerow is a hedgerow, regardless of where it is located. For those who are knowledgeable about landscapes and their regional variations and contrasts, this is ludicrous, especially since it introduces, in practice, a strong tendency towards uniformity.
13 All farm maps describing the specific measures and elements are digitalized and brought together in a programme that allows for quick adaptations, for higher-level perspectives and for greatly reduced transaction costs.
14 Traditionally, a simple protective cage was put over the nests or a stick was placed near it for recognition. However, predators increasingly came to know this practice and simply went wherever a stick or a cage could be seen in the meadows. Now, bird watchers are trying out a GPS device that registers the coordinates of each nest. When working in the field, the farmer or, even more importantly, the contracted machine operator is automatically warned, through GPS, when he comes close to a nest.
15 Historically, reformism is a term juxtaposed with revolution. It is a pejorative term. It has been assumed that only after a regime shift (i.e. a revolution) were real reforms possible. Any intent to do so before the political takeover of power was thought to be counterproductive – hence, the negative connotation given to 'reformism'. Only later, during the Chilean experience and within Euro communism, was it thought that evolving reforms might well become a vehicle for regime change. It is interesting that this same dilemma arises again in the debates about strategic niche management and socio-technical regimes.
16 Scientists such as Johan Bouma, an international expert in soil science and simultaneously member of the influential Scientific Council for Advice to Government (WRR), Lijbert Brussaard, one of the foremost experts on soil biology, animal scientists such as Jaap van Bruchem, Joan Reijs and Frank Verhoeven, and agronomists such as Jeroen Groot and Egbert Lantinga all played an important role here. The funding of a large multidisciplinary research programme by the Netherlands Scientific Research Organization (NWO) also played an important role.
17 Important episodes have been the so-called 'grassland experiment' (see Ploeg et al, 2006) and the first studies of the socio-economic impact of the 'environmental track' (see Ploeg et al, 2003).
18 This first design principle can be seen as contrasting with the approaches to transition now dominant. Where transition increasingly is thought of as an all-embracing adieu to the situation as it is (therefore implying a massive rupture),

this first principle represents, instead, a move that is rooted in, and starts from, the existing situation.
19 The range of ongoing and highly promising scientific work that has resulted from cooperation between Wageningen scientists and the NFW is far wider than the two examples that follow. Here I must also acknowledge the interesting work of Eddy Weeda (Weeda et al, 2004) on how to use existing biodiversity as a measure of the range of environmental qualities, and Marthijn Sonneveld's impressive 3MG research (Sonneveld, 2006) geared towards the development of new smart monitoring systems that operate at regional level (thus avoiding the high costs associated with monitoring each and every single unit).

Chapter 8
Tamed Hedgerows, a Global Cow and a 'Bug': The Creation and Demolition of Controllability

1 This chapter focuses on some of the many controversies that currently link peasants to their environment, some of which were discussed in a chapter I wrote for Lammert Jansma, the former scientific director of the Frisian Academy (Ploeg, 2007). The same controversies triggered heated debates within scientific circles (see WB, 2003). The chapter could not have been written without the help of Folkert Algra, Geale Atsma, Joop Atsma, Pieter van Geel, Foppe Nijboer, Joan Reijs, Arie Rip, Frank Verhoeven and Harm Evert Waalkens. In a slightly different way, Peter Munters of the Ministry of Agriculture was also of great help.
2 Many farmers believe that browsing on different trees and shrubs is helpful in preventing and curing several animal diseases. Local veterinarians support this view.
3 A special problem was that 'Brussels' was annoyed at the time with The Netherlands. During the 2002 to 2006 period, there was a kind of standstill. While The Netherlands failed to resolve some of the main issues (especially the manure issue), Brussels would not accept any (re-)discussion of other dossiers.
4 The cases were only indirectly related to the bramble issue; but having won these cases, the position of NFW vis-à-vis the different government services was much strengthened.
5 To really compensate for labour input and costs, three times the current payment would be needed. This relates, in part, to the high labour input needed to work lands that form part of, or are surrounded by, landscape and nature elements. But improving connectivity and flexibility (as outlined at the end of Chapter 7), in combination with newly designed farming styles, might also be very helpful (i.e. decrease the financial burden of nature management).
6 In rural sociology, the notion of the 'global chicken' is a well-known point of reference (Bonnano et al, 1994). It refers to the worldwide division of labour in the meat industry and to the interrelated movements of inputs and (parts of) outputs over the globe. The point I want to make here is that *'non-travelling'*, locally bounded issues, items, animals, artefacts, fields, etc. are also increasingly subjected to global parameters, relations and control.
7 A specific part is lost in the form of ammonia emission within the cowshed. This is estimated to be 11.75 per cent (which is of astonishing exactness).
8 It is amusing that in Italy an average of some 80kg per animal was calculated as being the nitrogen excretion. This is remarkable since it is to meet the same

European Nitrate Directive, and Italian dairy farming, mainly located in the Po Valley, is as intensive as Dutch farming. This implies that by importing Italian cows, the Dutch nitrogen surpluses could be considerably relieved – on paper at least.

9 The Manure Law imposes, or at least strongly favours, a specific farm development trajectory. This is in order to maximize milk yields (three cows each producing 10,000kg of milk also generate, according to the general formula, less nitrogen in the slurry than five cows each producing 6000kg of milk), which, in turn, implies increased concentrates, and increased fertilizer use in the fields in order to obtain high-energy, high-protein silage, and probably increased maize production in order to compensate the high protein levels (and to reduce urea). Thus, it tends towards further artificiality of agricultural production in the aftermath of which real nitrogen losses might well increase.

10 When everything is done exactly according to formal rules and procedures, a considerable slowdown will emerge. This was once a device used as an important bargaining mechanism for those who did not have the right to go on strike.

11 With a group of colleagues and a large group of farmers, I have been involved in the construction of such a way forward. The legal possibility to do so is derived from a concession of the minister at the request of parliament that farmers who *de facto* prove that they have a better method to stay within the legal norms will be allowed to do so. In the last section of this chapter I describe such a way out.

12 The variability of manure was one of the reasons why it more or less disappeared from the agenda of scientific research. An interesting exception is in Portela (1994).

13 Admittedly, this is an ugly expression. But then, some ugly tendencies occur in today's sciences as empirical fields.

14 These models introduce fixed (i.e. non-movable) interrelations between parts of nature, between people, and between people and nature. They deny, in synthesis, the dynamics of co-production. Following this reduction, social and natural realities are actively subordinated to these models: they have to 'behave' according to the assumed and imposed parameters.

15 Figure 8.4 shows that 'physical properties' indeed *escape* from the 'laws' formulated by science (see Figure 8.3). This evidently is due to co-production (Ploeg et al, 2006).

16 This is the basic difference with previous constellations. Regional farming styles were very much based on cultural repertoire that defined farming as it ought to be. However, the same repertoire allowed for deviations that often were considered to be legitimate and a vehicle for change. Furthermore, the application of the implied rules passed through debates, arguments and counter-arguments – it was, in short, a reflexive application. Third, it was a kind of self-control exerted by the community over itself. Current schemes, however, do not allow for deviations: they do not pass through more or less reflexive deliberation but through rigid protocols and procedures. They are instruments for control at a distance.

17 In practice, this comes down to injection technologies. Other modes, which are sometimes far superior, are *de facto* illegal. Their use is a legal offence. Many farmers have been subjected to high penalties.

18 These levels are related, through complex models, with the 50mg objective.

19 This is a serious problem in Italy: farmers from the northern Po Valley rent extensive lands in the Apennine Mountains for administrative purposes only (i.e. to match the legally required animal–land ratio). This land lies idle and can no longer be used by local shepherds. In The Netherlands, the same problem crops up in the

interrelation between the south and central eastern parts of the country. Farmers from the south rent land from the central–east areas for administrative reasons and for additional maize production, leading to regional economic impoverishment in these latter areas.
20 The first versions made farmers say: 'Wait a minute; we have far better manure, containing far less nitrogen, than suggested by your models.'
21 Note that the newly developed software indeed contributes to enlarge (or to restore) an important part of autonomy.
22 This is reflected in the strange 0.95 factor in the 'global algorithm' in Box 8.1.
23 It is important to note that this bug could only be introduced due to the strong coalition and cooperation of three groups: peasants such as those of the NFW co-operative and similar organizations; a small but strong group of scientists who aligned themselves with the first group; and a small but powerful group of members of parliament (MPs) who could mobilize enough support from fellow MPs to overrule at decisive moments high-ranking officials of the Ministry of Agriculture and several exponents of the associated expert system.
24 This graph shows how nitrogen excretion *escapes* from the general algorithm presented in Box 8.1. The decisive interrelations are again made 'movable'. This is due to the active and goal-oriented adjustment of co-production.

Chapter 9
Empire, Food and Farming: A Synthesis

1 I presented an outline of this chapter in a keynote address delivered to the Congress of the European Society of Rural Sociology, held in Keszthely, Hungary, in August 2005. The contents of this chapter were also discussed extensively at the European Society for Rural Sociology (ESRS) Summer School held in Gorizia, Italy, during the summer of 2006 and at the 2006 Congress of Greek Agricultural Economists. I am grateful to the following people for their comments on earlier drafts: Bettina Bock, Rutgerd Boelens, Gianluca Brunori, Jasper Eshuis, Harriet Friedmann, Norman Long, Egon Noe, Giorgio Osti, Eduardo Sevilla Guzman, Arie Rip and Scott Willis. I am grateful to Durk van der Ploeg, ICT specialist, for explaining many of the technical details and complications of ICT.
2 Similar difficulties arise when seeking to relate this theoretical concept to its theoretical antecedents. As a concept, Empire partly reflects centre–periphery relations as elaborated upon theoretically from the 1960s onwards. Empire also relates to globalization. Empire contains many elements of the authoritarian state, just as it is about multinational corporations. Empire is about formalization (Benvenuti, 1991). However, as a theoretical concept, Empire is much more than the addition of all these separate lines. It also reflects two major breakthroughs that in the meantime changed the world. The first was the development and widespread application of ICT, which revolutionized the administrative ordering of many, so far 'uncaptured', domains, phenomena and processes. The second consists of the worldwide liberalization of markets, which allows for the elaboration of linkages that until recently seemed impossible.
3 In my own intellectual journey, the Sociology of Technology and Science (STS) concept of 'socio-technical regime' (Rip and Kemp, 1998) has been very helpful in arriving at a better understanding of Empire. The same applies to the notion of mega-project as discussed by James Scott (1998). But I think that the concept of Empire is needed to grasp the *current expansion* of socio-technical regimes, as well

as the *generalization* of 'mega-projects'. The latter are not anymore the weird (albeit often deadly) projects of authoritarian states or of expert systems that have ample room to pursue their own fiction. Currently, nearly all domains of society are being subjected to huge 'mega-projects'. The work of Benvenuti (1975a, 1982) on TATE and formalization has also been an important tool for coming to grips with some aspects of Empire.

4   These and the following quotes are based on my translations of the Spanish version of Kamen's book.
5   I use the adjective 'structural' here to refer to certain characteristics that remain more or less stable over space and time. I am not implying any structural notion here.
6   I believe that these three features do not exhaustively define Empire. The feature of being 'void' and other features to be developed further on in this chapter are equally needed, I think. If this expanded list of features is taken into account and if the economy is explicitly included in the analysis, one arrives at conclusions that differ from those elaborated upon by Colás (2007). Colás limits his analysis too much to states only.
7   According to www.agriholland.nl/nieuws/artikel.html?id=56879 and 56774, respectively.
8   Preventive health control through widespread use of antibiotics is an example of shifts that allow one to go beyond the limitations of nature. It is now increasingly becoming clear that such practices are highly detrimental to public health.
9   I have expanded upon this quote somewhat in the square brackets.
10 The Italo–Dutch scholar Bruno Benvenuti was the first social scientist to identify, describe and theoretically elaborate upon this kind of 'techno-administrative' prescription and control. Later, the same phenomenon has also been analysed within the STS tradition (Benvenuti, 1982; Moors et al, 2004).
11 There is a growing tendency to delegate the design and implementation of regulatory schemes that apply to food processing to the food industries themselves. Once defined, they also oversee on-farm processing. Thus, highly unequal terms are created that are often detrimental to farmers as well as to small- and medium-sized food enterprises. It is interesting that this tendency is highly favoured, if not spurred on, by countries such as The Netherlands and the UK, while Mediterranean countries, Germany and France actively try to contain this tendency. In this respect, the European Union and especially the European Commission are important, if not decisive, battlegrounds.
12 Regulatory schemes might be used to exert extra-economic power over specific markets. A well-known case is 'raw milk'. Large dairy corporations have time and again lobbied in Brussels for regulations that explicitly forbid its use. This would eliminate much competition from small- and medium-sized enterprises (SMEs) that master the art of making excellent cheeses out of it.
13 In general terms, Colás (2007, p71) observes that 'Empires have fostered the administrative, communicative, legal and military infrastructure for both long-distance trade and local commerce'. He also observes that 'public–private partnerships' often are decisive in this respect (Colás, 2007, p73).
14 In this respect the four conditions developed by Scott (1998) remain highly relevant:

   1   an administrative ordering of nature and society;
   2   a high modernist ideology;
   3   an authoritarian state, 'willing and able to use the full weight of its coercive

power to bring the high modernist designs into being'; and
4 a prostrate civil society that lacks the capacity to resist.

15 Until recently, rules delineated fields of competence and expected outcomes. Those operating in a specific field (in a self-organizing space) had the *responsibility* of reaching the expected outcomes and were judged according to the level and qualities of the realized outcomes. The *way in which* they realized them was their own responsibility. What has changed, with the constitution of Empire, is that the 'way' in which things are done is highly specified through the application of new rules. The outcomes do not matter anymore; what counts is whether one has operated according to the rules and procedures. This strongly associates with the poor performance of many institutions of civil society (care, education, safety, etc.), while it simultaneously translates into an explosion of transactions costs.

16 Hernan de Soto (2000) has argued that the main characteristic of the informal economy resides in its lack of registered property rights. This is partly right, although I do not agree with de Soto that the introduction of such rights would generate effective capital for the poorer sector and thus contribute to the end of underdevelopment. It is the lack of civil and, especially, labour rights that is a far more decisive feature.

17 This new and distinctive feature applies also because Empire strengthens two other tendencies. The first is that of immediacy – that is, production (the immediate making of money) prevails over reproduction. The second concerns the fact that only those processes of conversion are favoured (or even allowed) which convert money into more money. Other useful conversions (e.g. labour into improved land or experimental space into a new market outlet) are – within the context defined and governed by Empire – increasingly impossible.

18 This already applies to the supermarket case. Supermarkets are not in the first place a series of physical infrastructural elements (as shops, storage and distribution centres, etc.). The latter might be sold and leased back. Their supply might also be outsourced to others. Supermarkets are, above all, a 'third level' structure: they offer space where products are on sale and a massive daily flow of consumers – that is, a meeting point of two flows. Consequently, this 'meeting point' is itself converted into commodity value, just as access to it increasingly is a commodity as well. Suppliers have to pay (directly or indirectly) for the possibility of using it; and consumers also pay, albeit mostly indirectly and without being aware of doing so.

19 Currently, the procedure is generally as follows: immediately after the takeover of other enterprises (see Figure 4.1), Empire resells the available buildings in order to lease them back. This enlarges the immediate cash flow and financial fluidity.

20 If the supply of asparagus is today mainly based on production in Peru, it may be shifted tomorrow to China – just as organic potatoes come today from Holland and tomorrow from Austria. And if Schiphol stops being interested in acting as a passage point, the harbour of Gdansk (or the airport of Krakow) could take over. The best metaphor for this 'mobility' is seen in the huge bulk carriers floating on the world's oceans that are loaded with grain, rice, soya, frozen rabbit meat, etc. Yesterday they routed to New York; today they receive orders to go to Hamburg or Rotterdam, tomorrow, perhaps, to China. The possibility of continually changing the 'timetable' (to gamble even with different 'parallel' timetables) offers Empire the power to condition world market developments and even more trends in different regional markets. One shipload of American potatoes is enough to make the price of potato chips drop considerably in The Netherlands.

21 If this means consumers getting ill, there is clearly a huge problem. The 'error indication' needs to be generated preferably before the assembly of the 'composed' food product. If not, considerable economic losses and damage will occur.
22 The state programme for nature conservation discussed in Chapter 8 has transaction and control costs that are as high as 27 per cent of the total budget.
23 What we are facing here is neo-liberalism – not just as ideology but as a material mechanism through which Empire organizes a considerable part of its expansion.
24 Precisely at this point Colás (2007, pp30, 166) is wrong when he refers to 'the *dual* processes of capitalist reproduction and state formation [that] make empire a thing of the past'.
25 In her 2006 publication, Harriet Friedmann stresses the centrality of 'complementary expectations' in the constitution of food regimes (Friedmann, 2006, p125).

Chapter 10
The Peasant Principle

1 Apart from having been involved in several episodes of peasant struggle, I have been engaged in political discussions both at national and supranational level. In The Netherlands, I have been a member, for nine years, of the Council for the Rural Areas (RLG) that directly advises the government and parliament on issues concerning agriculture, food, nature and the countryside. I have been chairman of the Working Group on Agriculture of the European Environmental and Agricultural Councils (EEAC) and participated in some debates within the European Commission (see, for example, Delors, 1994, and Prodi, 2002). In Italy I am a member of the scientific committee of Istituto di Servizi per il Mercato Agricolo Alimentare (ISMEA), which directly advises the minister of agriculture. The simultaneous involvement in grassroots initiatives and in formal and informal forums of policy formulation has allowed me to obtain insights into the mechanics of rural and agrarian policies and to gain a better understanding of how policies translate in the worlds of rural people. What I have learned, in particular, through this involvement is, first, that the agricultural sector is not simply moulded, in a unidirectional way, through political interventions and regulation – policies and evolving regulatory schemes (and, for that matter, agribusiness strategies) are moulded as much by the dynamics, contradictions and omnipresent resistance entailed in agriculture. Second, I have learned that at the level of policies there are many tendencies that tend to align with the peasantries of the third millennium. These two observations are important points of departure for the present chapter.
2 What I imply here is that other ordering principles are increasingly deactivated by and/or subordinated to Empire as an ordering principle (see Law, 1994, for a discussion of hierarchal relations or 'black boxing' between different modes of ordering). Of course, alongside hierarchal relations, differently patterned balances are also possible. Latour (1994) refers in this respect to translation, composition and delegation.
3 These key words coincide with the contours of the agrarian crisis introduced in Chapter 1. Disintegration applies to the once organic relation between farming and nature, precariousness to the actors involved, and redundancy to the relation with society at large. Farmers no longer matter; Empire tends to make them irrelevant.
4 It will be clear, I guess, that the concept used here, 'multitude *of responses*', is meant as a critique of Hardt and Negri (2000). In their work 'multitude' is basi-

cally void; it is without intentionality. Here I distance my analysis from their highly abstract concept: in the end, their concept of 'multitude' is as depersonalized as 'class' in many analyses of the past. In contrast, my use of 'multitude of responses' refers to specific fields of actions in which concrete responses are developed; it also refers to the real social actors who create, develop and implement these responses.

5   Long (2007) summarizes this point as follows: 'Discourses produce texts – written and spoken – and even non-verbal "texts" like the meanings embodied in infrastructure such as asphalt roads, dams and irrigation systems, and in farming styles and technologies'. Here the 'discourse' is embodied in the resource base, in the sheep, the fodder, etc., and, subsequently, it is the resource base that expresses and underlines the 'discourse'.

6   I talk of *citoyens* because later I introduce *citoyennité*, which, according to the French rural sociologist Jollivet (2001), is a distinct feature of the peasantry (and simultaneously a *claim*). I also introduce this concept because land invasions in Peru are typically characterized by the omnipresence of the national flag underlining 'that we also belong to the nation and therefore have rights as everyone else'. A similar symbol is related to MST: the camps they create (in order to initiate invasions) are always located alongside main and highly frequented roads 'so that everybody can see us'.

7   Dutch farmers demonstrate, in practice, that they are often far more efficient and productive in managing nature than large organizations that specialize in nature conservation. A famous logo for this is, in The Netherlands, the *'Jisperveld'*.

8   The Hazard Analysis and Critical Control Point is a regulatory scheme that centres on hygiene in food production, processing and distribution. See Whatmore and Stassart (2001) for a critical discussion.

9   Burawoy (2007, p7) stresses that 'it is our task [as committed social scientists] to explore with all the technical instruments at our disposal ... the conditions of existence and expanded reproduction of these actual existing utopias.'

10 This strongly contrasts with the previous modernization epoch in which, as Bauman (2004) has signalled, essentially local problems were countered with global solutions. A basic preoccupation remains: can all new global problems such as, for example, global heating be tackled in a decentralized way?

11 Looking only at the statistical data one might conclude that Brazil is an exception. Yet, Brazil is also a case where we encounter massive return migration from the cities to the countryside.

12 According to Martinez-Alier (2002), Netting not only 'praised the peasant economy as able to absorb population increases. ... Supported by careful fieldwork in several countries, he [also] added the argument that peasant agriculture was more energy efficient than industrial agriculture.' In an impressive historical account, this conclusion is supported by González de Molina and Guzmán Casado (2006). Ventura (1995) presents a comparative analysis that shows that cattle breeding organized in a peasant way is energetically superior to an entrepreneurial organization of breeding. Marsden (2003) gives a theoretical elaboration of the chronic 'unsustainability' of current agriculture and food processing.

13 The struggle for land not only concerns, of course, these 'fragments' but may positively affect the labour market as a whole. This was the case, as argued earlier, with respect to Norway (and many other countries), and currently is the case in countries such as Brazil, while China represents an opposite situation. In Europe, the multidimensional repeasantization that is unfolding might also augment or at least sustain levels of productive employment. In this way, agricultural systems can contribute to newly required levels of social and regional cohesion.

14 It is a bit ironic that Bernstein questions the ability of peasant agriculture to 'provision the world with adequate and healthy sources of nutrition' (Bernstein, 2007a, p458, note 9), and especially to feed 'the massive urban populations of the world' (Bernstein, 2007a, p458). If anything threatens such provision, then it is the new 'predatory' food empires. It is also telling that 'fragments' of the urban poor actively try to enter the peasant condition, either through movements such as the MST or through different forms of urban agriculture.

15 These meetings are especially interesting since there is no direct negotiation or bargaining. They address the underlying political issues that are present but far less visible and explicit during the tough negotiations of formal meetings. The Taormina meeting was particularly interesting since it took place on the eve of the Cancun negotiations and also because Italy is far more sensitive to development issues in Northern Africa (especially the Magreb) than other member states. The Taormina document shows that a reconsideration of *peasants* and their role in today's agriculture – although banned in the north-west of Europe – is omnipresent in other parts of Europe.

16 The underlying argument was that liberalization could not be understood as an *adieu* to agrarian and rural policies. According to the document, 'more market' necessarily implies 'more policy'. Accompanying policies basically condition and order markets in an indirect way – that is, by embedding them in socio-political and normative frameworks.

17 It should be emphasized that this right extends beyond narrow perceptions of time and place. It is important to prevent the loss of land (through erosion, reduction of soil fertility, degradation of water systems, terraces, etc.) and the associated agricultural potential. This is particularly urgent in less developed countries (LDCs) that are generally more susceptible to such threats and where such losses are frequently irreversible.

18 The emphasis is mine. See Tracy (1997), Buckwell et al (1997) and COPA (1998) for a discussion of the European Agricultural Model.

19 In a way, the Taormina document might be understood as an expression of a 'moral economy' reigning at the level of those who are responsible for agrarian and rural policies. The convergence between the 'moral economies' presented and discussed earlier in this book (see Box 3.1 in Chapter 3 and Box 7.1 in Chapter 7) and the Taormina paper is, in this respect, highly interesting.

20 Of the 25 EU member states (at this stage some of them still aspirant member), 21 supported the document. It was also supported by the delegations invited from Latin America, Africa and Asia. The four EU members that disagreed were the UK, The Netherlands, Denmark and Sweden. The first two argued that liberalization would imply, in the end, the elimination of *all* agrarian and rural policies. Denmark and Sweden differed for other reasons. It is also significant that the Taormina Meeting took place shortly before the World Trade Organization (WTO) meeting in Cancun during the same year.

21 As a matter of fact, this became highly manifest during the preparation of the (failed) Cancun negotiations within the WTO framework. At that time, Brazil had (for political reasons) two ministers for agriculture. The first, responsible for large commodity production and exports and closely linked to US agribusiness interests, tried to convince political circles in Europe to have Brazil produce most of Europe's food needs, thus allowing Europe to convert its agricultural sector into semi-natural landscape, into extended locations for new rural dwellers and into golf courses. The other, politically responsible for rural development, insisted that Brazil had to secure, above all, its own nutritional needs, which so far are inadequate.

22 An interesting effect is that total support might potentially even go beyond current levels. However, it will no longer be indiscriminate support that flows, in particular, to the large entrepreneurial farms (as is still the case with 'decoupled' support based on historical references).
23 This is precisely the 'profile' that characterizes people involved in endogenous processes of rural development (see Oostindie and Parrott, 2001; Oostindie et al, 2002; Ventura et al, 2007).
24 It is not excluded that the formation of such small- and medium-sized enterprises (SMEs) will equally be grassroots driven and supported, especially if not solely by regional and/or provincial policy institutions.
25 *Self-sufficiency* refers here to the availability of needed resources and the possibility of using them according to one's own insights and interests. Thus, *autonomy* is secured. Self-sufficiency strongly contrasts with Empire. Through its control over relevant connections, the latter tends to deny both self-sufficiency and autonomy. However, once self-sufficiency and autonomy are assessed, they represent alternatives to (and therefore function as material and symbolic critiques on) Empire. *Lien*, or *local rootedness*, is inherent in having an autonomous resource base and using it for the development of local solutions to global problems. Local rootedness represents the opposite of 'non-places'. *Citoyenneté* refers to the goal-oriented and knowledgeable practices of the actors involved; they emerge as *citizens* who actively exert agency.

References

Abramovay, R. (1992) *Paradigmas do Capitalismo Agrario em Questao*, Estudios Rurais, Editoria HUCITEC, ANPOCS, Editoria da UNICAMP, São Paolo, Rio de Janeiro, Campinas
Agrarisch Dagblad (2007) 30 March, p2
Aldridge, A. (2005) *The Market*, Polity Press, Cambridge, MA
Alexander, P. and Alexander, J. (2004) 'Setting prices, creating money, building markets: Notes on the politics of value in Jepara, Indonesia', in W. van Binsbergen and P. Geschiere (eds) *Commodification, Things, Agency, and Identities (The Social Life of Things Revisited)*, LIT, Leiden, The Netherlands
Alfa (Accountants en Adviseurs) (2005) *Cijfers die Spreken 2004, Analyse Melkveehouderij*, Alfa, Wageningen, The Netherlands
Alfa (2006) *Cijfers die Spreken, Melkveehouderij, Editie 2006*, Alfa, Wageningen, The Netherlands
Alfa (2007) *Cijfers die Spreken, Melkveehouderij*, Alfa, Wageningen, The Netherlands
Altieri, M. A. (1990) *Agro-Ecology and Small Farm Development*, CRC Press, Ann Arbor, MI
Altieri, M. A. (1999) 'The ecological role of biodiversity in agroecosystems', *Agriculture, Ecosystems and Environment*, vol 74, pp19–32
Altieri, M. A. (2002) 'Agroecology: The science of natural resource management for poor farmers in marginal environments', *Agriculture, Ecosystems and Environment*, vol 93, pp1–24
Antuma, S. J., Berentsen, P. B. M and Giesen, G. (1993) *Friese melkveehouderij, waarheen? Een verkenning van de Friese melkveehouderij in 2005; modelberekeningen voor diverse bedrijfsstijlen onder uiteenlopende scenario's*, Vakgroep Agrarische Bedrijfseconomie, LUW, Wageningen, The Netherlands
Appadurai, A. (1986) *The Social Life of Things: Commodities in Cultural Perspective*, Cambridge University Press, Cambridge, UK
Arce, A. and Long, N. (2000) *Anthropology, Development and Modernities,* Routledge Press, London, UK
Athias, G. (1999) 'MST transforma excluidos urbanos em militantes: Fazenda em Porto Feliz foi ocupada por desempregados e sem–teto', *O Estado de Sao Paolo*, São Paolo, segunda–feira, 15 March
Atsma, G., Benedictus, F., Verhoeven, F. and Stuiver, M. (2000) *De Sporen van Twee Milieucoöperaties*, VEL & VANLA, Drachten
Averbeke, W. van and Mohamed, S. S. (2006) 'Smallholder farming styles and development policy in South Africa: The case of Dzindi Irrigation Scheme', *Agrekon*, vol 45, no 2, June, pp136–157
Badstue, L. B. (2006) *Smallholder Seed Practices: Maize Seed Management in the Central Valley of Oaxaca, Mexico*, Wageningen University, Wageningen, The Netherlands
Bagnasco, A. (1988) *La Costruzione Sociale del Mercato, studi sullo sviluppo di piccole imprese in Italia*, Il Mulino, Bologna, Italy
Bakan, J. (2004) *The Corporation: La Patologica Ricerca del Profitto e del Potere* (a cura di Andrea Grechi), Fandango s.r.l., Roma, Italy

Bakker, E. de (2001) *De cynische verkleuring van legitimiteit en acceptatie: Een rechtssociologische studie naar de regulering van seizoenarbeid in de aspergeteelt van Zuidoost-Nederland*, Aksant, Amsterdam, The Netherlands

Banks, J. (2002) 'Direct marketing on the English–Welsh border', in J. D. van der Ploeg, A. Long and J. Banks (eds) *Living Countrysides: Rural Development Processes in Europe – The State of the Art*, Elsevier, Doetinchem, The Netherlands

Barlett, P. F. (1984) 'Microdynamics of debt, drought, and default in south Georgia', *American Journal of Agricultural Economics*, December, pp836–853

Barros Nock, M. (1997) 'Small farmers in the global economy: The case of the fruit and vegetable business in Mexico', PhD thesis, ISS, The Hague, The Netherlands

Bauman, Z. (2004) *Vite di Scarto [Wasted Lives: Modernity and Its Outcasts]*, Edizione Laterza, Roma/Bari, Italy

Beccatini, G. (1987) *Mercato e forze locali: Il distretto industriale*, Il Mulino, Bologna, Italy

Benedictus, M. de and Cosentino, V. (1979) *Economia dell'Azienda agrarian: toeria e metodi*, Il Mulino, Bologna, Italy

Benvenuti, B. (1975a) 'General systems theory and entrepreneurial autonomy in farming: Towards a new feudalism or towards democratic planning?', *Sociologia Ruralis*, no 1–2, pp46–61

Benvenuti, B. (1975b) 'Operatore agricolo e Potere', *Rivista di Economia Agraria*, vol XXX, no 3, pp489–521

Benvenuti, B. (1982) 'De technologisch administratieve taakomgeving (TATE) van landbouwbedrijven', *Marquetalia*, vol 5, pp111–136

Benvenuti, B. (1991) 'Towards the formalisation of professional knowledge in farming: Growing problems in agricultural extension', in *Proceedings for the International Workshop on Knowledge Systems and the Role of Extension*, Hohenheim, Germany

Benvenuti, B. (2005) 'Een beschouwing over endo-culturele gemeenschappen in 8 stappen', in J. D. van der Ploeg and H. Wiskerke (eds) *Het landbouwpolitieke gebeuren, Liber Amicorum voor Jaap Frouws*, Wageningen Universiteit, Wageningen, The Netherlands

Benvenuti, B. and Ploeg, J. D. van der (1985) 'Modelli di sviluppo aziendale agrario e loro importanza per l'agricoltura mediterranea', *La Questione Agraria*, no 17, pp85–105

Benvenuti, B., Antonello, S., Roest, C. de, Sauda, E. and Ploeg, J. D. van der (1988) *Produttore agricolo e potere; modernizzazione delle relazioni sociali ed economiche e fattori determinanti dell'imprenditorialita agricola*, CNR/IPRA, Roma, Italy

Bernstein, H. (1977) 'Notes on capital and peasantry', *Review of African Political Economy*, no 10, pp60–73

Bernstein, H. (1986) 'Capitalism and petty commodity production', *Social Analysis: Journal of Social and Cultural Practice*, no 20, pp11–28

Bernstein, H. (2001) 'The peasantry in global capitalism', in L. Panitch and C. Leys (eds) *Socialist Register, 2001: Working Classes, Global Realities*, Monthly Review Press, New York, NY

Bernstein, H. (2004) 'Changing before our very eyes: Agrarian questions and the politics of land in capitalism today', *Journal of Agrarian Change*, vol 4, no 1 and 2, pp190–225

Bernstein, H. (2006) 'From transition to globalization: Agrarian questions of capital and labour', Paper presented at the Conference on Land, Poverty, Social Justice and Development, ISS, The Hague, The Netherlands

Bernstein, H. (2007a) 'Is there an agrarian question in the 21st century?', *Canadian Journal of Development Studies*, vol 27, no 4, pp449–460

Bernstein, H. (2007b) 'Agrarian questions of capital and labour: Some theory about land reform (and a periodisation)', in L. Ntsebeza, L. and R. Hall (eds) *The Land Question in South Africa: The Challenge of Transformation and Redistribution*, Human Sciences Research Council Press, Cape Town, South Africa

Bernstein, H. and Byres, T. J. (2001) 'From peasant studies to agrarian change', *Journal of Agrarian Change*, vol 1, no 1, pp1–56

Bernstein, H. and Woodhouse, P. (2000) 'Whose environments, whose livelihoods?', in P. Woodhouse, H. Bernstein and D. Hulme (eds) *African Enclosures? The Social Dynamics of Wetlands in Drylands*, James Currey, Oxford, UK

Bernstein, H., Crow, B., Mackintosh, M. and Martin, C. (1990) *The Food Question: Profits versus People?*, Earthscan, London, UK

Berry, S. (1985) *Fathers Work for Their Sons: Accumulation, Mobility and Class Formation in an Extended Yoruba Community*, University of California Press, Berkeley, CA

Beuken, F. van den (2006) 'Flexibel requirementsmanagement crucial voor certificering', *Bits & Chips*, 2 November, pp54–55

Bieleman, J. (1992) *Geschiedenis van de landbouw in Nederland, 1500–1950*, Boom, Meppel, The Netherlands

Bock, B. (1998) *Vrouwen en vernieuwing van landbouw en platteland: De kloof tussen praktijk en beleid in Nederland en andere Europese landen*, Studies van Landbouw en Platteland, 27, LUW, Wageningen, The Netherlands

Bock, B. B. and Rooij, S. J. G. de (2000) *Social Exclusion of Smallholders and Women Smallholders in Dutch Agriculture: Final National Report for the EU Project – Causes and Mechanisms of Social Exclusion of Women Smallholders*, WUR, Wageningen, The Netherlands

Boeke, J. H. (1947) *The Evolution of The Netherland Indies Economy*, Tjeenk Willink, Haarlem, The Netherlands

Boer, J. de (2003) *Veldgids landschapselementen Noardlike Fryske Walden*, Landschapsbeheer Friesland, Beetsterzwaag, The Netherlands

Bolhuis, E. E. and Ploeg, J. D. van der (1985) *Boerenarbeid en stijlen van landbouwbeoefening*, Leiden Development Studies, Rijksuniversiteit Leiden, Leiden, The Netherlands

Bonnano, A., Busch, L., Friedland, W., Gouveia, L. and Mingione, E. (1994) *From Columbus to Conagra: The Globalization of Agriculture and Food*, University Press of Kansas, Lawrence, KS

Boonstra, W. J. (2002) 'Heterogeniteit als effect van liberalisering: Een studie naar bedrijfsstijlen in Australie, *TSL*, vol 17, no 1, pp21–35

Borras, S. (1997) *The Bibinka Strategy to Land Reform Implementation: Autonomous Peasant Mobilizations and State Reformists in the Philippines*, Research Paper, ISS, The Hague, The Netherlands

Boserup, E. (1970) *Evolution Agrarie et Pression Demographique*, Flammarion, Paris, France

Bouma, J. and Sonneveld, M. (2004) 'Waarden en normen in het mestbeleid: Doelgericht inzetten op innoverend vermogen', *Spil*, no 207–208, pp22–26

Bourdieu, P. (1986) 'The forms of capital', in J. G. Richardson (ed) *Handbook of Theory and Research for the Sociology of Education*, Greenword, New York, NY

Bourdieu, P. (2005) *The Social Structures of the Economy*, Polity, London, UK

Bové, J. (2003) *Un Contadino del Mondo*, Feltrinelli Editore, Milan, Italy

Brade-Birks, G. (1950) *Modern Farming: A Practical Illustrated Guide*, Waverley Book Company, London, UK

Branford, S. and Rocha, J. (2002) *Cutting the Wire: The Story of the Landless Movement in Brazil*, Latin American Bureau, London, UK

Braverman, H. (1974) *Labor and Monopoly Capital: The Degradation of Work in the 20th Century*, Monthly Review Press, New York, NY

Bray, F. (1986) *The Rice Economies: Technology and Development in Asian Societies*, Blackwell, Oxford, UK

Breeman, G. (2006) 'Cultivating trust: How do public policies become trusted?', PhD thesis, Leiden University, Leiden, The Netherlands

Brenner, R. P. (2001) 'The Low Countries in transition to capitalism', in P. Hoppenbrouwers and J. L. van Zanden (eds) *Peasants into farmers? The Transformation of the Rural Economy and Society in the Low Countries (Middle Ages–19th Century) in Light of the Brenner Debate*, CORN Publication Series 4, Turnhout, Belgium

Broek, H. P. van der (1998) *Labour, Networks and Lifestyles: Survival and Succession Strategies of Farm Households in the Basque Country*, WAU, Wageningen, The Netherlands

Broekhuizen, R. van and Ploeg, J. D. van der (1997) *Over de kwaliteit van plattelandsontwikkeling: Opstellen over doeleinden, sociaal–economische impact en mechanismen*, Circle for Rural European Studies, Studies van Landbouw en Platteland 24, Wageningen, The Netherlands

Broekhuizen, R. van and Ploeg, J. D. van der (1999) 'The malleability of agrarian and rural employment – the political challenges ahead', Paper for the EU seminar Prevention of Depopulation in Rural Areas, Joensuu, Finland, 2 October 1999

Broekhuizen, R. van, Klep, L., Oostindie H. and Ploeg, J. D. van der (eds) (1997) *Renewing the Countryside: An Atlas with Two Hundred Examples from Dutch Rural Society*, Misset, Doetinchem, The Netherlands

Brox, O. (2006) *The Political Economy of Rural Development: Modernisation without Centralisation?* (edited and introduced by J. Bryden and R. Storey), Eburon, Delft

Bruin, R. de and Ploeg, J. D. van der (1992a) *Maat Houden, bedrijfsstijlen en het beheer van natuur en landschap in de Noordelijke Friese Wouden en het Zuidelijk Westerkwartier*, BLB/LUW, Wageningen, Utrecht, The Netherlands

Bruin, R. de, Oostindie, H. and Ploeg, J. D. van der (1991) *Niet klein te krijgen: bedrijfsstijlen in de Gelderse Vallei*, LUW, Wageningen, The Netherlands

Bruin, R. de, Oostindie, H., Ploeg, J. D. van der and Roep, D. (1992) *Verbrede plattelandsontwikkeling in praktijk*, Studierapport Rijksplanologische Dienst 54, Vakgroep Agrarische Sociologie niet–westers, The Netherlands

Brun, J.-M. (1996) *Le défi alimentaire mondial: Des enjeux marchands à la gestion du bien public*, Solagral, Paris, France

Brunori, G. and Rossi, A. (2000) 'Synergy and coherence through collective action: Some insights from Tuscany', *Sociologia Ruralis*, vol 40, no 4, pp409–423

Brunori, G., Rossie, A. and Bugnoli, S. (2005) *Multifunctionality of Activities, Plurality of Identities and New Institutional Arrangements*, Multiagri project, Department of Agronomy and Agro-Ecosystems Management, University of Pisa, Pisa, Italy

Brush, S. B., Heath, J. C. and Huaman Z. (1981) 'Dynamics of Andean potato agriculture', *Economic Botany*, vol 35, no 1, pp70–88

Bryceson, D. F. and Jamal, V. (1997) *Farewell to Farms: De-agrarianisation and Employment in Africa*, African Studies Centre, Leiden, The Netherlands

Bryceson, D., Kay C. and Mooij, J. (2000) *Disappearing Peasantries? Rural Labour in Africa, Asia and Latin America*, Intermediate Technology Publications, London, UK

Bryden, J. M., Bell, C., Gilliatt, I., Hawkins, E. and MacKinnon, N. (1992) *Farm Household Adjustment in Western Europe 1987–1991*, Final report on the research programme on farm structures and pluriactivity, vol 1 and 2, ATR/92/14, European Commission, Brussels, Belgium

Buckwell, A., Blom, J., Commins, P., Hervieu, B., Hofreitner, M., Meyer, H. von, Rabinowicz, E., Sotte, F. and Sumpsi Vina, J. M. (1997) *Towards a Common Agricultural and Rural Policy for Europe*, Report of an Expert Group, DG VI/A1, European Commission, Brussels, Belgium
Burawoy, M. (2007) 'Sociology and the fate of society', *View Point*, January–July, www.geocities.com/husociology/michaelb.htm?200711
Bussi, E. (2002) *Agricoltura e Alimentazione: Impegni, risorse e regole per lo sviluppo*, Relazione al Convegno dell'Istituto Cervi, Reggio Emilia, Italy
Buttel, F. H. (2001) 'Some reflections on late twentieth century agrarian political economy', *Sociologia Ruralis*, vol 41, no 2, pp165–181
Byres, T. J. (1991) 'The agrarian question and differing forms of capitalist transition: An essay with reference to Asia', in J. Breman and S. Mundle (eds) *Rural Transformations in Asia*, Oxford University Press, Delhi, India, pp3–76
Cabello Norder, L. A. (2004) *Politicas de Assentamento e Localidade; os desafios da reconstitucao do trabalho rural no Brasil*, PhD thesis, Wageningen University, Wageningen, The Netherlands
Camagni, R. (2002) 'Competitività territoriale, milieux locali e apprendimento collettivo: Una contro riflessione critica', in R. Camagni and R. Capello (eds) *Apprendimento Colletivo e Competitività Territoriale*, Franco Angeli, Milano, Italy
Castells, M. (1996) *The Rise of the Network Society: Volume I, The Information Age: Economy, Society and Culture*, Blackwell, Oxford, UK
Caron, P. and Cotty, T. le (2006) 'A review of the different concepts of multifunctionality and their evolution', *European Series of Multifunctionality*, no 10, pp1–179
Charvet, J. P. (1987) *Le desordre alimentaire mondial: Surplus et penuries, le scandale*, Hatier, Paris, France
Charvet, J. P. (2005) *Transrural Initiatives*, Harmattan, Paris, France
Chayanov, A.V. (1966) *The Theory of Peasant Economy* (edited by D. Thorner et al), Manchester University Press, Manchester, UK
Chomsky, N. (2005) *Democrazie e Impero; interviste su USA, Europa, Medio Oriente, America Latina*, Datanews Editrice, Roma, Italy
CIDA (Comite Interamericano de Desarrollo Agricola) (1966) *Tenencia de la tierra y desarollo socio–economico del sector agricola: Peru*, Washington, DC
CIDA (1973) 'Bodennutzung und Betriebsfuhrung in einer Latifundio–landwirtschaft', in E. Feder (ed) *Gewalt und Ausbeutung, Lateinamerikas Landwirtschaft*, Hofmann und Campe Verlag, Hamburg, Germany
Cloke, P., Marsden, T. and Mooney, P. H. (2006) *Handbook of Rural Studies*, Sage Publications, London, UK
Colás, A. (2007) *Empire*, Polity Press, Cambridge, MA
Coldiretti (Movimento Giovanile) (1999) *Nuova Impresa, idee ed evoluzione dei giovani agricoltori in Italia*. Edizione Tellus, Roma, Italy
Columella (1977) *L'arte dell'agricoltura*, reprint, Einaudi Editore, Torino, Italy
Commandeur, M. (2003) *Styles of Pig Farming: A Techno–Sociological Inquiry of Processes and Constructions in Twente and the Achterhoek*, WUR, Wageningen, The Netherlands
Commissione Internazionale per il Futuro dell'Alimentazione e dell'Agricoltura (2006a) *Manifesto sul Futuro dei Semi*, ARSIA, Regione Toscana, Florence, Italy
Commissione Internazionale per il Futuro dell'Alimentazione e dell'Agricoltura (2006b) *Manifesto sul Futuro del Cibo*, ARSIA, Regione Toscana, Florence, Italy
COPA/COCEGA (1998) *The European Model of Agriculture: The Way Ahead* [Pr(98)12F2, 2 April 1998], Brussels, Belgium

Cork Declaration (1996) *'A Living Countryside': Conclusions of the European Conference on Rural Development*, Cork, Ireland, 7–9 November 1996

Countryside Council (Raad voor het Landelijk Gebied) (1997) *Ten Points for the Future: Advice on the Policy Agenda for the Rural Area in the Twenty-First Century*, RLG Publication, 97/2a, Amersfoort

Crozier, M (1964) *The Bureaucratic Phenomenon*, University of Chicago Press, Chicago, IL

Cruz Villegas, J. (1982) *Catac Ccaos: Origen y evolucion historica de Catacaos*, CIPCA, Piura, Peru

Darré, J. P. (1985) *La parole et la technique: L'univers de pensée des éleveurs du Ternois*, Editions L'Harmattan, Paris, France

Delors, J. (1994) *En quete d'Europe; les carrefours de la science et de la culture*, Editions Apogee, Collection Politique Europeenne, Rennes, France

Depoele, L. van (1996) 'European rural development policy', in W. Heijman, H. Hetsen and J. Frouws (eds) *Rural Reconstruction in a Market Economy*, Mansholt Studies 5, Wageningen Agricultural University, Wageningen, The Netherlands

Diez Hurtado, A. (1998) *Comunes y Haciendas: Procesos de Comunalizaion en la Sierra de Piura (siglos XVIII al XX)*, Fondo Editorial CBC, Lima, Peru

Dijk, G. van (1990) *Is de tijd rijp voor milieucoöperaties?* NCR, Rijswijk, The Netherlands

Dijk, G. van (2005) *Als 'de markt' faalt; inleiding tot coöperatie*, SDU Uitgevers bv., Den Haag, The Netherlands

Djurfeldt, G. (1996) 'Defining and operationalizing family farming from a sociological perspective', *Sociologia Ruralis*, vol 36, no 3, pp340–351

Djurfeldt, G. (1999) 'Essentially non-peasant? Some critical comments on post-modernist discourse on the peasantry', *Sociologia Ruralis*, vol 39, no 2, pp262–270

Domínguez Garcia, L., Fernandez, X. S., Alonso Mielgo, A., Ramon Mauleón, J., Ramos Truchero, G. and Renting, H. (2006) 'Catching up with Europe: Rural development policies and practices in Spain', in D. O'Connor, H. Renting, M. Gorman and J. Kinsella (eds) *Driving Rural Development in Europe – The Role of Policy: Case Studies from Seven EU Countries*, Royal van Gorcum, Assen, The Netherlands

Dries, A. van der (2002) *The Art of Irrigation: The Development, Stagnation and Redesign of Farmer-Managed Irrigation Systems in Northern Portugal*, Circle for Rural European Studies, Wageningen University, Wageningen, The Netherlands

Dupuis, M. E. and Goodman, D. (2005) 'Should we go 'home' to eat? Toward reflexive politics of localism', *Journal of Rural Studies*, vol 21, pp359–371

DVL (1998) *Verzeichnis der Regional–Initiativen: 230 Beispiele zur Nachhaltigen Entwicklung*, Deutscher Verband für Landschaftspflege, Ansbach, Germany

EC (European Commission) Directorate-General for Agriculture and Rural Development, G5 (2006) *Agricultural Trade Policy Analysis: Agricultural Commodity Markets – Past Developments and Outlook*, Brussels, Belgium

Ecologiste (Edition française de *The Ecologist*) (2004) 'La resistance des paysans, Afrique, Asie, Amerique Latine, Europe', *Ecologiste*, vol 5, no 3, pp1–64

Eizner, N. (1985) *Les Paradoxes de l'Agriculture Française; essai d'analyse a partir des Etats Généraux de Développement Agricole, avril 1982–fevrier 1983*, Harmattan, Paris, France

Ellis, F. (1988/1993) *Peasant Economics: Farm Households and Agrarian Development*, Wye Studies in Agricultural and Rural Development, Cambridge University Press, Cambridge, UK

Ellis, F. (2000a) 'The determinants of rural livelihood diversification in developing countries', *Journal of Agricultural Economics*, vol 51, pp289–302

Ellis, F. (2000b) *Rural Livelihoods and Diversity in Developing Countries*, Oxford University Press, Oxford, UK

Enriquez, L. J. (2003) *Economic Reform and Repeasantization in Post-1990 Cuba*, University of Texas Press, Austin, TX

Eshuis, J. (2006) *Kostbaar vertrouwen; een studie naar proceskosten en procesvertrouwen in beleid voor agrarisch natuurbeheer*, Eburon, Delft, The Netherlands

Eshuis J. and Stuiver M. (2004) 'Creating situated knowledge through joint learning processes among dairy farmers and scientists in a mineral project in the Netherlands', *Agriculture and Human Values*, vol 22, no 2, pp137–148

Eshuis, J., Stuiver, M., Verhoeven, F. and Ploeg, J. D. van der (2001) *Goede mest stinkt niet: Een studie over drijfmest, ervaringskennis en het terugdringen van mineralenverliezen in de melkveehouderij*, Studies van Landbouw en Platteland, 31, Rurale Sociologie, WUR, Wageningen, The Netherlands

Ettema, M., Nooij, A., Dijk, G. van, Ploeg, J. D. van der and Broekhuizen, R. van (1995) *De toekomst: Een bespreking van de derde Boerderij–enquête voor het Nationaal Landbouwdebat*, Misset Uitgeverij bv, Doetinchem, The Netherlands

Feder, E. (1971) *The Rape of the Peasantry: Latin America's Landholding System*, Anchor Books, New York

Feder, E. (1973) *Gewalt und Ausbeutung, Lateinamerikas Landwirtschaft*, Hoffmann und Campe Verlag, Hamburg, Germany

Feder, E. (1977) *Strawberry Imperialism*, Research Report Series, Institute of Social Studies, The Hague, The Netherlands

Feder, E. (1977, 1978) 'Campesinistas y descampesinistas: tres enfoques divergentes (no incompatibles) sobre la destrucción del campesinado', *Comercio Exterior*, vol 27, no 12, and vol 28, no 1, México

Figueroa, A. (1986) 'Accumulacion, control de excedentes y desarrollo en la sierra', in Universidad Nacional Agraria 'La Molina' y Centro de Estudios Rurales Andinos 'Bartolome de las Casas' (ed) *Estrategias para el desarrolllo de la sierra*, Centro Bartolome de las Casas, Cusco, Peru

Fischler, F. (1996) 'Europe and its rural areas in the year 2000: Integrated rural development as a challenge for policy making', Opening speech presented at the European Conference on Rural Development: Rural Europe–Future Perspectives, Cork, Ireland

Fischler, F. (1998) 'Food and the environment: Agriculture's contribution to a sustainable society', in WUR *Compendium van een driedaagse confrontatie tussen wetenschap, samenleving en cultuur, 16, 17 en 18 april te Wageningen*, WAU, Wageningen, The Netherlands

Flora, C. B. (2005) 'Book review: Seeds of transition – essays on novelty production, niches and regimes in agriculture, edited by J. Wiskerke and J. D. van der Ploeg', *Journal of Environmental Quality*, vol 34, pp400–401

Flören, R. H. (2002) *Crown Princes in the Clay: An Empirical Study on the Tackling of Succession Challenges in Dutch Family Farms*, Royal van Gorcum, Assen, The Netherlands

Fort, A., Boucher, S. and Riesco, G. (2001) *La pequena agricultura piurana: Evidencias sobre ingreso, credito y asistencia tecnica*, Universidad del Pacifico/CIPCA, Lima/Piura, Peru

Foucault, M. (1972) *Les mots et les choses: Une archeologie des sciences humaines*, Gallimard, Mayenne, France

Foucault, M. (1975) *Surveiller et punir: Naissance de la prison*, Gallimard, Paris, France

Franks, J. and McGloin, A. (2006) *Co-operative Management of the Agricultural Environment*, SAFRD, Newcastle upon Tyne, UK

Franzini, G. (2004) *Il crac Parmalat, Storia del crollo dell'impero del latte*, Editore Riuniti, Roma, Italy

Friedmann, H., (1980) 'Household production and the national economy: Concepts for the analysis of agrarian formations', *Journal of Peasant Studies*, vol 7, pp158–184

Friedmann, H. (1993) 'The political economy of food: A global crisis', *New Left Review*, vol 1, p197

Friedmann, H. (2004) 'Feeding the Empire: The pathologies of globalized agriculture', in Miliband, R. (ed) *The Socialist Register*, Merlin Press, London, UK, pp124–143

Friedmann, H. (2006) 'Focusing on agriculture: A comment on Henry Bernstein's "Is there an agrarian question in the 21st century?"', *Canadian Journal of Development Studies*, vol xxvii, no 4, pp461–465

Frouws, J. (1993) *Mest en macht: een politiek–sociologische studie naar belangenbehartiging en beleidsvorming inzake de mestproblematiek in Nederland vanaf 1970*, Studies van Landbouw en Platteland no. 11, LUW, Wageningen, The Netherlands

Frouws, J. and Ploeg, J. D. van der (1974) *Materiaal voor een Kritiek op Agrarische Socialogie en Voorlichtingskunde*, Boerengroep/Studium Generale, Wageningen, The Netherlands

Frouws, J., Oerlemans, N., Ettema, M., Hees, E., Broekhuizen, R. van, Ploeg, J. D. van der (1996) *Naar de geest of naar de letter: een onderzoek naar knellende regelgeving in de agrarische sector*, Studies van Landbouw en Platteland, 19, LUW, Wageningen, The Netherlands

Galjart, B. (2003) 'Sociaal kapitaal, vertrouwen en ontwikkeling', *Sociologische Gids*, vol 50, no 1, pp26–50

García-Sayán, D. (1982) *Tomas de Tierras en el Peru*, DESCO, Lima, Peru

Gates, M. (1993) *In Default: Peasants, the Debt Crisis, and the Agricultural Challenge in Mexico*, Westview Press, Boulder, CO

Geels, F. W (2002) 'Technological transitions as evolutionary reconfiguration processes: A multi-level perspective and a case study', *Research Policy*, vol 31, pp1257–1274

Geertz, C. (1963) *Agricultural Involution*, University of California Press, Berkeley, CA

Gerritsen, P. R. W. (2002) *Diversity at Stake: A Farmers' Perspective on Biodiversity and Conservation in Western Mexico*, Circle for Rural European Studies, Wageningen University, Wageningen, The Netherlands

Gerritsen, P. R. W. and Morales, J. (ed) (2007) *Respuestas Locales Frente a la Globalización Económica. Productos regionales de la Costa Sur Jalisco*, Universidad de Guadalajara, ITESO/RASA, Guadalajara, Mexico

Gerritsen, P. R. W., Villalvazo, V. L., Figueroa, P. B., Cruz, G. S. and Morales, J. H. (2005) 'Productos regionales y sustentabilidad: Experiencias de la Costa Sur de Jalisco', Paper presented at the 5th Mexican Congress of Rural Studies, Oaxaca, Mexico, 25–28 May 2005

Gibbon, P. and Neocosmos, M. (1985) 'Some problems in the political economy of "African socialism"', in H. Bernstein and B. K. Campbell (eds) *Contradictions of Accumulation in Africa*, Sage, Beverley Hills, CA

Goede, R. G. M. de, Brussaard, L. and Akkermans, A. D. L. (2003) 'On-farm impact of cattle slurry manure management on biological soil quality', *NJAS*, vol 51, no 1–2, pp103–134

Goede, R. G. M. de, Vliet, P. C. J. van, Stelt, B. van der, Verhoeven, F. P. M., Temminghoff, E. J. M., Bloem, J., Dimmers, W. J., Jagers op Akkerhuis, G. A. J. M., Brussaard, L. and Riemsdijk, W. H. van (2004) *Verantwoorde Toepassing van Rundermest in Graslandbodems*, SV-411, SKB, Gouda, The Netherlands

González Chávez, H. (1994) *El empresario agricola en el jugoso negocio de las frutas y hortalizas de México*, PhD thesis, WAU, Wageningen, The Netherlands

González de Molina, M. and Guzmán Casado, G. (2006) *Tras los pasos de la insustentabilidad: agricultura y medio ambiente en perspectiva histórica* (s.XVIII–XX), Icaria Editorial, Barcelona, Spain

Goodman, D. (1999) 'Agro-food studies in the "Age of Ecology": Nature, corporeality, bio-politics', *Sociologia Ruralis*, vol 39, no 1, pp17–38

Goodman, D. (2004) 'Rural Europe redux? Reflections on alternative agro-food networks and paradigm change', *Sociologia Ruralis*, vol 44, no 1, pp3–16

Goodman, D. and Watts, M. J. (1997) *Globalising Food: Agrarian Questions and Global Restructuring*, Routledge, London, UK

Gorgoni, M. (1980) 'Il contadino tra azienda e mercato del lavoro: Un modello teorico', *Rivista di Economia Agraria*, vol XXXV, no 4, pp683–718

Gorgoni, M. (1987) 'Review of Jan Douwe van der Ploeg: La ristrutturazione del lavoro agricolo', *Questione Agraria*, vol 27, pp187–190

Gorlach, K. and Mooney, P. (2004) *European Union Expansion: The Impacts of Integration on Social Relations and Social Movements in Rural Poland*, Cornell University Mellon Sawyer Seminar, Ithaca, NY

Gouldner, A. (1978) 'The concept of functional autonomy', in P. Worsley (ed) *Modern Sociology*, 2nd edition, Penguin, New York, NY

Griffin, K., Rahman, A. Z. and Ickowitz, A. (2002) 'Poverty and the distribution of land', *Journal of Agrarian Change*, vol 2, no 3, pp279–330

Groen, A. F., Groot, K. de, Ploeg, J. D. van der and Roep, D. (1993) *Stijlvol fokken, een orienterende studie naar de relatie tussen sociaal–economische verscheidenheid en bedrijfsspecifieke fokdoeldefinitie*, Bedrijfsstijlenstudie no 9, Vakgroep Veefokkerij en Vakgroep Rurale Sociologie, Landbouwuniversiteit, Wageningen, The Netherlands

Groot, J. C. J., Rossing, W. A. H., Lantinga, E. A. and Keulen, H. van (2003) 'Exploring the potential for improved internal nutrient cycling in dairy farming systems using an eco-mathematical model', *NJAS – Wageningen Journal of Life Sciences*, vol 51, pp165–194

Groot, J. C. J., Stuiver, M. and Brussaard, L. (2004) 'New opportunities and demands for decision support in eco-technological agricultural practices', *Grassland Science in Europe*, vol 9, pp1202–1204

Groot, J. C. J., Rossing, W. A. H. and Lantinga, E. A. (2006a) 'Evolution of farm management, nitrogen efficiency and economic performance of dairy farms reducing external inputs', *Livestock Production Science*, vol 100, pp99–110

Groot, J. C. J., Rossing, W. A. H., Jellema, A. and Ittersum M. K. van (2006b) 'Landscape design and agricultural land-use allocation using Pareto-based multi-objective differential evolution', in A. Voinov, A. J. Jakeman and A. E. Rizzoli (eds) *Proceedings of the iEMSs Third Biennial Meeting, Summit on Environmental Modelling and Software*. International Environmental Modelling and Software Society, Burlington, VT

Groot, J. C. J., Rossing, W. A. J., Jellema, A., Stobbelaar, D. J., Renting, H. and Ittersum, M. K. van (2007a) 'Exploring multi-scale trade-offs between nature conservation, agricultural profits and landscape quality – a methodology to support discussions on land-use perspectives', *Agriculture Ecosystems and Environment*, vol 120, pp58–69

Groot, J. C. J., Ploeg, J. D. van der, Verhoeven, F. P. M. and Lantinga, E. A. (2007b) 'Interpretation of results from on-farm experiments: Manure–nitrogen recovery on grassland as affected by manure quality and application technique, 1, an agronomic analysis', *NJAS*, vol 54, no 3, pp235–254

Gudeman, S. (1978) *The Demise of a Rural Economy, from Subsistence to Capitalism in a Latin American Village*, Routledge and Kegan Paul, London, UK

Guzman Casado, G. I., González de Molina, M. and Sevilla Guzmán, E. (2000) *Introduccion a la Agroecologia Como Desarrollo Rural Sostenible*, Ediciones Mundi-Prensa, Madrid, Spain

Guzman-Flores, E. (1995) *The Political Organization of Sugarcane Production in Western Mexico*, Agricultural University, Wageningen, The Netherlands

Haan, H. de (1993) *In the Shadow of the Tree: Kinship, Property and Inheritance among Farm Families*, Het Spinhuis, Amsterdam, The Netherlands

Haar, G. van der (2001) *Gaining Ground, Land Reform and the Constitution of Community in the Tojolabal Highlands of Chiapas, Mexico*, Thela Latin America Series, Amsterdam, The Netherlands

Hagedorn, K. (2002) *Environmental Co-operation and Institutional Change: Theories and Policies for European Agriculture*, Elgar, Cheltenham, UK

Halamska, M. (2004) 'A different end of the peasants', *Polish Sociological Review*, vol 3, no 147, pp205–268

Hammond, J. L. (1999) 'Law and disorder: The Brazilian landless farmworkers' movement', *Bulletin of Latin American Research*, vol 18, no 4, pp469–489

Hanlon, J. (2004) *The Land Debate in Mozambique: Will Foreign Investors, the Urban Elite, Advanced Peasant or Family Farmers Drive Rural Development?*, Research paper commissioned by Oxfam GB – Regional Management Centre for Southern Africa, Maputo

Hann, C. (2003) *The Postsocialist Agrarian Question*, LIT Verlag, Munster

Hansen, M., Lannoye, P., Pons, S. and Gilles-Eric, S. (2001) *La guerre au vivant: Organismes génétiquement modifies & autres mystifications scientifiques*, Agone Editeur, Marseille, France

Hardt, M. and Negri, A. (2000) *Empire*, Harvard University Press, Cambridge, MA

Harriss, J. (1982) *Rural Development: Theories of Peasant Economy and Agrarian Change*, Hutchinson University Library, London, UK

Harriss, J. (1997) *The Making of Rural Development: Actors, Arenas and Paradigms*, Paper for the anniversary symposium of the Department of Rural Sociology of the Agricultural University, LU, Wageningen, The Netherlands

Harriss, J. (2002) *Depoliticizing Development: The World Bank and Social Capital*, Anthem, London, UK

Hassink, J. (1996) 'Voorspellen van het stikstofleverend vermogen van graslandgronden', in J. W. G. M. Loonen and W. E. M. Bach-de Wit, *Stikstof in Beeld: Naar Een Nieuw Bemestingsadvies op Grasland*, DLO, Wageningen, The Netherlands

Hayami, Y. and Ruttan, V. W. (1985) *Agricultural Development: An International Perspective* (revised and expanded edition), John Hopkins, Baltimore and London

Hebinck, P. and Averbeke, W. van (2007) 'Livelihoods and landscapes: People, resources and land use', in P. Hebinck and P. C. Lent (eds) *Livelihoods and Landscapes: The People of Guquka and Koloni and Their Resources*, Brill, Leiden, The Netherlands

Hebinck, P. and Monde, N. (2007) 'Production of crops in arable fields and home gardens', in P. Hebinck and P. C. Lent (eds) *Livelihoods and Landscapes: The People of Guquka and Koloni and Their Resources*, Brill, Leiden, The Netherlands

Hees, E. (2000) *Trekkers naast de trap, een zoektocht naar de dynamiek in de relatie tussen boer en overheid*, PhD thesis, Wageningen Universiteit, Wageningen, The Netherlands

Hees, E., Rooij, S. de and Renting, H. (1994) *Naar lokale zelfregulering, samenwerk-*

ingsverbanden voor integratie van landbouw, milieu natuur en landschap, Studies van Landbouw en Platteland 14, LUW, Wageningen, The Netherlands

Heijman, W. (2005) *Boeren in het landschap, een studie naar de kosten van agrarisch natuurbeheer in de noordelijke Friese wouden*, WUR, Wageningen, The Netherlands

Heijman, W., Hubregtse M. H. and Ophem, J. A. C. van (2002) 'Regional economic impact of non-standard activities on farms: Method and application to the province of Zeeland in The Netherlands', in J. D. van der Ploeg, A. Long and J. Banks (eds) *Living Countrysides*, Elsevier, Doetinchem, The Netherlands

Held, D., McGrew, A., Goldblatt, D. and Perraton, J. (1999) *Global Transformations: Politics, Economics and Culture*, Polity, Cambridge, MA

Hemme, T., Deeken, E. and Ramanovich, M. (2004) *IFCN Dairy Report*, IFCN, The Hague, The Netherlands

Hervieu, M. B. (2005) *La multifunctionalite et l'agriculture*, INRA, Paris, France

Heynig, K. (1982) 'Principales enfoques sobre la economía campesina', *Revista de la CEPAL*, no 16, pp115–143

Hofstee, E. W. (1985a) *Groningen van Grasland naar Bouwland, 1750–1930*, Pudoc, Wageningen, The Netherlands

Hofstee, E. W. (1985b) 'Differentiële sociologie in kort bestek, schets van de differentiële sociologie en haar functie in het concrete sociaal–wetenschappelijke onderzoek', in E. W. Hofstee, H. M. Jolles and I. Gadourek (1985) *Differentiële Sociologie in Discussie*, VUGA, Amsterdam, The Netherlands

Holloway, J. (2002) *Cambiar el Mundo sin Tomar el Poder: El Significado de la Revolución Hoy*, El Viejo Topo, Madrid, Spain

Hoog, K. de and Vinkers, J. (2000) *De beleving van armoede in agrarische gezinsbedrijven*, Wetenschapswinkel, nr 165, WUR, Wageningen, The Netherlands

Hoogma, R., Kemp, R., Schot, J. and Truffer, B. (2002) *Experimenting for Sustainable Transport: The Approach of Strategic Niche Management*, Spon Press, London, UK

Hoppenbrouwers, P. and Zanden, J. L. van (2001) *Peasants into Farmers? The Transformation of Rural Economy and Society in the Low Countries (Middle Ages – 19th Century) in Light of the Brenner Debate*, Corn Publication Series, Brepols, Turnhout, Belgium

Horlings, I. (1996) *Duurzaam produceren met beleid: Innovatiegroepen in de Nederlandse landbouw*, LUW, Wageningen, The Netherlands

Howe, S. (2002) *Empire: A Very Short Introduction*, Oxford University Press, Oxford, UK

Huylenbroeck, G. van and Durand, G. (2003) *Multifunctional Agriculture: A New Paradigm for European Agriculture and Rural Development*, Ashgate, Aldershot, UK

Huijsmans, J. F. M., Hol, J. M. G., Smits, M. C. J., Verwijs, B. R., Meer, H. G. van der, Rutgers, B. and Verhoeven, F. P. M. (2004) *Ammoniakemissie bij bovengronds breedwerpige mesttoediening*, Report 136, Agrotechnology and Food Innovations B.V, WUR, Wageningen, The Netherlands

Huizer, G. (1973) *Peasant Rebellion in Latin America*, The Pelican Latin American Library, Penguin Books Ltd, Harmandsworth, UK

IATP (1998) *Marketing Sustainable Agriculture: Case Studies and Analysis from Europe*, Institute for Agriculture and Trade Policy, Minneapolis, MN

Ikerd, J. (2000a) *Sustainable Agriculture: A Positive Alternative to Industrial Agriculture*, http://webmissouri.edu/ikerdj/papers/Ks-hrtld.htm

Ikerd, J. (2000b) *Sustainable Farming and Rural Community Development*, http://webmissouri.edu/ikerdj/papers/ND-NFCD.html

Ikerd, J. (2000c) *Sustainable Agriculture as a Rural Economic Strategy*, http://webmissouri.edu/ikerdj/papers/sa-cdst.htm

Immink, V. M. and Kroon, S. M. A. van der (2006) *Wat je vers haalt is lekker: Thuisverkoop op het platteland*, rapport 227, LEI/WUR, Wageningen, The Netherlands

ISMEA (2005), 'La competitività dell' agroalimentare italiano, check-up 2005', *Politica Agricola Internazionale*, Anno IV, no 1–3, pp71–95

Jacobs, D. (1999) *Het Kennisoffensief: Slim Concurreren in de Kenniseconomie*, Samsom, Deventer, The Netherlands

Janvry, A. de (2000) 'La logica delle aziende contadine e le strategie di sostegno allo sviluppo rurale', *La Questione Agraria*, no 4, pp7–38

Joannides, J., Bergan, S., Ritchie, M., Waterhouse, B. and Ukaga, O. (2001) *Renewing the Countryside*, Minnesota, Institute for Agriculture and Trade Policy, Minneapolis, MN

Johnson, H. (2004) 'Subsistence and control: The persistence of the peasantry in the developing world', *Undercurrent*, vol 1, no 1, pp55–65

Jollivet, M. (1988) *Pour une agriculture diversifiée: Arguments, questions, recherches*, Harmattan, Paris, France

Jollivet, M. (2001) *Pour une science sociale à travers champs: Paysannerie, ruralité, capitalisme (France XXe siecle)*, AP editions, Paris, France

Kamen, H. (2003) *Imperio: La forja de Espana como Potencia Mundial*, Aguilar, Madrid, Spain

Karel, E. H. (2005) *De maakbare boer; streekverbetering als instrument van het Nederlandse landbouwbeleid, 1953–1970*, Historia Agriculturae, NAHI, Groningen, The Netherlands

Kautsky, K. (1899/1970) *La question agraire, etude sur les tendences de l'agriculture moderne*, Maspéro, Paris, France

Kayser, B. (1995) 'The future of the countryside', in J. D. van der Ploeg and G. van Dijk, *Beyond Modernization: The Impact of Endogenous Rural Development*, Royal van Gorcum, Assen, The Netherlands

Kearney, M. (1996) *Reconceptualizing the Peasantry: Anthropology in Global Perspective*, Westview Press, Boulder, CO

Keat, R. (2000) *Cultural Goods and the Limits of the Market*, Palgrave, London, UK

Kemp, R., Schot, J. and Hoogma, R. (1998) 'Regime shifts to sustainability through processes of niche formation: The approach of strategic niche management', *Technology Analysis and Strategic Management* vol 10, pp175–196

Kemp, R., Rip, A. and Schot, J. (2001) 'Constructing transition paths through the management of niches', in R. Garud and P. Karnoe (eds) *Path Dependence and Creation*, Lawrence Erlbaumm Associates, London, UK

Kessel, J. van (1990) 'Produktieritueel en technisch betoog bij de Andesvolkeren', *DerdeWereld*, vol 1990, no 1–2, pp77–97

Kimball, S. T. and Arensberg, C. M. (1965) *Keeping the Name on the Land*, Harcourt, Brace and World, New York, NY

Kinsella, J., Wilson, S., Jong, F. de and Renting, H. (2000) 'Pluriactivity as a livelihood strategy in Irish farm households and its role in rural development', *Sociologia Ruralis*, vol 40, no 4, pp481–496

Kinsella, J., Bogue, P., Mannion, J. and Wilson, S. (2002) 'Cost reduction for small-scale dairy farms in County Clare', in J. D. van der Ploeg, A. Long and J. Banks (eds) *Living Countrysides: Rural Development Processes in Europe – The State of Art*, Elsevier, Doetinchem, The Netherlands

Knickel, K. (2002) 'Energy crops in Mecklenburg–Vorpommern: The rural development potential of crop diversification and processing in Germany', in J. D. van der Ploeg, A. Long and J. Banks (eds) *Living Countrysides: Rural Development Processes in Europe – The State of Art*, Elsevier, Doetinchem, The Netherlands

Knickel, K. (2006) 'Agrarwende – agriculture at a turning point: Rural development practices and policies in Germany', in D. O'Connor, H. Renting, M. Gorman and J. Kinsella (eds) *Driving Rural Development in Europe – The Role of Policy: Case Studies from Seven EU Countries*, Royal van Gorcum, Assen, The Netherlands

Knickel, K. and Hof, S. (2002) 'Direct retailing in Germany: Farmers markets in Frankfurt', in J. D. van der Ploeg, A. Long and J. Banks (eds) *Living Countrysides: Rural Development Processes in Europe – The State of the Art*, Elsevier, Doetinchem, The Netherlands

Knorr-Cetina, K. D. (1981) 'The micro-sociological challenge of the macro-sociological: Towards a reconstruction of social theory and methodology', in K. D. Knorr-Cetina and A.V. Cicourel (eds) *Advances in Social Theory and Methodology: Towards an Integration of Micro- and Macro-Sociologies*, Routledge and Kegan Paul, Boston, MA

Koningsveld, H. (1987) 'Klassieke landbouwwetenschap, een wetenschapsfilosofische beschouwing', in H. Koningsveld, J. Mertens and S. Lijmbach (eds) *Landbouw, Landbouwwetenschap en Samenleving: Filosofische Opstellen*, Mededelingen van de vakgroepen voor sociologie, no 20, Landbouwuniversiteit, Wageningen, The Netherlands

Kop, P. van de, Sautier, D. and Gerz, A. (2006) *Origin-Based Products: Lessons for Pro-Poor Market Development*, Royal Tropical Institute/CIRAD, Amsterdam, The Netherlands

Korten, D. C. (2001) *When Corporations Rule the World*, second edition, Kumarian Press Inc/Berrett-Koehler Publishers Inc, San Francisco, CA

Lacroix, A. (1981) *Transformations du proces de travail agricole, incidences de l'industrialisation sur les conditions de travail paysannes*, INRA, Grenoble, France

Lang, T. and Heasman, M. (2004) *Food Wars: The Global Battle for Mouths, Minds and Markets*, Earthscan, London/Sterling VA

Lanner, S. (1996) *Der Stolz der Bauern; die Entwicklung des ländlichen Raumes: Gefahren und Chancen*, Ibera & Molden Verlag/European University Press, Vienna, Austria

Latour, B. (1994) 'On technical mediation – philosophy, sociology, genealogy', *Common Knowledge*, vol 3, no 2, pp29–64

Laurent, C. and Remy, J. (1998) 'Agricultural holdings: Hindsight and foresight', *Etudes et Recherches des Systemes Agraires et Developpement*, no 31, pp415–430

Lauwere, C. de, Verhaar, K. and Drost, H. (2002) *Het Mysterie van het Ondernemerschap: Boeren en tuinders op zoek naar nieuwe wegen in een dynamische maatschappij*, IMAG, Wageningen, The Netherlands

Law, J. (1994) *Organizing Modernity*, Blackwell Publishers. Oxford, UK

Leeuwis, C. (1993) *Of Computers, Myths and Modelling: The Social Construction of Diversity, Knowledge, Information and Communication Technologies in Dutch Horticulture and Agricultural Extension*, LUW, Wageningen, The Netherlands

LEI (2005) *Landbouw Economisch Bericht (LEB)*, LEI, Den Haag, The Netherlands

Lenin, V. I. (1961) 'The agrarian question and the "critics of Marx"', in *Collected Works*, vol V, Progress Publishers, Moscow

Lenin, V. I. (1964) *The Development of Capitalism in Russia*, Progress Publishers, Moscow

Lin, N. (1999) 'Building a network theory of social capital', *Connections*, vol 22, no 1, pp28–51

Llambi, L. (1988) 'Small modern farmers: Neither peasants nor fully-fledged capitalists?', *Journal of Peasant Studies*, vol 5, no 3, pp350–372

Llambi, L. (1994) 'Comparative advantages and disadvantages', in P. McMichael (ed) *Latin American Nontraditional Fruit and Vegetable Exports in The Global Restructuring of Agro-Food Systems*, Cornell University Press, Ithaca and London

LNV (2005) *Perspectieven voor de Agrarische Sector in Nederland*, Achtergrondsrapport bij 'Kiezen voor landbouw', LNV, Den Haag, The Netherlands

Long, N. (1977) *An Introduction to the Sociology of Rural Development*, Tavistock, London, UK

Long, N. (1985) 'Creating space for change: A perspective on the sociology of development', *Sociologia Ruralis*, vol XXIV, no 3/4, pp168–184

Long, N. (2001) *Development Sociology: Actor Perspectives*, Routledge, London, UK

Long, N. (2007) 'Resistance, agency and counter-work: A theoretical positioning', in W. Wright and G. Middendorf (eds) *Food Fights*, Penn State University Press, Pennsylvania

Long, N. and Long, A. (1992) *Battlefields of Knowledge: The Interlocking of Theory and Practice in Social Research and Development*, Routledge, London, UK

Long, N. and Ploeg, J. D. van der (1994) 'Heterogeneity, actor and structure: Towards a reconstitution of the concept of structure', in D. Booth (ed) *Rethinking Social Development: Theory, Research and Practice*, Longman Scientific and Technical, Harlow, UK

Long, N. and Roberts, B. (2005) 'Changing rural scenarios and research agendas in Latin America in the new century', in F. Buttel and P. McMichael (eds) *New Directions in the Sociology of Global Development, Research in Rural Sociology and Development*, vol 11, pp57–90

Long, N., Ploeg, J. D. van der, Curtin, C. and Box, L. (1986) *The Commoditization Debate: Labour Process, Strategy and Social Network*, Papers of the departments of sociology, 17, LUW, Wageningen, The Netherlands

MacIntyre, A. (1981) *After Virtue*, Duckworth, London, UK

Mak, G. (1996) *Hoe God verdween uit Jorwerd; een Nederlands dorp in de twintigste eeuw*, Uitgeverij Atlas, Amsterdam, The Netherlands

Mariátegui, J. C. (1925) *Siete Ensayos de Interpretación de la Realidad Peruana*, Amauta, Lima, Peru

Marsden, T. K. (1991) 'Theoretical issues in the continuity of petty commodity production', in S. Whatmore, P. Lowe and T. Marsden (eds) *Rural Enterprise: Shifting Perspectives on Small-Scale Production*, David Fulton Publishers, London, UK

Marsden, T. (1998) 'Agriculture beyond the treadmill? Issues for policy, theory and research practice', *Progress in Human Geography*, vol 22, no 2, pp265–275

Marsden, T. (2003) *The Condition of Rural Sustainability*, Royal van Gorcum, Assen, The Netherlands

Marsden, T. and Murdoch, J. (2006) *Between the Local and the Global: Confronting Complexity in the Contemporary Agri-Food Sector*, Elsevier, Amsterdam, The Netherlands

Marsden, T., Banks, J. and Bristow, G. (2000a) 'Food supply chains approaches: Exploring their role in rural development', *Sociologia Ruralis*, vol 40, no 4, pp424–438

Marsden, T., Flynn, A. and Harrison, M. (2000b) *Consuming Interests: The Social Provision of Food*, UCL Press, London, UK

Martinez-Alier, J. (2002) *The Environmentalism of the Poor*, Edward Elgar, Cheltenham, UK

Marx, K. (1867/1970) *Het Kapitaal; een kritische beschouwing over de economie* (translation I. Lipschits), De Haan, Bussum, The Netherlands

McMichael, P. (ed) (1994) *The Global Restructuring of Agro-Food Systems*, Cornell University Press, Ithaca, NY
McMichael, P. (2007) 'Feeding the world: Agriculture, development and ecology', *The Socialist Register*, vol 2007, pp1–25
Melhum, H., Moene, K. and Torvik, R. (2006) 'Institutions and resource curse', *Economic Journal*, vol 116, pp1–20
Mendras, H. (1967) *La fin des paysans – innovations et changement dans l'agriculture Francaise*, Futuribles/SEDEIS, Paris, France
Mendras, H. (1970) *The Vanishing Peasant: Innovation and Change in French Agriculture*, Cambridge University Press, Cambridge, UK
Mendras, H. (1976) *Sociétés paysannes, éléments pur une théorie de la paysannerie*, Armand Colin, Paris, France
Menghi, A. (2002) 'I prezzi al consumo crescono di piu di quelli alla produzione', *Unalat Informe*, no 60, pp23–25
Meulen, H. S. van der (2000) *Circuits in de Landbouwvoedselketen: Verscheidenheid en samenhang in de productie en vermarkting van rundvlees in Midden-Italie*, PhD thesis, Wageningen University, Wageningen, The Netherlands
Miele, M. (2001) *Creating Sustainability: The Social Construction of the Market for Organic Products*, WUR, Wageningen, The Netherlands
Milone, P. (2004) *Agricoltura in transizione: la forza dei piccoli passi; un analisi neo–istituzionale delle innovazioni contadine*, PhD thesis, Wageningen University, Wageningen, The Netherlands
Milone, P. and Ventura, F. (2000) 'Theory and practice of multi-product farms: Farm butcheries in Umbria', *Sociologia Ruralis*, vol 40, no 4, pp452–465
Moerman, M. (1968) *Agricultural Change and Peasant Choice in a Thai Village*, University of California Press, Berkeley, CA
Montoya, R. (1986) *El factor etnico y el desarrollo andino: Estrategias para el desarrollo de la sierra*, Centro Bartolome de las Casas, Cusco, Peru
Moors, E. H., Rip, A. and Wiskerke, J. (2004) 'The dynamics of innovation: A multilevel co-evolutionary perspective', in J. S. C. Wiskerke and J. D. van der Ploeg (eds) *Seeds of Transition: Essays on Novelty Production, Niches and Regimes in Agriculture*, Royal van Gorcum, Assen
Moquot, G. (1988) 'Alcuni risultati nella ricerca applicata dei laboratory dell'INRA specializzati nell'industria lattiera, Atti del Convegno sulle biotecnologie in caseificio (Lodi, 8/9 Ottobre 1987), *L'Industria del latte*, April/June
Morgan, K. and Sonnino, R. (2006) 'Empowering consumers: The creative procurement of school meals in Italy and the UK', *International Journal of Consumer Studies*, vol 31, no 1, pp19–25
Mourik, R. M. (2004) *Did Water Kill the Cows? The Distribution and Democratisation of Risk, Responsibility and Liability in a Dutch Agricultural Controversy on Water Pollution and Cattle Sickness*, Pallas Publications, University of Maastricht, Maastricht, The Netherlands
MPAF (Ministero delle Politiche Agricole e Forestale) (2003) *La Poverta in Agricultura*, Eurispes, Roma, Italy
Murdoch, J. (2006) *Post-Structuralist Geography: A Guide to Relational Space*, SAGE Publications, London, UK
Negri, A. (2003) *Cinque Lezioni di Metodo su Moltitudine e Imperio*, Rubettino, Soveria Mannelli, Italy
Negri, A. (2006) *Movimenti nell'Impero, passagi e paesaggi*, Rafaello Cortina Editore, Milano, Italy
Netting, R. Mc. (1993) *Smallholders, Householders: Farm Families and the Ecology of*

Intensive, Sustainable Agriculture, Stanford University Press, Stanford, UK
NFW (2004) *Intententieverklaring en werkprogramma*, NLTO, Drachten, The Netherlands
Nooteboom, G. (2003) *A Matter of Style: Social Security and Livelihood in Upland East Java*, Nijmegen University, Nijmegen, The Netherlands
North, D. C. (1990) *Institutions, Institutional Change and Economic Performance*, Cambridge University Press, New York/Cambridge, UK
NRLO (Nationale Raad voor Landbouwkundig Onderzoek) (1997) *Input–Output Relaties en de Besluitvorming van Boeren*, rappart 97/21, NRLO, Den Haag, The Netherlands
O'Connor, D., Renting, H., Gorman, M. and Kinsella, J. (2006) *Driving Rural Development: Policy and Practice in Seven EU countries*, Royal van Gorcum, Assen, The Netherlands
OECD (Organisation for Economic Co-operation and Development) (1996) *Co-operative approaches to sustainable agriculture* (COM/AGR/CA/ENV/EPOC(96)131), OECD, Paris, France
OECD (2000) *Multifunctionality – Towards an Analytical Framework*, www.oecd.org, AGR/CA/APM(2000)3/FINAL, Paris, France
Ontita, E. G. (2007) *Creativity in Everyday Practice: Resources and Livelihoods in Nyamira, Kenya*, Wageningen University, Wageningen, The Netherlands
Oosterveer, P. (2005) *Global Food Governance*, Wageningen University, Wageningen, The Netherlands
Oostindie, H. and Broekhuizen, R. van (2004) *Landbouw en platteland in de Wolden: een studie naar agrarische ontwikkeling, berbrede landbouw en nieuwe bedrijvigheid in voormalige boerderijen in de gemeente Wolden*, Wageningen Universiteit, dep. maatschappijwetenschappen, Wageningen, The Netherlands
Oostindie, H. and Parrott, N. (2001) 'Farmers' attitudes to rural development: Results of a transnational survey', Working paper, Impact Programme, www.rural-impact.net
Oostindie, H., Ploeg, J. D. van der and Renting, H. (2002) 'Farmers' experiences with and views on rural development practices and processes: Outcomes of a transnational European survey', in J. D. van der Ploeg, A. Long and J. Banks (eds) *Living Countrysides: Rural Development Processes in Europe – The State of the Art*, Elsevier, Doetinchem, The Netherlands
Osmont, S. (2004) *Il Capitale*, Rizzoli romanzo, Milano, Italy
Osti, G. (1991) *Gli innovatori della periferia, la figura sociale dell'innovatore nell'agricoltura di montagna*, Reverdito Edizioni, Torino, Italy
Ostrom, E. (1990) *Governing the Commons: The Evolution of Institutions for Collective Action*, Cambridge University Press, New York, NY
Ostrom, E. (1992) *Crafting Institutions for Self-Governing Irrigation Systems*, ICP Press, San Francisco, CA
Otsuki, K. (2007) *Paradise in a Brazil Nut Cemetery: Sustainability Discourses and Social Action in Pará, the Brazilian Amazon*, PhD thesis, Wageningen University, Wageningen, The Netherlands
Ottati, G. dei (1995) *Tra mercato e communita: aspetti concettuali e ricerche empiriche sul distretto industriale*, Franco Angeli, Milano, Italy
Owen, W. F. (1966) 'The double developmental squeeze on agriculture', *The American Economic Review*, vol LVI, pp43–67
Paige, J. (1975) *Agrarian Revolution: Social Movements and Export Agriculture in the Underdeveloped World*, The Free Press, New York, NY
Palerm, A. (1980) 'Antropologos y campesinos: origines y transformaciones',

Antropologia y Marxismo, Nueva Imagen, Mexico
Paz, R. (1999) 'Campesinado, globalización y desarrollo: Una perspectiva diferente', *Revista Europea de Estudios Rurales Latinoamericanos y del Caribe*, no 66, Amsterdam, The Netherlands
Paz, R. (2004) 'Mercantilización de la pequeña producción caprina: Desaparicion o permanencia; studio de caso de la principal Cuenca lecheara de Argentina', in F. Forni (ed) *Caminos de Solidaridad de la Economia Argentina*, Ediciones CICCUS, Buenos Aires, Argentina
Paz, R. (2006a) 'El campesinado en el agro argentino: Repensando el debate teórico o un intento de reconceptualización?', *Revista Europea de Estudios Latinoamericanos y del Caribe*, no 81, pp3–23
Paz, R. (2006b) 'Desparición o permanencia de los campesinos ocupantes en el noroeste argentino? Evolución y crecimiento en la última década', *Canadian Journal of Latin American and Caribbean Studies*, vol 31, no 61, pp169–197
Pearse, A. (1975) *The Latin American Peasant*, Frank Cass, London, UK
Peppelenbos, L. (2005) *The Chilean Miracle: Patrimonialism in a Modern Free Market Democracy*, PhD thesis, Wageningen University, Wageningen, The Netherlands
Pérez-Vitoria, S. (2005) *Les paysans sont de retour*, essai, Actes Sud, Arles, France
Pernet, F. (1982) *Resistances paysannes*, Presses Universitaires de Grenoble, Grenoble, France
Platteau, J. P. (1992) *Land Reform and Structural Adjustment in Sub-Saharan Africa: Controversias and Guidelines*, FAO, Roma, Italy
Ploeg, J. D. van der (1977) *De Gestolen Toekomst: Imperialisme, Landhervorming en Boerenstrijd in Peru*, De Uytbuyt, Wageningen, The Netherlands
Ploeg, J. D. van der (1987a) *De Verwetenschappelijking van de Landbouwbeoefening*, Mededelingen van de vakgroepen voor sociologie, 21, Lanndbouwuniversiteit, Wageningen, The Netherlands
Ploeg, J. D. van der (1987b) *La Ristrutturazione del Lavora Agricolo*, Presentazioni di Giuseppe Barbero, postilla di Bruno Benvenuti, Ricerche e Studi Socio-economici, La Reda, Roma, Italy
Ploeg, J. D. van der (1990a) *Labour, Markets, and Agricultural Production*, Westview Special Studies in Agriculture Science and Policy, Westview Press, Boulder/San Francisco/Oxford
Ploeg, J. D. van der (1990b) 'Autarky and technical change in rice production in Guinea Bissau: On the importance of commoditisation and decommoditisation as interrelated processes', in M. Haswell and D. Hunt (eds) *Rural Households in Emerging Societies: Technology and Change in Sub-Saharan Africa*, Berg Publisher Ltd, Oxford, Hamburg and New York
Ploeg, J. D. van der (1990c) 'Modelli differenziali di crescita aziendale agricola: Ossia il legame fra "senso" e "strutturazione"', *Rivista di Economia Agraria*, no 2, pp171–199
Ploeg, J. D. van der (1993) 'On potatoes and metaphors', in M. Hobart (ed) *An Anthropological Critique of Development: The Growth of Ignorance*, Routledge, London and New York
Ploeg, J. D. van der (1994) 'Agrarisch natuurbeheer: Nieuwe perspectieven', *Gorteria, Tijdschrift voor onderzoek aan de wilde flora*, vol 20, no 2/3, pp36–41
Ploeg, J. D. van der (1995) 'Voorbodes van plattelandsvernieuwing', *Agrarisch Dagblad*, 12 April, p11
Ploeg, J. D. van der (1997a) 'Om de plaats van arbeid: kansen voor een derde weg in het debat over wereldvoedselproductie?', *Spil*, no 4/5, pp52–60

Ploeg, J. D. van der (1997b) 'On rurality, rural development and rural sociology', in H. de Haan and N. Long (eds) *Images and Realities of Rural Life: Wageningen Perspectives on Rural Transformations*, Royal van Gorcum, Assen, The Netherlands

Ploeg, J. D. van der (1998) *Landhervorming: Onvoltooid Verleden en Toekomstige Tijd*, Diesrede ter gelegenheid van het 80 jarig bestaan van de Landbouw Universiteit Wageningen, Wageningen, The Netherlands

Ploeg, J. D. van der (1999), *De Virtuele Boer*, Royal van Gorcum, Assen, The Netherlands

Ploeg, J. D. van der (2000) 'Revitalizing agriculture: Farming economically as starting ground for rural development', *Sociologia Ruralis*, vol 40, no 4, pp497–511

Ploeg, J. D. van der (2002) *Kleurrijk Platteland: Zicht op een Nieuwe Land – en Tuinbouw*, LTO/Royal van Gorcum, Assen, The Netherlands

Ploeg, J. D. van der (2003a) *The Virtual Farmer: Past, Present and Future of the Dutch Peasantry*, Royal van Gorcum, Assen, The Netherlands

Ploeg, J. D. van der (2003b) 'I contadini fra passato e futuro', in M. Pacetti, P. Bedogni and A. Boldrini (eds) *Agricoltura e societa contadina all'esordio degli anni 2000*, Istituto Alcide Cervi, Reggio Emilia, Italy, pp53–76

Ploeg, J. D. van der (2005a) 'L'innovazione istituzionale e tecnoligica a sostegno dei cambiamenti in atto in agricoltura e per lo sviluppo rurale', *Politica Agricola Internazionale (PAGRI)*, vol IV, no 1, 2, 3, pp25–35

Ploeg, J. D. van der (2005b) 'Landbouwbeleid: de kameleon van Europa', *Socialisme & Democratie*, Maandblad van de Wiardi Beckman Stichting, vol 62, no 10, pp25–31

Ploeg, J. D. van der (2006a) 'Agricultural production in crisis', in P. T. Cloke, T. Marsden and P. H. Mooney (eds) *Handbook of Rural Studies*, Sage, London, UK, pp258–277

Ploeg, J. D. van der (2006b) *Oltre la Modernizzazione: Processi di Sviluppo Rurale in Europa*, Rubbettino Editore, Cosenza, Italy

Ploeg, J. D. van der (2006c) 'O modo de produção camponês revisitado', in S. Schneider (ed) *A Diversidade da Agricultura Familiar*, UFRGS Editora, Porto Alegre, Brazil

Ploeg, J. D. van der (2006d) *El Futuro Robado: Tierra, Agua y Lucha Campesina*, Instituto de Estudios Peruanos, Lima, Peru

Ploeg, J. D. van der (2007) 'The mystery of the local in times of Empire', in J. van Ophem and C. Verhaar (eds) *On the Mysteries of Research: Essays in Various Fields of Humaniora*, Fryske Akademy, Leeuwarden, The Netherlands

Ploeg, J. D. van der and Bolhuis, E. E. (1983) *Scelte Tecniche e Incorporamento delle aziende Zootecniche nelle Strutture Esterne: Una Indagine Nella Realta Emiliana*, Quaderni di studio, Universita di Parma, Italy

Ploeg, J. D. van der and Dijk, G. van (1995) *Beyond Modernization: The Impact of Endogenous Development*, Royal van Gorcum, Assen, The Netherlands

Ploeg, J. D. van der and Frouws, J. (1999) 'On power and weakness, capacity and impotence: Rigidity and flexibility in food chains', *International Planning Studies*, vol 4, no 3, pp333–347

Ploeg, J. D. van der and Long, A. (1994) *Born from Within: Practices and Perspectives of Endogenous Development*, Royal van Gorcum, Assen, The Netherlands

Ploeg, J. D. van der and Renting, H. (2004) 'Behind the "redux": A rejoinder to David Goodman', *Sociologia Ruralis*, vol 44, no 2, pp231–242

Ploeg, J. D. van der and Rooij, S. J. G. de (1999) 'Agriculture in central and Eastern Europe: Industrialization or repeasantization?', in *Proceedings of the Research Conference on Rural Development in Central and Eastern Europe*, 6 December, Podbanska, Slovakia

Ploeg, J. D. van der, Saccomandi, V. and Roep, D. (1990) 'Differentiele groeipatronen in de landbouw: Het verband tussen zingeving en structurering', *TSL*, vol 5, no 2, pp108–132
Ploeg, J. D. van der, Ettema, M. and Roex, J. (1994) *De Crisis: Een Bespreking van de Eerste Boerderij–enquete voor het Nationaal Landbouwdebat*, Misset, Doetinchem, The Netherlands
Ploeg, J. D. van der, Roex, J. and Koole, B. (1996) *Bedrijfsstijlen en Kengetallen: Zicht op Informatie*, DOBI report no 3, Landbouw Economisch Instituut, Den Haag, The Netherlands
Ploeg, J. D. van der, Renting, H., Brunori, G., Knickel, K., Mannion, J., Marsden, T., Roest, K. de, Sevilla Guzman, E. and Ventura, F. (2000) 'Rural development: From practices and policies towards theory', *Sociologia Ruralis*, vol 40, no 4, pp391–408
Ploeg, J. D. van der, Frouws, J. and Renting, H. (2002a) 'Self-regulation as new response to over-regulation', in J. D. van der Ploeg, A. Long and J. Banks (eds) *Living Countrysides: Rural Development Processes in Europe – The State of Art*, Elsevier, Doetinchem, The Netherlands
Ploeg, J. D. van der, Roep, D., Renting, H., Banks, J., Alonso Mielgo, A., Gorman, M., Knickel, K., Schaefer, B. and Ventura, F. (2002b) 'The socio-economic impact of rural development processes within Europe', in J. D. van der Ploeg, A. Long and J. Banks (eds) *Living Countrysides: Rural Development Processes in Europe – The State of Art*, Elsevier, Doetinchem, The Netherlands
Ploeg, J. D. van der, Long, A. and Banks, J. (2002c) *Living Countrysides: Rural Development Processes in Europe – The State of the Art*, Elsevier, Doetinchem, The Netherlands
Ploeg, J. D. van der, Long, A. and Banks, J. (2002d) 'Rural development: The state of the art', in J. D. van der Ploeg, A. Long and J. Banks (eds) *Living Countrysides: Rural Development Processes in Europe – The State of Art*, Elsevier, Doetinchem, The Netherlands
Ploeg, J. D. van der, Verhoeven, F., Oostindie, H. and Groot, J. (2003) *Wat smyt it op; een verkennende analyse van bedrijfseconomische en landbouwkundige gegevens van Vel & Vanla bedrijven*, WUR/NLTO noord, Wageningen, The Netherlands
Ploeg, J. D. van der, Benvenuti, B., Bussi, E., Losi, G., Piagnagnoli, C. and Roest, C. de (2004a) *Latte Vivo: Il Lungo Viaggio del Latte dai Campi alla Tavola – Prospettive Dopo il Parmacrack*, Diabasis, Reggio Emilia, Italy
Ploeg, J. D. van der, Bouma, J., Rip, A., Rijkenberg, F., Ventura, F. and Wiskerke, J. (2004b) 'On regimes, novelties, niches and co-production', in J. S. C. Wiskerke and J. D. van der Ploeg (eds) *Seeds of Transition: Essays on Novelty Production, Niches and Regimes in Agriculture*, Royal van Gorcum, Assen
Ploeg, J. D. van der, Verschuren, P., Verhoeven, F. and Pepels, J. (2006) 'Dealing with novelties: A grassland experiment reconsidered', *Journal of Environmental Policy and Planning*, vol 8, no 3, pp199–218
Ploeg, J. D. van der, Groot, J. C. J., Verhoeven, F. P. M. and Lantinga, E. A. (2007) 'Interpretation of results from on-farm experiments: Manure–nitrogen recovery on grassland as affected by manure quality and application technique, 2 – a sociological analysis', *NJAS*, vol 54–3, pp255–268
Polanyi, K. (1957) *The Great Transformation: The Political and Economic Origins of Our Time*, Beacon Press, Boston, MA
Pollin, R., Epstein, G., Heintz, J. and Ndikumana, L. (2007) *An Employment-Targeted Economic Program for South Africa*, Edward Elgar, Cheltenham, UK
Portela, E. (1994) 'Manuring in Barroso: A crucial farming practice', in J. D. van der

Ploeg and A. Long (eds) *Born from Within: Practice and Perspectives of Endogenous Rural Development*, Van Gorcum, Assen, The Netherlands
Portela, J. and Caldas, J. C. (2003) *Portugal Chão*, Celta Editora, Oeiras, Portugal
Presidency (2003) *The European Union and Developing Countries after Cancún: Common Objectives for Agricultural Policies, Food Security and Rural Development*, Working Document of the Italian Presidency for the Informal Meeting of Agricultural Ministers, held in Taormina, Sicily, 20–24 September 2003, Ministry of Agriculture, Roma
Prins, B.(2006) 'Waarde quotum in 2010 nihil', *Nieuwe Oogst, ledenblad van LTO Noord*, editie Oost, vol 2, no 4, 4 November, p1
Prodi, R. (2002) 'Foreword', in J. D. van der Ploeg, A. Long and J. Banks (eds) *Living Countrysides: Rural Development Processes in Europe – The State of the Art*, Elsevier, Doetinchem, The Netherlands
Prodi, R. (2004) 'La sfida contadina', *La Stampa, Cultura e Spettacoli*, giovedì 23 a giovedì 1 aprile 2004, p23
Putnam, R. (1993) *Making Democracy Work: Civic Traditions in Modern Italy*, Princeton University Press, Princeton, NJ
Rabbinge, R. (2001) 'Megatrends in landbouwontwikkeling en ruimtelijk beleid: Premissen, taboes, mythes, paradoxen en dilemma's', *Spil*, no 173–174, pp18–21
Ragin, C. C. (1989) *The Comparative Method: Moving Beyond Qualitative and Quantitative Strategies*, University of California Press, Berkeley, CA
Ranger, T. (1985) *Peasant Consciousness and Guerrilla Warfare in Zimbabwe: A Comparative Study*, Currey, London, UK
Rassegna Stampa Italiana dal Ministero delle Politiche Agricole e Forestali (2005) 'Domenica 6 di Marzo: Rassegna Stampa' Domenica 6, 7 March, ANSA 05-03-05, 19:17
Raup, P. M. (1978) 'Some questions of value and scale in American agriculture', *American Journal of Agricultural Economics*, May, pp303–308
Reijntjes, C., Haverkort, B. and Waters-Bay, A. (1992) *Farming for the Future: An Introduction to Low External Input and Sustainable Agriculture*, ILEA/MacMillan, Leusden/London
Reijs, J. (2007) *Improving Slurry by Diet Adjustments: A Novelty to Reduce N Losses from Grassland Based Dairy Farms*, PhD thesis, Wageningen University, Wageningen, The Netherlands
Reijs, J. and Verhoeven, F. P. M. (2006) 'Handreiking is nuttig managementinstrument', *Nieuwe Oogst*, vol 1, no 8, pp16–17
Reijs, J., Verhoeven, F. P. M., Bruchem, J. van, Ploeg, J. D. van der and Lantinga, E. A. (2004) 'The nutrient management project of the VEL and VANLA environmental co-operatives', in J. S. C. Wiskerke and J. D. van der Ploeg (eds) *Seeds of Transition: Essays on Novelty Production, Niches and Regimes in Agriculture*, Royal van Gorcum, Assen, The Netherlands
Reijs, J., Meijer, W. H., Bakker, E. J. and Lantinga, E. A (2003) 'Explorative research into quality of slurry manure from dairy farms with different feeding strategies', *NJAS – Wageningen Journal of Life Sciences*, vol 51, pp67–89
Reijs, J., Sonneveld, M. P. W., Pol, A. van der, Visser, M. de and Lantinga, E. A. (2005) *Nitrogen Utilisation of Cattle Slurry in Field and Pot Experiments Originating from Different Diets*, Wageningen University, Wageningen, The Netherlands
Reinhardt, N. and Barlett, P. (1990) 'The persistence of family farms in United States agriculture, *Rural Sociology*, vol 55, no 3, pp203–225
Remmers, G. (1998) *Con cojones y maestría: un estudio sociológico–agronómico acerca del desarrollo rural endógeno y procesos de localización en la Sierra de la Contraviesa*

(España), Wageningen Studies on Heterogeneity and Relocalization, 2, CERES, LUW, Wageningen, The Netherlands

Renting, H. and Ploeg, J. D. van der (2001) 'Reconnecting nature, farming and society: Environmental cooperatives in The Netherlands as institutional arrangements for creating coherence', *Journal of Environmental Policy and Planning*, vol 3, no 2, pp85–102

La Repubblica (2007) 'Imprese e Mercati: Parmalat, la grande fuga delle banche: ecco gli istituti che hanno venduto 200 millioni di bond prima del crac', *La Repubblica*, mercoledi 25 di aprile, p38

Revesz, B. (1989) *Agro y Campesinado*, CIPCA, Piura, Peru

Revesz, B., Aldana Rivera, S., Hurtado Galvan, L. and Requna, J. (1997) *Piura: Region y Sociedad, derrotero bibliografico para el desarollo; Archivos de Historia Andina 22*, CIPCA, CBC, Piura, Cusco, Peru

Richards, P. (1985) *Indigenous Agricultural Revolution: Ecology and Food Production in West Africa*, Unwin Hyman, London, UK

Rip, A. (2006) *Interlocking Socio-Technical Worlds*, Paper presented at the STeHPS Colloquium, 14 June 2006, University of Twente, Enschede (also presented and discussed at the European University Institute, Florence and Wageningen University, Wageningen), The Netherlands

Rip, A. and Kemp, R. (1998) 'Technological change', in S. Rayner and E. L. Malone (eds) *Human Choice and Climate Change*, vol 2, Battelle Press, Columbus, OH, pp327–399

Rip, A. and Schot, J. W. (2001) 'Identifying loci for influencing the dynamics of technological development', in J. Williamson and P. Sørensen (eds) *Social Shaping of Technology*, Edward Elgar, London, UK

Ritzer, G. (1993) *The McDonaldization of Society: An Investigation Into the Changing Character of Contemporary Social Life*, Pine Forge Press, London, UK

Ritzer, G. (2004) *The Globalization of Nothing*, Sage, London, UK

RLG (Raad voor het Landelijk Gebied) (2001) *Agribusiness: Steeds Meer Business, Steeds Minder Agri* (advies 01/5), RLG, Amersfoort, The Netherlands

Robertson, S. J. W. (1912) *A Free Farmer in a Free State: A Study of Rural Life and Industry and Agricultural Politics in an Agricultural Country*, Heinemann, London, UK

Roep, D. (2000) *Vernieuwend werken; sporen van vermogen en onvermogen (een sociomateriele studie over verniewuing in de landbouw uitgewerkt voor de westelijke veenweidegebieden)*, Studies van Landbouw en Platteland 28, Circle for Rural European Studies, Wageningen University, Wageningen, The Netherlands

Roep, D. (2002) 'Value of quality and region: The Waddengroup Foundation', in J. D. van der Ploeg, A. Long and J. Banks (eds) *Living Countrysides: Rural Development Processes in Europe – The State of Art*, Elsevier, Doetinchem, The Netherlands

Roep, D. and Wiskerke, H. (2004) 'Reflecting on novelty production and niche management in agriculture', in H. Wiskerke and J. D. van der Ploeg (eds) *Seeds of Transition*, Royal van Gorcum, Assen, The Netherlands

Roep, D., Ploeg, J. D. van der and Wiskerke, H. (2003) 'Managing technical–institutional design processes: Some strategic lessons from environmental co-operatives in the Netherlands', *NJAS*, vol 51, no 1–2, pp195–216

Roest, K. de (2000) *The Production of Parmigiano-Reggiano Cheese: The Force of an Artisanal System in an Industrialised World*, Royal van Gorcum, Assen, The Netherlands

Roex, J. and Ploeg, J. D. van der (1993) *Op zoek naar het verbredingsaanbod; een statistische analyse*, RPD, Den Haag, The Netherlands

Rogers, E. M. and Shoemaker, F. (1971) *Communiation of Innovations: A Cross-*

Cultural Approach, The Free Press, New York/Collier–MacMillan Ltd, London

Rooij, S. de (1992) *Werk van de Tweede Soort: Boerinnen in de Melkveehouderij*, Van Gorcum, Assen, The Netherlands

Rooij, S. de (2005) 'Environmental cooperatives: A farming strategy with potential', *Compas Magazine*, no 8, ETC Leusden, pp5–10

Rooij, S. de, Brouwer, E. and Broekhuizen, R. van (1995) *Agrarische Vrouwen en bedrijfsontwikkeling*, LUW/WLTO, Wageningen, The Netherlands

Roos, N. (2006) 'Weg met requirements, interview met Durk van der Ploeg', *Bits & Chips*, 2 November, pp24–26

Ross, M. (1999) 'The political economy of the resource curse', *World Politics*, vol 51, no 2, pp297–232

Sabourin, E. (2006) 'Praticas sociais, políticas públicas e valores humanos', in S. Schneider (ed) *A Diversidade da Agricultura Familiar*, UFRGS Editora, Porto Alegre, Italy

Saccomandi, V. (1990) 'Presentazione', in J. D. van der Ploeg (ed) *Lo Sviluppo Tecnologico in Agricoltura: Il Caso della Zootecnia*, INEA: Studi e Ricerche, Il Mulino, Bologna, Italy

Saccomandi, V. (1991) *Istituzioni di Economia del Mercato dei Prodotti Agricoli*, REDA, Roma, Italy

Saccomandi, V. (1998) *Agricultural Market Economics: A Neo-Institutional Analysis of Exchange, Circulation and Distribution of Agricultural Products*, Royal van Gorcum, Assen, The Netherlands

Sachs, J. and Warner, A. M. (2001) 'Natural resources and economic development: The curse of natural resources', *European Economics Review*, vol 45, pp827–838

Salamon, S. (1985) 'Ethnic communities and the structure of agriculture', *Rural Sociology*, vol 50, no 3, pp323–340

Salazar, C. (1996) *A Sentimental Economy: Commodity and Community in Rural Ireland*, Berghahn Books, Providence, Rhode Island

Salter, W. E. G. (1966) *Productivity and Technical Change*, Cambridge University Press, New York, NY

Sandt, J. van de (2007) *Behind the Mask of Recognition: Defending Autonomy and Communal Resource Management in Indigenous Resguardos*, Colombia, Universiteit van Amsterdam, Amsterdam, The Netherlands

Saraceno, E. (1996) *Jobs, Equal Opportunities and Entrepreneurship in Rural Areas*, Paper presented to the European Conference on Rural Development: Rural Europe – Future Perspectives, Cork, Ireland

SARE (2001) *The New American Farmer: Profiles of Agricultural Innovation*, USDA Sustainable Agriculture Research and Education (SARE) programme, Sustainable Agriculture Network (SAN), Waldorf, MD

Scettri, R. (ed) (2001) *Novità in Campagna: Innovatori Agricoli nel sud Italia*, ACLI Terra/IREF, Roma, Italy

Schaminee, J., Stortelder, A. and Weeda, E. (2004) *Streekeigen Natuur op de Grens van Zand, Klei en Veen*, Alterra, Wageningen, The Netherlands

Schejtman, A. (1980) 'Economía campesina: Lógica interna, articulación y persistencia', *Revista de la CEPAL*, no 11, pp121–140

Schlosser, E. (2001) *Fast Food Nation: The Dark Side of the All-American Meal*, Houghton Mifflin Company, Boston/New York

Schmitter, P. (2001) *What Is There to Legitimize in the European Union, and How Might This Be Accomplished?* Paper presented at the workshop Linking Political Science and the Law – the Provision of Common Goods, Held at the Max Planck Projectgruppe Recht der Gemeinschaftsgueter, Bonn, January 2001

Schmitzberger, I., Wrbka, T., Steurer, B., Aschenbrenner, G., Peterseil, J. and Zechmeister, H. G. (2005) 'How farming styles influence biodiversity maintenance in Austrian agricultural landscapes', *Agriculture Ecosystems and Environment*, vol 108, pp274–290

Schnabel, P. (2001) *Waarom blijven boeren? Over voortgang en beëindiging van het boerenbedrijf*, Sociaal Cultureel Planbureau, Den Haag, The Netherlands

Schneider, S. (2003) *A Pluriatividade na Agricultura Familiar*, UFRGS Editora, Porte Alegre, Brazil

Schneider, S. (2006) *A Diversidade da Agricultura Familiar*, UFRGS Editora, Porto Alegre, Brazil

Schoorlemmer, H. B., Munneke, F. J. and Broker, M. J. E. (2006) *Verbreding Onder de Loep: Potenties van Multifunctionele Landbouw*, PPO, WUR, Lelystad, The Netherlands

Schuite, H. (2000) *Pioneers in Agriculture: A Study on Direct Sales and on Farm Transformation in the Province Gelderland*, Rural Sociology Group, Wageningen University, Wageningen, The Netherlands

Schultz, T. W. (1964) *Transforming Traditional Agriculture*, Yale University Press, New Haven, CT

Schüren, U. (2003) 'Reconceptualizing the post-peasantry: Household strategies in Mexican Ejidos', *Revista Europea de Estudios Latinoamericanos y del Caribe*, vol 75, pp47–63

Scorza, M. (1974) *Garabombo el Invisible*, Uitgeverij Contact, Amsterdam, The Netherlands

Scott, J. C. (1976) *The Moral Economy of the Peasant*, Yale University Press, New Haven, NJ

Scott, J. C. (1985) *Weapons of the Weak: Everyday Forms of Peasant Resistance*, Yale University Press, New Haven and London

Scott, J. C. (1998) *Seeing Like a State: How Certain Schemes to Improve the Human Condition Have Failed*, Yale University Press, New Haven and London

Scottish Office, Land Reform Policy Group (1998) *Identifying the Problems and Identifying the Solutions*, Scottish Office, Edinburgh, UK

Sender, J. and Johnston, D. (2004) 'Searching for a weapon of mass production in rural Africa: Unconvincing arguments for land reform', *Journal of Agrarian Change*, vol 4, no 1 and 2, pp142–164

Servizi Commerciali Allevatori (2005) *Vendita Diretta di Latte Crudo Sfuso: Dal Produttore als Consumatore*, Società Cooperativa a r.l., Erba

Sevilla Guzman, E. (2006) *Desde del Pensamiento Social Agrario: perspectivas agroecologicas del instituto de sociologia y estudios campesinos*, Servicio de Publicaciones, Universidad de Cordoba, Cordoba, Spain

Sevilla Guzman, E. (2007) *De la Sociologia Rural a la Agroecologia: Perspectivas Agroecologicas*, Icaria Editorial, Barcelona, Spain

Sevilla Guzman, E. and Gonzalez de Molina, M. (1990) 'Ecosociologia: Elementos teoricos para el analisis de la coevolucion social y ecologica en la agricultura', *Revista Espanola de Investigaciones Sociologicas*, no 52, pp7–45

Sevilla Guzman, E. and Martínez-Alier, J. (2006) 'New rural social movements and agroecology', in P. Cloke, T. Marsden and P. Mooney (eds) *Handbook on Rural Studies*, Sage Publications, London, UK

Shanin, T. (1971) *Peasants and Peasant Societies*, Penguin Books, Harmondsworth, UK

Shanin, T. (1972) *The Awkward Class: Political Sociology of Peasantry in a Developing Society: Russia 1910–1925*, Clarendon Press, Oxford, UK

Shanin, T. (1990) *Defining Peasants*, Basil Blackwell, London, UK

Sivini, G. (2007) *Resistance to Modernization in Africa: Journey among Peasants and Nomads*, Transaction Publishers, Rutgers, NJ

Slee, B., Gibbon, D. and Taylor, J. (2006) *Habitus and Style of Farming in Explaining the Adoption of Environmental Sustainability-Enhancing Behaviour*, Countryside and Community Research Unit, University of Gloucestershire, Cheltenham, UK

Slicher van Bath, B. (1960) *De agrarische geschiedenis van West–Europa, 500–1850*, Het Spectrum, Utrecht/Antwerpen, The Netherlands

Slicher van Bath, B. H. (1978) 'Over boerenvrijheid (inaugurele rede Groningen, 1948)', in B. H. Slicher van Bath and A. C. van Oss (eds) *Geschiedenis van Maatschappij en Cultuur*, Basisboeken Ambo, Baarn, The Netherlands

Smeding, F. W. (2001) *Steps Towards Food Web Management on Farms*, Wageningen University, Wageningen, The Netherlands

Smit, J. (2004) *Het drama Ahold*, Uitgeverij Balans, Amsterdam, The Netherlands

Sonneveld, M. P. W. (2004) *Impressions of Interactions: Land as a Dynamic Result of Co-Production between Man and Nature*, PhD thesis, Wageningen University, Wageningen, The Netherlands

Sonneveld, M. P. W. (ed) (2006) *Effectiviteit van het 'Alternatieve Spoor' in de Noordelijke Friese Wouden*, Tussenrapportage 2006, WUR, Wageningen, The Netherlands

Sonneveld, M. P. W. and Bouma, H. (2003) 'Effects of combinations of land use history and nitrogen application on nitrate concentration in the groundwater', *NJAS*, vol 51, no 1–2, pp135–146

Sonneveld, M. P. W. and Bouma, J. (ed) (2004) *Onderzoek op het bedrijf Spruit*, Tussenrapportage voor 2004, WUR–Bodemkunde & Geologie, Intern Rapport 2004–043, Wageningen, The Netherlands

Soto, H. de (2000) *Het Mysterie van het Kapitaal: Waarom het Kapitalisme zo'n Succes is in het Westen Maar Faalt in de Rest van de Wereld*, Het Spectrum, Utrecht, The Netherlands

Souza Martins, S. de (2003) *Travessias: A Vivencia da Reforma Agraria nos Assentamentos* UFRGS Editora, Porte Alegre, Brazil

Speerstra, H. (1999) *It Wrede Paradys, Libbensferhalen fan Fryske folksferhuzers*, Friese Pers Boekerij, Leeuwarden/Ljouwert

SRA (2006) *Benchmark Melkveehouderij 2005*, SRA, Nieuwegein, The Netherlands

Stassart, P. and Engelen, G. van (eds) (1999) *Van de grond tot in je mond: 101 pistes voor een kwaliteitsvoeding*, Vredeseilanden–Coopibo and Fondation Universitaire Luxembourgeoise, Brussels, Belgium

Steenhuijsen Piters, B. de (1995) *Diversity of Fields and Farmers: Explaining Yield Variations in Northern Cameroon*, Agricultural University, Wageningen, The Netherlands

Stiglitz, J. (2002) *Globalization and Its Discontents*, Penguin Books, London, UK

Stiglitz, J. (2003) *The Roaring Nineties: Seeds of Destruction*, Allan Lane, Penguin Group, London, UK

Strange, M. (1985) *Family Farming: A New Economic Vision*, University of Nebraska Press, Lincoln and London, and Institute for Food and Development Policy, San Francisco, CA

Straten, R. van (2006) 'Requirements: Niet voor software alleen', *Bits & Chips*, 2 November, pp52–53

Stuiver, M. (2006) 'Highlighting the retro side of innovation and its potential for regime change in agriculture', in J. Murdoch and T. Marsden (eds) *Between the Local and the Global: Confronting Complexity in the Contemporary Agri-Food Sector – Research in Rural Sociology and Development*, volume 12, Elsevier, Amsterdam, The Netherlands

Stuiver, M. and Wiskerke, J. S. C. (2004) 'The VEL & VANLA environmental co-operatives as a niche for sustainable development', in J. S. C. Wiskerke and J. D. van der Ploeg (eds) *Seeds of Transition: Essays on Novelty Creation, Niches and Regimes in Agriculture*, Royal van Gorcum, Assen, The Netherlands

Stuiver, M., Ploeg, J. D. van der and Leeuwis, C. (2003) 'The VEL and VANLA cooperatives as field laboratories', *NJAS*, vol 51, no 1–2, pp27–40

Stuiver, M., Leeuwis, C. and Ploeg, J. D. van der (2004) 'The power of experience: Farmers' knowledge and sustainable innovations in agriculture', in J. S. C. Wiskerke and J. D. van der Ploeg (eds) *Seeds of Transition: Essays on Novelty Creation, Niches and Regimes in Agriculture*, Royal van Gorcum, Assen, The Netherlands

Swagemakers, P. (2002) *Verschil Maken: Novelproductie en de Contouren van een Streekcooperatie*, Circle for Rural European Studies/ Leerstoelgroep Rurale Sociologie, Wageningen, The Netherlands

Swagemakers, P., Wiskerke, H. and Ploeg, J. D. van der (2007) 'Linking birds, fields and farmers', Working Document 26, Rural Sociology Group, Wageningen University, Wageningen

Tepicht, J. (1973) *Marxisme et agriculture: Le paysan polonais*, Armand Collin, Paris

Thiel, H. van (2006) 'Requirementsmanagement staat of valt met communicatie', *Bits & Chips*, 2 November, pp50–51

Thiesenhuisen, W.C. (1995) *Broken Promises: Agrarian Reform and the Latin American Campesino*, Westview Press, Boulder, CO

Toledo, V. M. (1981) 'Intercambio ecológico e intercambio económico en el proceso productivo primario', in E. Leff (ed) *Biosociología y Articulación de las Ciencias*, UNAM, Mexico City, Mexico

Toledo, V. M. (1990) 'The ecological rationality of peasant production', in M. Altieri and S. Hecht (eds) *Agroecology and Small Farm Development*, CRC Press, Ann Arbor, MI

Toledo, V. (1992) 'La racionalidad ecologica de la produccion campesina', in E. Sevilla Guzman and M. Gonzalez de Molina (eds) *Ecologia, Campesinado e Historia*, Las Ediciones de la Piqueta, Madrid, Spain

Toledo, V. M. (1994) *La Apropiación Campesina de la Naturaleza: Un análisis Etnoecológico*, PhD thesis, Facultad de Ciencias, UNAM, Mexico

Toledo, V. M. (1995) *Campesinidad, Agroindustrialidad, Sostenibilidad: Los Fundamentos Ecológicos e Históricos del Desarrollo Rural*, Cuadernos de Trabajo 3, Grupo Interamericano para el Desarrollo Sostenible de la Agricultura y los Recursos Naturales, Mexico

Toledo, V. (2000) *La Paz en Chiapas, Ecologia, Luchas Indigenas y Modernidad Alternativa*, Ediciones Quinto Sol, Mexico.

Tönnies, F. (1887) *Gemeinschaft und Gesellschaft: Grundbegriffe der reinen Soziologie*, Fues, Leipzig, Germany

Torres, G. (1994) *The Force of Irony: Studying the Everyday Life of Tomato Workers in Western Mexico*, Wageningen University, Wageningen, The Netherlands

Tracy, M. (1997) *Agricultural Policy in the European Union, Agricultural Policy Studies*, European Commission, Brussels

Tudge, C. (2004) *So Shall We Reap: What's Gone Wrong with the World's Food – and How to Fix It*, Penguin Books, New York, NY

Twist, M. van and Veeneman, W. (eds) (1999) *Marktwerking op weg: Over concurrentiebevordering in infrastructuurgebonden sectoren*, Lemma, Utrecht, The Netherlands

Unalat (2002), *Unalat Informe*, vol 60, October, p25

UNRISD (1998) *Outline for a Programme on Grassroots Movements and Initiatives for Land Reform in Developing Countries*, UNRISD, Geneva

Ullrich, O. (1979) *Weltniveau*, Rotbuch Verlag, Berlin, Germany

Uvin, P. (1994) *The International Organization of Hunger*, Kegan Paul International, London, UK

Vaeren, P. van der (2000) *Perdidos en la Selva; un estudio del proceso de re-arraigo y de desarrollo de la Comunidad – Cooperativa Unión Maya Itza, formada por campesinos guatemaltecos, antiguos refugiados, reasentados en el Departamento de El Petén, Guatemala*, PhD thesis, Wageningen University, Wageningen, The Netherlands

Valentini, D. (2006) 'La spesa? Si fa dal contadino', *La Repubblica*, Venerdi 20 Gennaio 2006, pIX

Veenhuizen, R. van (2006) *Cities Farming for the Future: Urban Agriculture for Green and Productive Cities*, RUAF Foundation, IDRC and IIRR, Leusden, The Netherlands

Venema, G., Pager, J., Doorneweert, B. and Klooster, A. van der (2006) *Verbreding onder de loep: Monitoring economische positie van agrarische bedrijven met verbreding in recreatie, huisveerkoop en zorg*, LEI, Den Haag, The Netherlands

Ventura, F. (1995) 'Styles of beef cattle breeding and resource use efficiency in Umbria', in J. D. van der Ploeg and G. van Dijk (eds) *Beyond Modernization: The Impact of Endogenous Rural Development*, Van Gorcum, Assen, The Netherlands

Ventura, F. (2001) *Organizzarsi per Sopravvivere: Un analisi neo-istituzionale dello sviluppo endogeno nell'agricoltura Umbra*, PhD thesis, Wageningen University, Wageningen, The Netherlands

Ventura, F. and Meulen, H. van der (1994) *La Costruzione della Qualita: Produzione, Commercializzazione e Consumo della Carne Bovina in Umbria*, CESAR, Assisi, Italy

Ventura, F. and Milone, P. (2004) 'Novelty as redefinition of farm boundaries', in H. Wiskerke and J. D. van der Ploeg (eds) *Seeds of Transition: Essays on Novelty Production, Niches and Regimes in Agriculture*, Royal van Gorcum, Assen, The Netherlands

Ventura, F. and Milone, P. (2005a) *Innovatività Contadina e Sviluppo Rurale: Un'analisi neo-istitutionale del cambiamento in agricoltura in tre regioni del Sud Italia*, Franco Angeli, Milano, Italy

Ventura, F. and Milone, P. (2005b) *Traiettorie di Sviluppo: Il sostegno a modelli di sviluppo endogeno: Dall'esperienza del distretto viti-vinicolo di Montefalco alla valorizzazione dell'area della Valnerina*, CESAR, Assisi, Italy

Ventura, F. and Milone, P. (2007) *I Contadini del Terzo Millennio*, Franco Angeli, Milano, Italy

Ventura, F., Milone, P. and Ploeg, J. D. van der (2007) *Qualità della vita fuori città*, AMP Editore, Perugia, Italy

Verhoeven, F. P. M., Reijs, J. W. and Ploeg, J. D. van der (2003) 'Re-balancing soil–plant–animal interactions: Towards reduction of nitrogen losses', *NJAS*, vol 51, no 1–2, pp147–164

Volkskrant (2006) 'Het hoofdkantoor gaat er als eerste aan: een bedrijf in stukjes hakken scheelt kosten en creëert helderheid', *Volkskrant*, 16 August, p2

Vries, W. de (1995) *Pluri-activiteit in de Nederlandse landbouw*, Studies van Landbouw en Platteland, 17, LUW, Wageningen, The Netherlands

Ward, N. (1993) 'The agricultural treadmill and the rural environment in the post-productivist era', *Sociologia Ruralis*, vol 33, no 3–4, pp348–364

Warman, A. (1976) *Y venimos a contradecir, los campesinos de Morelos y el Estado Nacional*, Ediciones de la Casa Chata, Mexico

Wartena, D. (2006) *Styles of Making a Living and Ecological Change on the Fon and*

Adja Plateaux in South Benin, ca 1600–1900, PhD thesis, Wageningen University, Wageningen, The Netherlands
WB (Wageningen Blad) (2003) 'Bovengronds mest uitrijden wel of juist niet beter voor milieu? Sociologen en dieronderzoekers betwisten elkaars conclusies', *WB 20*, 19 June, p11
Weeda, E., Swagemakers, P., Bijlsma, R. J. and Spruit, H. (2004) *Boerendiversiteit voor biodiversiteit: Een inventarisatie van de spontane plantengroei op vijf natuurvriendelijke bedrijven*, Alterra (rapport 973), Wageningen, The Netherlands
Weis, T. (2007) *The Global Food Economy: The Battle for the Future of Farming*, Zed Books, London, UK
Wertheim, W. F. (1971) *Evolutie en Revolutie: De Golfslag der Emancipatie*, Van Gennep, Amsterdam, The Netherlands
Whatmore, S. and Stassart, P. (2001) 'Metabolizing risk: The assemblage of alternative meat networks in Belgium', Paper presented to Workshop on International Perspectives on Alternative Agro-Food Networks: Quality, Embeddedness, Bio-Politics, University of California, Santa Cruz, CA
Wielenga, E. (2001) *Netwerken als levend weefsel; een studie naar kennis, leiderschap en de rol van de overheid in de Nederlandse landbouw sinds 1945*, PhD thesis, Wageningen University, Wageningen, The Netherlands
Wijffels, H. (2004) 'Durf het anders te doen', in J. Proost and F. Verhoeven (eds) *Zo werkt het in de praktijk*, WUR, Wageningen, The Netherlands
Willis, S. and Campbell, H. (2004) 'The chestnut economy: The praxis of neo-peasantry in rural France', *Sociologia Ruralis*, vol 44, no 3, pp317–332
Wilson, S., Mannion, J. and Kinsella, J. (2002) 'The contribution of part-time farming to living countrysides in Ireland', in J. D. van der Ploeg, A. Long and J. Banks (eds) *Living Countrysides: Rural Development Processes in Europe – The State of Art*, Elsevier, Doetinchem, The Netherlands
Wiskerke, H. (1997) *Zeeuwse akkerbouw tussen verandering en continuiteit: En sociologische studie naar diversiteit in landbouwbeoefening, technologieontwikkeling en plattelandsvernieuwing*, Studies van Landbouw en Platteland, 25, LUW, Wageningen, The Netherlands
Wiskerke, H. (2001) 'Rural development and multifunctional agriculture: Topics for a new socio-economic research agenda', *Tijdschrift voor Sociaalwetenschappelijk onderzoek van de landbouw*, vol 16, no 2, pp144–19
Wiskerke, J. S. C. (2002) 'On promising niches and constraining sociotechnical regimes: The case of Dutch wheat and bread', *Environment and Planning A*, vol 35, pp429–448
Wiskerke, J. S. C. and Ploeg, J. D. van der (2004) *Seeds of Transition: Essays on Novelty Production, Niches and Regimes in Agriculture*, Royal van Gorcum, Assen, The Netherlands
Wiskerke, J. S. C., Bock, B. B., Stuiver, M. and Renting, H. (2003a) 'Environmental co-operatives as a new mode of rural governance', *NJAS*, vol 51, no 1–2, pp9–26
Wiskerke, J. S. C., Verhoeven, F. P. M., Brussaard, L., Wienk, J. and Struik, P. (eds) (2003b) 'Rethinking environmental management in Dutch dairy farming: a multi-disciplinary farmer-driven approach', Special issue of *NJAS – Wageningen Journal of Life Sciences*, vol 51
Wit, C. T. de (1992) 'Resource use efficiency in agriculture', *Agricultural Systems*, vol 40, pp125–151
Wolf, E. (1955) 'Types of Latin American peasantry: A preliminary discussion', *American Anthropologist*, vol 57, no 3, pp452–471
Wolf, E. (1966) *Peasants*, Englewood Cliffs, Prentice-Hall, New Jersey

Wolf, E. (1969) *Peasant Wars of the Twentieth Century*, Harper and Row, New York, NY
Wolleswinkel, A. P., Roep, D., Calker, K. J. van, Rooij, S. J. G. de and Verhoeven, F. P. M. (2004) *Atlas van innoverende melkveehouders, Veelbelovende vertrekpunten bij het verduurzamen van de melkveehouderij*, WUR, Wageningen, The Netherlands
Wrigley, N. and Lowe, M. S. (1996) (eds) *Retailing, Consumption and Capital: Towards a New Retail Geography*, Longman, Harlow, UK
WRR (Wetenschappelijke Raad voor het Regeringsbeleid) (2003) *Naar Nieuwe Wegen in het Milieubeleid*, SDU Uitgevers, Den Haag, The Netherlands
Yotopoulos, P. A. (1974) 'Rationality, efficiency and organizational behaviour through the production function: Darkly', *Food Research Institute Studies*, vol XIII, no 3, pp263–273
Zamosc, L. (1990) *Peasant Struggles and Agrarian Reform*, Latin American Issues monograph no 8, Allegheny College, Meadville, PA
Zanden, J. L. van (1985) *De Economische Ontwikkeling van de Nederlandse Landbouw in de Negentiende Eeuw, 1800–1914*, AAG Bijdragen, Landbouwuniversiteit, Wageningen, The Netherlands
Zhang, X., Xing, L., Fan, S. and Luo, X. (2007) *Resource Curse and Regional Development in China*, IFPRI, Washington, DC
Ziegler, J. (2006) *L'impero della Vergogna*, Marco Tropea Editore, Milano, Italy
Zijderveld, A. C. (1999) *The Waning of the Welfare State*, Transaction, Piscataway, New Jersey
Zuiderwijk, A. (1998) *Farming Gently, Farming Fast: Migration, Incorporation and Agricultural Change in the Mandara Mountains of Northern Cameroon*, CLM, Leiden, The Netherlands

Index

access
   commoditization of 87–88, 246–247
   credit 70
   food empire expansion 93
   skill-oriented technology 176
accountability 49
acid rain 185
additives 107, 108
agency 21–22, 36, 204, 234, 258, 274
agrarian crisis 182, 278–282
Agricomex SRL 108–109
Ahold 87, 91, 102
ammonia 109, 185, 218–219
artificialization 114–115, 257
asparagus 76, 77, 239
assembly-line systems 255, 256, 258
autonomy
   cooperation 34
   imperial conquest 234
   labour processes 26–27
   maintaining 36
   market relations 23, 27–30
   peasant and entrepreneurial contrasts 113–116
   Peasant Principle 261, 272, 286
   relative 32, 169–170
   repeasantization 7, 156, 157
   striving for 14, 23, 32, 156, 157, 181–209

backwardness 19, 25, 46, 47, 121, 123
Bajo Piura, Peru 53–85
Bernstein, Henry 280–282
bio-energy 176–178
biophysics of production 132–133
black milk 98
Bondi intervention 95, 100
borderlines 3, 36–38, 156, 256
boundaries 237, 238, 252–253
brambles 212, 213, 250
Brazil 277
broadening activities 158, 159, 160, 179

bubble economy 91
Buenos Aires, Peru 56
bugs 226–230, 268, 278

Cameroon 125
capital
   agrarian crisis 279–280, 281
   control 240
   cultural 122, 279–280
   distantiation 50, 51
   ecological 114, 278, 280
   Empire 77–78, 96–97
   human 117
   mobilization of 92
   modernization 127
   virtual 97
   working 70
   see also social capital
capitalist farming 2, 3, 9, 12, 38, 120
carbon enriched manure 167–168
Catacaos, Peru 12, 53–85, 266, 267, 269
cattle 134–135, 214–218
cattle–manure–soil–fodder balance 193, 194
centrality of community 191
centrality of labour 43–44
centrality of nature 278
centralization
   bio-energy production 177–178
   control 196, 224–225, 263
   regressive 109–111
   value 87, 92, 104
   see also Empire
centrifugal filtering systems 170–171
certification 177, 228
chain management 248
cheese milk 97–98
chicken 237
Chira Valley, Peru 69–70
clever geography 71, 72, 78–79
coercion 54, 100, 101, 239, 245, 248, 255, 256

collateral moves 205
commoditization 42, 87–88, 116, 246–247, 264, 269
communal units of production (UCPs) 55, 57–59, 60, 61, 63–65, 80, 81–82
communication 175–176
community level 64, 80–81, 82–83, 190, 191
competition
　bio-energy 177
　dairy production 104, 106, 141–142, 143
　Empire 92, 100, 108, 109
　entrepreneurial farming 17
　repeasantization 151, 153, 177, 178
connectivity 3–5, 206–207
conquest 130, 133–134, 141, 233–234, 235, 237–238
　see also takeovers
constellations of farming 1–5, 8–10
constructed milk 102–103
consultations 189–190
consumers
　changing tastes 285
　knowledgeable 174
　milk prices 98–99, 100, 101
　producer relations 279
consumption milk 97, 98, 99, 102, 104, 174
containment 122–125
contaminated food 108
continuity 36, 122, 147
continuous degradation 106
contracts 188–189, 219
control
　creation and demolition 211–231
　Empire 233, 239, 240, 243, 245, 246
　markets 252, 253
　see also hierarchy; power
conversion mechanisms 51, 169, 269–271
cooperation 23, 34, 82–83, 154, 184, 188, 282
co-operatives
　loss of 80, 81–82
　repeasantization 55, 56, 57, 59
　territorial 13, 181–209
　co-production 23, 24–26, 31, 34, 38, 114–115, 209
corporate farming see capitalist farming
corporations 243–244

costs
　calculated 147
　repeasantization 66, 159
　skill-oriented technology 267
　transaction 92, 184
cotton production 57, 63, 257
counterproductivity 115, 240, 271
craftsmanship 116–119, 154, 170, 173, 174
　see also skill-oriented technologies
credit 33, 62, 64, 65, 70
crisis 10–11, 145, 182, 278–282
cultural level
　capital 122, 279–280
　novelty production 201
　repertoires 27–28, 66–67
　see also peasant principle

dairy production
　controllability 227–230
　emerging technologies 173–176
　food empires 87–111
　global cow 215
　market relations 41
　nitrogen excretion 231
　novelties 193–196
　peasant and entrepreneur contrast 12, 113–126, 131–133, 139–140, 141–142, 143–149
　peasant principle 275
　quota systems 134, 142, 246
　territorial co-operatives 13, 186–187
deactivation 1, 7–8, 9, 12, 144, 149
de-agrarianization 7
debate 249–250
debt 90, 95, 101, 131
decentralized bio-energy production 177
decomposition of food 247
deepening activities 158, 159, 179
defiance 265
　see also resistance
degradation 106–109, 131–133, 218
degrees of peasantness 29–30, 36–38, 137, 138
de-peasantization 35
dependency
　autonomy 14, 115–116, 205
　dissimilarities 38
　Empire 100, 259, 264
　market 27, 40, 41–42, 44, 45, 50, 62
　new patterns of 101

payment periods 97
peasant principle 278
pluriactivity 33
reducing 23, 31–32
repeasantization 7, 62
scale enlargement 126
standardization 220
worldwide threat of 39
deprivation 14, 31, 39, 263
deregulation 242
developed world 20, 39–40
developing world 39–40, 124
deviation 125–128, 192, 218, 220, 221, 223–224
see also novelties
De Wolden, The Netherlands 160, 167
differentiation 36–38
diminishing returns, law of 46–47
disconnection
  economics and production 249
  Empire 4, 5–6
  food empires 102, 107
  nature and farming 114, 115, 184
  repeasantization 78
disintegration 262, 263
dissimilarities 38–39
distantiation 49–52, 121, 140
distrust 27, 28, 182, 201, 272
  see also trust
diversification 68, 146, 152–153, 154, 158–159, 275
doggedness 272–273
drainage 78, 263, 264
dreams 254

ecological capital 114, 278, 280
economic level
  coercion 54, 100, 101
  Empire 242–243
  entrepreneurial farming 128–136
  environment relations reversal 201
  farming economically 145–146
  farm resources 28–29
  food empires 81
  informal 242, 277
  moral economies 125–126, 140–142, 191
  nature mutual exclusivity 208
  politico 22, 128, 233
  quality of life 164
  repeasantization 159

squeeze 128–129, 130
standardization 224
virtual 20, 242
education 283
effective reformism 202–203
El Dorados 235, 238, 253–254
electricity production 168, 176–177
emancipation 11, 26, 45, 62, 184, 262, 273, 283
embedded software 251
emerging technology 167–178
Emilia Romagna, Italy 113–126, 143
Empire
  agrarian crisis 11
  description of 3–4
  food empires 69–72, 87–111, 236, 241
  food and farming 233–259
  future of 286
  industrialization 6
  peasant principle 262–264, 276, 277–278
  rise of 69–80
  science 220–226, 254–255
employment 57, 58, 60, 123, 124
  see also labour
energy production 168, 176–177, 178
enlargement 119–121, 122–125
  see also expansion; scale
entrepreneurial condition 113, 128, 144
entrepreneurial farming
  borderlines 37–38
  deactivation 8
  description 1–2, 3, 17, 19
  future 285
  industrialization 9
  peasant comparisons 113–149
  standardization 220
  value added 156–157
environmental level 11, 14, 201
European level
  autonomy 14
  distantiation 51
  food empires 80, 87–111, 129
  labour intensification 47–48
  market relations 41
  milk production 98–99, 142
  policies 282–287
  repeasantization 151–180
  social wealth decline 135
  value added 129
expansion 88–93, 148–149, 235,

237–238, 241, 255–256
see also enlargement; scale
expert systems
  Empire 255
  entrepreneurial condition 128
  moving away from 185
  multifunctionality 121
  novelty production 196–197
  regulations 205–206
  repeasantization 154, 155
  science 221–222, 226
externalization 115
extraction processes see drainage
extra-economic forces 100, 101, 243

family capital 50
farmers' freedom 32, 46
farming
  economically 13, 145–146, 153, 154, 159
  gently 125, 138, 190
  roughly 125, 126, 138
farm-specific data 216, 227, 228
fast farming 125
fencing hedgerows 211–212
fertilizers 68, 217, 218, 254
field guides 198–201, 205
filtering processes 170–172
financial level 89–90, 213, 226
  see also capital; income
fixed iso-curves 207, 208
flexibility 176, 177, 207
fodder 193, 194
food 6, 11, 233–259, 286, 286–287
food empires 69–72, 87–111, 236, 241
free markets 256
free trade 142
fresh blue milk 101–105, 111, 257
Friesland, The Netherlands 136
fruit and vegetable production 76, 93
future aspects 258, 274, 285, 286

gas production 167–168
generalization 221, 233, 240–241
gentle farming 125, 138, 190
Germany 99
global cow 214–218, 222, 230, 239
globalization 142–149, 233
global level
  agrarian crisis 11
  approaches 229

dependency 39
food empires 88–93, 110
markets 110, 255
outsourcing 247
standardization 220
  see also international level
good manure 146, 186, 192–194, 198, 203, 204, 218–219, 227
governance 184–185, 202, 233
Granarolo 102, 104
grassland management 265–267
growth
  food industry 109–111, 238
  peasant and entrepreneur contrast 123
  see also enlargement; expansion
Guinea Bissau 125

haciendas 55, 57, 79–80, 257, 269
harvesting, new modes of 100–101
health 282
heat production 168, 177
hedgerows 182, 183, 186, 198–201, 211–231, 250
herbicides 265
heterogeneity 136–140, 193, 204, 223, 265
hierarchy 219, 238–240, 241, 252, 253, 256
high tech enterprises 139–140, 146
highways 245
historical level
  agrarian entrepreneurship 125–128
  North Frisian Woodlands 185–191
  peasantry 261
hit-and-run approaches 70, 79, 144–145
Holstein cattle 131
hope 274
human capital 117

ICT see information and communication technologies
identity 104, 106–107, 269
Imperial see Empire
imports of milk 99, 102
impoverishment 134–135
income
  differences in 39
  peasant and entrepreneurial contrasts 118–119, 124
  peasant principle 276–277
  pluriactivity 33

repeasantization 160
support for farmers 286
Taormina policy document 283
see also labour
industrialization 1, 5–6, 8, 9, 11
inequality 21, 97, 98, 101
informal economies 242, 277
information and communication technology (ICT) 240, 247–251
information networks 175
infrastructure 93, 94, 245, 264
innovation 178–179, 192
see also novelties
institutional level
  cooperation 34, 82–83
  distantiation 51–52
  distrust 28
  Empire 76–77
  markets 252–253
  reliability 264
  repeasantization 155
  territorial co-operatives 181–209, 185
  trust 90–91
intrinsic backwardness *see* backwardness
insubordination 270, 278
integration 140, 203
integrity 6
intensification
  labour 19, 45–49, 60, 76, 119–121, 137, 138
  production and repeasantization 62–63
  spurred 63–65
intergenerational transfer 50, 51–52
international level 141, 256
see also global level
internet 269
investment 79, 89, 141, 144, 145
invisibility 269
involution 46, 62, 124
Iowa Mission 59–60
irrigation 57, 69, 70
Italy 40, 87–111, 113–126, 160–162, 237, 238, 257, 277

knowledge 116, 174, 176, 203, 226, 279–280

labour
  agrarian crisis 281, 282
  autonomy 156
  co-production 26–27
  cotton production 57
  division of 64, 127
  equality 67
  evolution in Latin America 273
  intensification 19, 45–49, 60, 76, 119–121, 137, 138
  mobilization of 270
  peasant mode of farming 35, 43–44, 45
  quality of 117, 118
  repeasantization 67, 156
  see also employment; income; man–land ratios
land
  agrarian crisis 281–282
  control over 53
  reform 12, 19, 59–60, 61, 281
  repeasantization 55, 56–57
landless peasants 59, 277
Landscape IMAGES 206–207
landscapes 184, 186, 188, 190, 198–201, 206–209
Latin America 53–85, 273
latte fresco blu *see* fresh blue milk
law of diminishing returns 46–47
leasing milk 134
legal level
  chain management 248
  connectivity 206
  fresh milk 105
  manure law 198, 216–217, 226–227, 228
  slurry 219, 225
level three *see* third level
liberalization 92, 142–149
literature shortcomings 20–23
livelihoods 274, 283
living milk 174–176
loans 62
local level
  control 218–219
  Empire impacts 233
  novelty production 192
  repeasantization 179
  repertoires 28
  technology 168–169, 175
logic of farming 117–119
longitudinal studies 12
lookalike products 106–109, 216, 269, 279

lookalike sustainability 218
low-cost farms 139–140, 146
low external-input agriculture 66–67, 278

machinery 75, 107–108, 186, 219, 251
maize production 176, 177
management of nutrient accountancy systems (MINAS) 186, 227
man–land ratios 53–54, 58, 59–60
manure
  global cow 215, 216–217
  new technology 167–168
  state level 218–219, 224, 226–228
  territorial co-operatives 186, 192–194, 198, 203, 204
marginal areas 160–161, 162, 283
marginalization 19, 39, 60, 62, 80
market level
  assembly-line fusion 255, 256, 258
  autonomy 23, 27–30
  commoditization 246, 264
  conversion processes 269–271
  distantiation 49, 50, 52
  emerging technology 169–170
  entrepreneurial condition 128
  food empires 89, 101
  global 110, 255
  liberalization 92
  moral economy 140–141
  peasant and entrepreneurial contrasts 117
  production relations 44, 62
  reciprocity 48–49
  relations with 39–42
  repeasantization 62, 65, 152–153
  state and institutions 252–253
  territorial co-operatives 182
  virtual realities 20
meadows 222, 223
means to do the job 66–67
mechanical technologies 170, 172, 173, 251
mercantile–industrial food regimes 256, 257
microfiltration 101–105, 108–109, 237
migrant labour 33
migration 277
milk production *see* dairy production
MINAS *see* management of nutrient accountancy systems
mission statements 189, 190–191

misunderstood changes 19
mobilization
  capital 92
  peasant principle 277
  resources 28–29, 152
modeling approaches 197, 207–209, 221–226
modernization 17–18, 22, 125–128, 193
modes of farming 1–5, 8–10, 113–149, 152
monopolization 155, 239, 243, 255, 286
moral economies 125–126, 140–142, 191
mortgaging 89, 90, 92
movability 206–209
movements of milk 93–95
multifunctionality 121–122, 151, 160–162, 165
multilevel distantiation 49–52
multi-occupation households 66
multiple degradation 131–133
multi-product farms 153, 160
municipal quality of life 160–162
mystery of farming 13

natural disasters 70
nature
  autonomy and progress 14
  centrality of 278
  co-production 24
  economy mutual exclusivity 208
  hedgerows 211–231
  peasant and entrepreneurial contrasts 114–115
  regrounding of agriculture 153
  repeasantization 157
  territorial co–operatives 184, 186, 188, 198–201
nested markets 280
The Netherlands
  alternative technology 167
  controllability 211–231
  dairy production 13, 115, 134, 139–140, 141–142, 148, 246
  degrees of peasantness 37
  distantiation 49–50, 51
  food industry growth 238
  labour intensification 47
  market dependency 40, 41–42
  pluriactivity 33
  repeasantization 160, 167–168, 169, 179

survival 30
territorial co-operatives 181–182, 185–209
networks
  control 239
  Empire 236, 252
  information 175
  railways and corporations 243
  repeasantization 173, 174
  social capital 163–166
  territorial co-operatives 185
  three-tiered 93–96
new rural areas 161, 162
NFW *see* North Frisian Woodlands, The Netherlands
NGOs *see* non-governmental organizations
nitrate levels 224
nitrogen 187, 214–217, 222, 223, 227–230
Noardlike Fryske Walden *see* North Frisian Woodlands, The Netherlands
non-agency 234
non-economic coercion 100, 101
non-governmental organizations (NGOs) 287
non-identities 104, 269
non-importables 286
non-origins 107
non-places 6, 71, 103–104, 107, 233, 258, 269
North Frisian Woodlands (NFW), The Netherlands 13, 181–182, 185–209, 211–231, 257
Norway 276
novelties 66, 67–68, 192–201, 220, 221
  see also deviation
nutrition 227

oil-exporting countries 276
olive harvests 270
olive oil 170–173
on-farm processing 153, 154, 155
order 240–241
organizational level 34, 64–65, 67, 82–83, 252, 253
outstretched hands approach 198, 199, 227, 229
ownership 55, 57, 245

paprika 241

parallel enterprises 95, 100
  see also third level
Pareto optimization 207, 208
Parmalat, Italy 87–111, 237
Parmesan cheese 12, 97–98, 113–126, 143, 173–174, 247
Parmigiano-Reggiano *see* Parmesan cheese
pasteurization 173
payments
  access 246–247
  milk production 97, 98
peasant condition 23–42, 274
peasant farming
  borderlines 37–38
  description 1, 3, 9, 23, 35, 42–45
  distantiation 52
  entrepreneur comparisons 113–149
  future 285
  globalization and liberalization 143, 144
  standardization 220
peasant principle 138, 261–287
peasant studies 18–23, 191
performance 91, 190, 228, 276
Peru 12, 40, 53–85, 257, 266, 267, 269
physical infrastructure 93, 94
physical quality of life 164
pluriactivity 32–33, 154, 158, 159
policy level 8, 101, 152, 179–180, 212–213, 249, 282–287
political level 81, 128–136, 203–204
politico-economic level 22, 128, 233
poverty
  Empire 242, 259, 263, 264
  European level 51
  income 39, 124
  pluriactivity 33
  repeasantization 70, 71, 80
  sharing of 60, 62
power
  countervailing 80, 83, 201, 262, 267, 277
  Empire 234, 236, 243, 253
  inequality 21, 97, 98, 101
  resistance 272
  see also control, hierarchy
precariousness 263, 268
prices
  economic squeeze 128–129
  fluctuations 284

food 110, 258
  globalization and liberalization 142, 143, 149
  milk 93–95, 98–99, 100–101, 102, 104
  scale and labour intensification 120–121
principles of peasant struggles 61
  see also shared values
printing technology 240
processing 88, 153, 154, 155, 286
producers 279
production
  Empire 238
  energy 168, 176–177, 178
  entrepreneurial farming 135
  labour intensification 45
  market relations 44, 62
  peasant and entrepreneur contrast 122–125, 131–133
  repeasantization 65–66
  value added 42–43, 44–45, 268
  worldwide outsourcing 247
  see also co-production; dairy production
production repeasantization 62–63, 66–67
profitability 92, 95, 102, 169, 240, 244, 252, 269
progressive reformism 202–203
pumping stations 69, 70

quality
  dairy production 104, 105, 174
  labour 117, 118
quality of life 160–166
quota systems 134, 142, 246

railway systems 243–244
real economy 130, 242
reciprocity 48–49, 50, 270
recomposition of food 247
reconstituting the peasantry 271–273
redistributive growth 109–111
redundancy 263, 264, 268
regional cooperation 184, 188
regressive centralization 109–111
regressive modes of wealth 242
regrounding of agriculture upon nature 153
regulation
  acid rain 185

controllability 211
deregulation 242
Empire 255, 256
food production and ICT 248, 249, 250–251
hedgerows 211–213
hierarchy 239–240
novelty production 198
peasant and entrepreneur contrast 140
rural life 179
science 220–221
self-regulation 279
shifting interfaces 205
standardization 219–220
territorial co-operatives 182–184
regulations, expert systems 205–206
relative autonomy 32, 169–170
repeasantization
  description 1, 6–7, 8, 35
  European level 151–180
  future 285
  imperial food regimes 257
  Latin America 53–85
  peasant principle 278, 286
reproduction 26, 44, 45, 137
requirements of food 247
re-regionalization of production and consumption 284–285
resistance 262, 265–271, 271–272
resources
  autonomy 14, 113, 156
  co-production 25–26
  defending 36
  dissimilarities 38
  distantiation 50, 51
  economic relations 28–30
  Empire 235–236, 262–263
  food empires 72–73, 76, 77–78, 79, 83–85, 101, 236
  mobilization 28–29, 152
  peasant mode of farming 43
  peasant principle 278
  remoulding 197
  scarcity of 76, 78, 142–143
  search for new 254
  self-controlled base 25–27
  strengthening of 23, 31
retro-innovation 265
rice production 70
rough farming 125, 126, 138
rupture 122, 192

rural development 151–180, 283, 285
rural governance 184–185

scale 2, 118, 119–121, 126, 130, 135, 143, 148–149
  see also enlargement; expansion
scarcity of resources 76, 78, 142–143
Schultz thesis 122
science 211, 220–226, 253–255
self-consumption 65–66
self-controlled resource base 25–27, 36
self-regulation 181, 184, 279
self-sacrifice 67
self-sufficiency 30
services 164, 165
session days 198, 204
settler–colonial food regimes 256, 257
shared values 61, 163, 189, 190–191, 269, 282–283
silage 132, 177, 227
simplicity 250
skill-oriented technologies 170, 171, 172, 173, 174, 176, 267
  see also craftsmanship
slurry 186, 193, 204, 215, 218–219, 224, 225
social level
  capital 163–166, 189–190, 279, 280
  ICT 251
  order 264
  patterning of time 122
  quality of life 164
  struggles 26, 178–180
social wealth
  agrarian crisis 282
  food empires 95, 104, 111
  peasant and entrepreneur contrast 114, 122, 129, 130, 135
  see also value added
society 11, 157, 261
socio-institutional relations 228
socio-political struggles 12, 26, 59, 60–61, 65, 178–180, 265
socio-technical level 234
socio-technical networks 69–80, 93, 94, 174
soil–crop–animal–manure cycle 227
soils 193, 194, 222, 223
sovereignty 282, 286
Spanish Empire 235–236, 253–254
spatial level 62, 80

specialization 121–122, 127
specificity 39, 172, 173, 217, 250
spurred intensification 63–65
squeeze on agriculture 130, 149, 151, 195, 220
stagnation 47
standardization 214–218, 219–220, 223–224
state level
  Empire 218–220, 236
  entrepreneurial condition 128
  markets and institutions 252–253
  territorial co-operatives 182–184
  see also regulations
strategic niche management 192, 201–204
structural features of Empire 236–241
struggles *see* socio-political struggles
styles of farming 26, 136–140
  see also heterogeneity
subjectivity 274
subordination 21
subsidiarity 184, 241, 287
substitution 78
suburbia 161–162
sunk costs 73, 79
superiority 83, 124, 146, 191, 228, 276, 287
supermarkets 244
survey commissions 201
survival 23, 30, 38
sustainability 214, 217, 218, 264

takeovers 89, 104, 126, 129, 233, 263
  see also conquest
tamed hedgerows 211–231
Taormina policy document 282–284
technical ceilings 46, 122
technico-administrative level 126–127
techno-institutional design 168
technology
  entrepreneurial farming 119
  heterogeneity 137, 138
  modernization 127
  new 167–178, 254, 268
tenancy 51
territorial co-operatives 13, 181–209
third level 95, 100, 101, 106, 245–247, 253, 258, 259
three-tiered networks 93–96
throwaway products 132

tied sales mechanisms 107–108
time dimension 2, 12, 122, 131–132, 138
tomatoes 107–108
transaction costs 92, 184
transitional processes 9–10
trust 28, 90–91, 99, 102
   see also distrust

UCPs *see* communal units of production
underdeveloped world 20
universities 253, 255
urban areas 8, 37, 272
urea 215–216
utopian dimension 271

value
   evaporation of 258–259
   food empires 87–88, 90, 95, 96–101, 104, 236
value added
   centralization 111
   economic squeeze 129
   Empire 264
   enlarging or containing 119, 122–125
   food industry 238
   novelty production 195
   production of 42–43, 44–45, 268
   repeasantization 156–157, 158, 159
   skill-oriented technology 176
   see also social wealth
values, shared 61, 163, 189, 190–191, 269, 282–283
virtual capital 97
virtual economy 242
virtual realities 19–20
virtual sustainability 217, 264
visibility 269
volatility 284, 285, 286

water resources 72–73, 76, 78
wealth 70–71, 96, 100–101, 104, 111, 238, 242
web of interconnected novelties 197, 198, 199, 200
working capital, lack of 70

yield 54, 62, 63–64, 131

Zwiggelte, The Netherlands 167–168, 169, 179

Lightning Source UK Ltd.
Milton Keynes UK
UKOW05f0501010617
302424UK00002B/34/P